Ultra-Sensitive PIN and Avalanche Photodiode Receivers

Online at: https://doi.org/10.1088/978-0-7503-5437-0

IOP Series in Advances in Optics, Photonics and Optoelectronics

SERIES EDITOR

Professor Rajpal S Sirohi Consultant Scientist

About the Editor

Rajpal S Sirohi is currently working as a faculty member in the Department of Physics, Alabama A&M University, Huntsville, AL, USA. Prior to this, he was a consultant scientist at the Indian Institute of Science, Bangalore, and before that he was Chair Professor in the Department of Physics, Tezpur University, Assam. During 2000–11 he was an academic administrator, being vice-chancellor to a couple of universities and the director of the Indian Institute of Technology, Delhi. He is the recipient of many international and national awards and the author of more than 400 papers. Dr Sirohi is involved with research concerning optical metrology, optical instrumentation, holography, and speckle phenomena.

About the series

Optics, photonics, and optoelectronics are enabling technologies in many branches of science, engineering, medicine, and agriculture. These technologies have reshaped our outlook and our ways of interacting with each other, and have brought people closer together. They help us to understand many phenomena better and provide deeper insight into the functioning of nature. Further, these technologies themselves are evolving at a rapid rate. Their applications encompass very large spatial scales, from nanometers to the astronomical scale, and a very large temporal range, from picoseconds to billions of years. This series on advances in optics, photonics, and optoelectronics aims to cover topics that are of interest to both academia and industry. Some of the topics to be covered by the books in this series include biophotonics and medical imaging, devices, electromagnetics, fiber optics, information storage, instrumentation, light sources, CCD and CMOS imagers, metamaterials, optical metrology, optical networks, photovoltaics, free-form optics and its evaluation, singular optics, cryptography, and sensors.

About IOP ebooks

The authors are encouraged to take advantage of the features made possible by electronic publication to enhance the reader experience through the use of color, animation, and video and by incorporating supplementary files in their work.

Do you have an idea for a book you'd like to explore?

For further information and details of submitting book proposals, see iopscience.org/books or contact Ashley Gasque at Ashley.gasque@iop.org.

A full list of titles published in this series can be found here: https://iopscience.iop.org/bookListInfo/series-on-advances-in-optics-photonics-and-optoelectronics.

Ultra-Sensitive PIN and Avalanche Photodiode Receivers

Horst Zimmermann
Institute of Electrodynamics, Microwave and Circuit Engineering, TU Wien,
Vienna, Austria

IOP Publishing, Bristol, UK

ISBN 978-0-7503-5437-0 (ebook)
ISBN 978-0-7503-5435-6 (print)
ISBN 978-0-7503-5438-7 (myPrint)
ISBN 978-0-7503-5436-3 (mobi)

DOI 10.1088/978-0-7503-5437-0

Version: 20231001

IOP ebooks

British Library Cataloguing-in-Publication Data: A catalogue record for this book is available from the British Library.

Published by IOP Publishing, wholly owned by The Institute of Physics, London

IOP Publishing, No.2 The Distillery, Glassfields, Avon Street, Bristol, BS2 0GR, UK

US Office: IOP Publishing, Inc., 190 North Independence Mall West, Suite 601, Philadelphia, PA 19106, USA

Contents

Preface

The era of photodiode ICs (photodiode integrated circuits, PDICs) started in the 1990s with applications in audio compact disk (CD) players, CD-ROM and digital versatile disk (DVD) systems when bandwidths and transimpedance of optical sensors consisting of discrete photodiodes/photodiode arrays and amplifier ICs came to a limit. Since that time, rapid progress was made in bandwidth/data rate and quantum efficiency/responsivity of integrated photodiodes and sensor ICs. Integrated photodiodes allowed not only increased bandwidth and transimpedance gains. They were essential for image sensors with high and very high numbers of pixels. Integrated photodiodes let to better immunity to electromagnetic interference, improved the reliability and, last but not least, reduced the costs of many optical sensors.

The first quantum leap in PDIC performance came from integrated PIN photodiodes. Then avalanche photodiodes (APDs) in the linear mode introduced the second step of performance gain of PDICs. The next quantum leap in integrated photodetectors was due to APDs operated in the Geiger mode, where detection of single photons is possible. But after about a decade of research and development on single-photon avalanche diode (SPAD) sensors, it is time to reconsider whether dark counts, afterpulsing and optical crosstalk of SPADs are avoidable.

In fact, there is a disruptive way to avoid these parasitic properties of SPADs and even to beat the sensitivity of SPAD receivers. This book will describe this way by focusing on ultra-sensitive PIN and avalanche photodiode receiver (USPAR) ICs for data communication. After more than 20 years of exploitation of the PIN-photodiode CMOS technology leading to efficient APD and SPAD receivers, this book describes how ultra-sensitive PIN photodiode receivers close the loop back to PIN photodiode ICs.

The advanced low-capacitance PIN photodiodes investigated in USPAR and described in this book will also allow the improvement of conventional (resistive-feedback) transimpedance amplifiers and many different optical sensors with respect to noise reduction and sensitivity enhancement.

The author would like to thank Ashley Gasque from IOP Publishing for supporting the idea for this book. He also has to thank Dr Kerstin Schneider-Hornstein, Dr Bernhard Goll, Christoph Gasser, Simon Laube, and Saman Kohneh Poushi from our institute for excellent work in the USPAR project as well as the former group members Dr Bernhard Steindl, Dr Dinka Milovancev, Dr Tomislav Jukic, Dr Paul Brandl, Wolfgang Gaberl and Dr Reinhard Enne for important and excellent contribution to our work on integrated PIN photodiodes, finger photodiodes, APDs and SPADs as well as on sophisticated circuits. In addition, he would like to thank Wolfgang Einbrodt, Dr Konrad Bach, Detlef Sommer, Dr Alexander Zimmer and Dr Daniel Gäbler from XFAB Semiconductor Foundries for the long cooperation and for enabling the huge progress with silicon PDICs. Last but not least, funding of the project P 34 649-N from the Austrian Science Foundation (FWF) is acknowledged.

Vienna, Austria
Horst Zimmermann

Author biography

Horst Zimmermann

Dr Horst Zimmermann, received the diploma in Physics in 1984 from the University of Bayreuth, Germany, and the Dr-Ing. degree from the University Erlangen-Nürnberg working at the Fraunhofer Institute for Integrated Circuits (IIS-B), Erlangen, Germany in 1991. Then, Dr Zimmermann was an Alexander-von-Humboldt Research-Fellow at Duke University, Durham, NC, working on diffusion in Si, GaAs, and InP until 1992. In 1993, he joined the Chair for Semiconductor Electronics at Kiel University, Kiel, Germany, where he lectured on optoelectronics and worked on optoelectronic integration. Since 2000 he has been full professor for electronic circuit engineering at Vienna University of Technology, Vienna, Austria. His main interests are in design and characterization of analog and nanometer CMOS circuits as well as optoelectronic integrated CMOS and BiCMOS circuits, in optical wireless communication, in single-photon detection and in electronic-photonic integration. He is author of the Springer books *Integrated Silicon Optoelectronics* and *Silicon Optoelectronic Integrated Circuits* as well as coauthor of *Highly Sensitive Optical Receivers*, *Optical Communication over Plastic Optical Fibers*, *Analog Filters in Nanometer CMOS*, *Comparators in Nanometer CMOS Technology*, *Optoelectronic Circuits in Nanometer CMOS Technology* and of the IOP book *Single-Photon Detection for Data Communication and Quantum Systems*. In addition, he is author and co-author of more than 600 publications. In 2002 he became Senior Member IEEE. He was primary guest editor of the November/December 2014 issue of IEEE Journal of Selected Topics in Quantum Electronics on Optical Detectors: Technology and Applications.

Symbols

ADC	Analog-to-digital converter
α	Optical absorption coefficient (μm^{-1})
α_n	Impact ionization coefficient of electrons (cm^{-1})
α_n	Impact ionization coefficient of holes (cm^{-1})
APD	Avalanche photodiode
APP	Afterpulsing probability
APS	Active pixel sensor
B	Bit rate (bits/s)
BER	Bit error ratio
c	Speed of light in a medium (cm s^{-1})
c_0	Speed of light in vacuum (cm s^{-1})
C_{int}	Integration capacitance (As V^{-1})
CIS	CMOS image sensor
CMOS	Complementary metal oxide semiconductor
CPU	Central processing unit
CW	Continuous wave
D	Diffusion coefficient (cm^2 s^{-1})
D_n	Diffusion coefficient of electrons (cm^2 s^{-1})
D_p	Diffusion coefficient of holes (cm^2 s^{-1})
DAC	Digital-to-analog converter
DCR	Dark count rate (s^{-1})
DR	Data rate (Mb s^{-1})
ϵ	Permittivity
ϵ_0	Vacuum permittivity
ϵ_r	Relative permittivity
η_d	Quantum efficiency (of a photodiode)
η_{int}	Integration efficiency
$\mathbf{E}$	Electric field (V cm^{-1})
f	Frequency
FEC	Feedward error correction
FET	Field-effect transistor
g_m	Transconductance (S = AV^{-1})
h	Planck constant (J s)
$h\nu$	Photon energy (eV)
I	Current (A)
I_{ph}	Photocurrent (A)
i_n	Noise curent (A)
j	Current density (A cm^{-2})
k_B	Boltzmann's constant (JK^{-1})
k_BT	Thermal energy (eV)
L	Length (μm)
L_n	Diffusion length of electrons (μm)
L_p	Diffusion length of holes (μm)
λ	Wavelength (m)
μ	Mobility (cm^2 V^{-1} s^{-1})
μ_n	Electron mobility (cm^2 V^{-1} s^{-1})
μ_p	Hole mobility (cm^2 V^{-1} s^{-1})

M	Avalanche gain
MOS	Metal oxide semiconductor
ν	Frequency of light (Hz)
n	Density of free electrons (cm^{-3})
N_A	Acceptor concentration (cm^{-3})
N_D	Donor concentration (cm^{-3})
N_{PH}	Number of photons
NIR	Near infra-red
NRZ	Non-return to zero
OCT	Optical crosstalk
OCTP	Optical crosstalk probability
p	Density of free holes (cm^{-3})
P_{opt}	Incident optical power (W)
PAM	Pulse amplitude modulation
PCB	Printed circuit board
PD	Photodiode
PDP	Photon detection probability
PRBS	Pseudo random bit sequence
Ψ	Potential (V)
q	Electron charge (As)
Q	Charge (As)
Q_{av}	Avalanche charge (As)
QIS	Quantum image sensor
ρ	Charge density (As cm^{-3})
R	Resistor ($V/A = \Omega$)
R_o	Responsivity (A W^{-1})
RAM	Randomn access memory
RS	Reset
RZ	Return to zero
S	Sensitivity (dBm)
SCR	Space-charge region
SF	Source follower
SIPM	Silicon photomultiplier
SPAD	Single-photon avalanche diode
STI	Shallow trench isolation
t	Time (s)
τ	Lifetime (s)
τ_n	Electron lifetime (s)
τ_p	Hole lifetime (s)
Θ	Angle (°)
Θ_i	Incidence angle (°)
T	Absolute temperature (K)
TIA	Transimpedance amplifier
U	Voltage (V)
U_{BE}	Base–emitter voltage (V)
U_D	Built-in voltage (V)
U_{DS}	Drain–source voltage (V)
U_{GS}	Gate–source voltage (V)
U_T	Thermal voltage $k_B T/q$ (V)
U_{th}	Thermal generation/recombination rate (cm^{-3} s^{-1})

U_{Th}	Threshold voltage (V)
USPAR	Ultra-sensitive PIN and avalanche photodiode receiver
v	Carrier velocity (cm s^{-1})
v_s	Saturation velocity (cm s^{-1})
v_n	Noise voltage (V)
V_{BD}	Breakdown voltage (V)
V_{EX}	Excess bias voltage of SPAD (V)
W	Width of space-charge region (μm)

IOP Publishing

Ultra-Sensitive PIN and Avalanche Photodiode Receivers

Horst Zimmermann

Chapter 1

Motivation

In this chapter the history of improving the sensitivities of integrated optical receivers with pin-photodiodes and avalanche photodiodes (APDs) in the linear mode is summarized. Then the further sensitivity enhancement with single-photon avalanche diodes (SPADs) including the difficulties and disadvantages is described. As a recent development, quantum image sensors, which are special image sensors down to single-photon resolution, are introduced. Furthermore, we throw a glance on the principle of the disruptive idea of ultra-sensitive optical receivers without SPADs.

1.1 PIN-photodiode and APD receivers

The number of publications on optical receivers with integrated photodiodes increased a lot at the turn of the last millenium. But before we have a look at certain receiver optoelectronic integrated circuits (OEICs), i.e. onto receiver chips with monolithically integrated photodiodes, we have to define the sensitivity S, which is used to characterize their noise.

$$S = 10 \log \frac{P_{\text{opt}}}{1\ \text{mW}}\ \text{dBm} \tag{1.1}$$

The sensitivity of an optical receiver is the average optical input power P_{opt}, which is necessary to achieve a certain bit error ratio (BER), for instance 10^{-9}. 'Average' means the mean value of the optical power in a logical '1' and of the optical power in a logical '0'[1]. The BER must be mentioned, for which the sensitivity is given. The wavelength of the light also has to be specified. The sensitivity S is often expressed in dBm as it can be obtained according to equation (1.1). As examples: $S = -30$ dBm corresponds to an average optical input power of 1 μW and $S = -60$ dBm is valid for 1 nW.

[1] Usually there are as many '1's as '0's used in a bit sequence.

doi:10.1088/978-0-7503-5437-0ch1

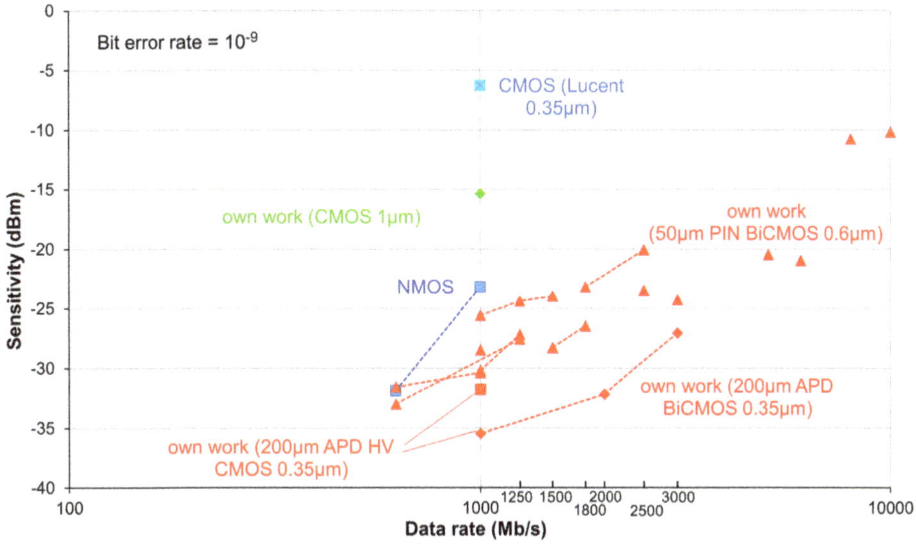

Figure 1.1. Sensitivities of integrated PIN-photodiode and APD receivers.

The sensitivities of the most interesting receiver OEICs are compared in figure 1.1. A receiver OEIC using a p-finger-inside-n-well photodiode in 0.35 μm Lucent CMOS [1] achieved a sensitivity of −6.3 dBm at 1 Gb s^{-1} with 850 nm light. With a lateral PIN photodiode in 1.0 μm NMOS technology a sensitivity of −31.9 dBm at 622 Mb s^{-1} and of −23.2 dBm at 1 Gb s^{-1} both with 30 V photodiode bias voltage with 850 nm light was obtained [2]. At 10 V photodiode bias, the sensitivity reduced to −23.2 dBm for 622 Mb s^{-1}. A silicon-on-insulator (SOI) layer with 2 μm thickness was used to fabricate an OEIC in 0.13 μm CMOS [3]. At 850 nm, it reached sensitivities of −19 dBm, −16.6 dBm, −15.4 dBm and −10.9 dBm at 1 Gb s^{-1}, 2 Gb s^{-1}, 3.125 Gb s^{-1} and 5 Gb s^{-1}, respectively.

Within the frame of the German lead project 'Optical Storage Systems', my group integrated a PIN photodiode in 1.0 μm CMOS and investigated BiCMOS opto-electronic integrated circuits (OEICs) [4]. In this project, one of our PIN-photodiode OEIC receivers achieved a data rate of 1 Gb s^{-1} with a sensitivity of −15.4 dBm with a supply voltage of 5 V, whereby the voltage across the PIN photodiode was only 3 V [5]. After my start at TU Wien, we improved the circuits and the measurement technique as well as the OEIC performance considerably when working in 0.6 μm (Bi)CMOS. The red symbols in figure 1.1 belong to such more advanced receivers and are valid for BER = 10^{-9}. The integrated optical receivers with PIN photo-diodes represented with red triangles in figure 1.1 had 50 μm square-shaped detectors [6]. Describing all these receivers in detail exeeds the scope of this book considerably and the reader is referred to [6]. Just two of them should be mentioned: in [7], the sensitivity of an OEIC with 50 μm PIN photodiode at 1 Gb s^{-1} with 660 nm light for BER = 10^{-9} was −30.4 dBm, which equals an average optical power of 0.912 μW corresponding to 3034 photons in average, i.e. to 6068 photons in a logical '1'. In [8], at 622 Mb s^{-1} with 660 nm the sensitivity again with 50 μm PIN photodiode was −30.7 dBm, corresponding to 4552 photons in average and 9104 photons in a logical ‘1’.

Here only the most interesting and newest receivers are summarized: a receiver OEIC with 390 μm-diameter PIN photodiode exhibited a sensitivity of −24.3 dBm at 1.25 Gb s^{-1} with 675 nm light. Its sensitivity is worse than the −30 dBm of [7] and [8] because the larger photodiode with its larger capacitance causes more electronic noise [6]. The OEICs with integrated APDs included in figure 1.1 were designed as receivers for optical wireless communication, where the incident light power per μm^2 is small. They therefore had an APD diameter of 200 μm. The sensitivity of the best APD OEIC in 0.35 μm high-voltage (HV) CMOS was −35.5 dBm at 1 Gb s^{-1} [9]. From this average optical power, a photon number of 1918 in a logical ‘1’ follows. Finally, an OEIC with a 800 μm-diameter APD was realized in 0.35 μm BiCMOS [10]. Its APD was operated at an avalanche gain of 22 and its sensitivity was −33 dBm at 1 Gb s^{-1} and −29.3 dBm at 2 Gb s^{-1}. When we consider the larger receiver noise (compared to the 50 μm-PIN-photodiode OEICs), the APDs improve the sensitivity by roughly 10 dB (a factor of 10). To make it clear: all these receiver OEICs implemented transimpedance amplifiers (TIAs) with resistive feedback (RF).

1.2 SPAD receivers

Conventional light-emitting diodes (LEDs) and semiconductor lasers are Poissonian light sources. Therefore, when attenuating the light intensity the number of photons emitted within a time interval T fluctuates. The probability $p_m(k)$ that k photons hit the detector in T, when m photons are arriving in average in T (this determines the average optical power $\langle P_{opt}\rangle$ and the sensitivity), can be obtained from the Poisson distribution [11]:

$$p_m(k) = \frac{m^k}{k!}e^{-m}. \tag{1.2}$$

For an ideal detection efficiency of 100%, in average 21 photons (m = 21) are necessary in a ‘1’-bit for BER = 10^{-9}, because the probability that a ‘0’ ($k = 0$) is detected has to be less than 10^{-9}. Feedforward error correction (FEC) nowadays allows BER = 2×10^{-3} enabling that only seven photons are necessary in average. But we have to be aware that an optical transmitter with an extinction ratio of infinity is necessary, meaning that for a ‘0’ no photon must be emitted (in practice the extinction ratio should be at least about 200).

The average optical power is given by ($h\nu$ is the photon energy, bit rate (data rate) $B = 1/T$, T = duration of a bit):

$$\langle P_{opt}\rangle = \frac{mh\nu B}{2}. \tag{1.3}$$

The sensitivity (see equation (1.1)) defines the so-called quantum limit representing the physical limit for the sensitivity of optical receivers operated together with Poissonian transmitters. Figure 1.2 compares the sensitivities corresponding to the quantum limit for BER = 10^{-9} and BER = 2×10^{-3}. There is a difference of about

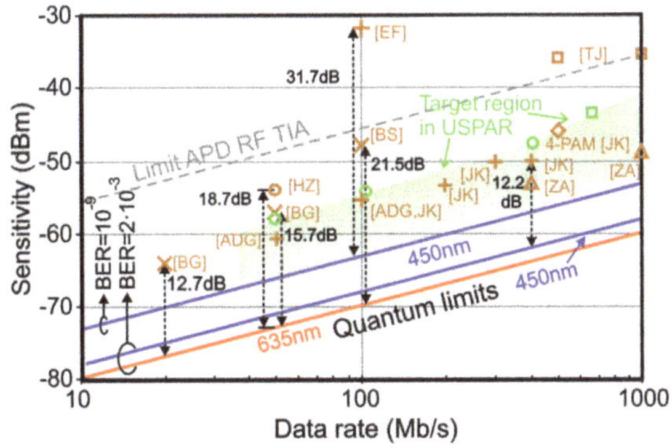

Figure 1.2. Sensitivities of SPAD receivers and quantum limit. The data points [ADG], [BG], [HZ], [BS], [JK] and [ZA] are valid for BER = 2×10^{-3} ([ADG]: [25], [BG]: [18], [HZ]: [14], [BS]: [15], [JK]: [19] and [ZA]: [20]). [ZA] represents SiPM receivers. The green symbols are the starting points of USPAR receivers with PIN photodiodes (green circles) and linear-mode APD receivers (green square). The gray dashed line shows the limit of RF TIAs with 200 μm-linear-mode APDs in 0.35 μm (Bi)CMOS for BER = 10^{-9}. The USPAR target region is also included for comparison. The squared symbols represent receivers with APDs in the linear mode.

5 dB between the quantum limits for these two BERs. It should be noted that the quantum limit shifts with the wavelength λ (because of the photon (light) frequency ν in equation (1.3), $\lambda = c/\nu$, c = light velocity). Figure 1.2 contains these quantum limits.

Due to the very high gain of SPADs exceeding 10^6, the electronic noise of amplifiers (shot noise, thermal noise) and excess noise of linear-mode APDs is not limiting anymore. A SPAD can already be connected to a digital gate. Therefore, SPAD receivers represent digital optical receivers. In principle, they do not need (analog) amplifiers like receivers with PIN photodiodes or APDs. But it will become clear in the following that in practice a large circuit overhead will be needed because of dark counts, afterpulses and optical crosstalk to achieve an acceptable bit error performance [12].

Several SPAD receivers were described in the literature. Figure 1.2 compares the sensitivities of SPAD receivers as they were reported until the start of the USPAR project in 2021. They achieve better sensitivities than linear-mode APD receivers, for which the estimated sensitivity limit valid for TIAs with RF is included. The first SPAD receiver with an array of 32 × 32 SPADs and a complex signal processing circuit [13] was optimized for a large dynamic range and achieved a sensitivity of −31.7 dBm at 100 Mb s^{-1} and 450 nm. Two 4-SPAD receivers in 0.35 μm PIN-photodiode CMOS with active quenching circuits were introduced by the group of the author and achieved sensitivities of −55.7 dBm at 50 Mb s^{-1} [14] and −48.5 dBm at 100 Mb s^{-1} [15], both at 635nm. To allow an easy comparison with the PIN and APD receivers mentioned above and the USPAR receivers, which will be described in this book, the photon number in a logical '1' corresponding to the average optical power of −55.7 dBm is 354. Four SPADs with 3 to 4 photon detections in a '1'-bit to keep the error rate due to dark count rate (DCR), afterpulsing probability (APP) and crosstalk below the error correction limit of 2×10^{-3} [16, 17] restricted the sensitivity. With one gated SPAD in 0.35 μm PIN-photodiode CMOS using 5 sub-bits, the detection of 2 photons in a '1'-bit was sufficient, leading to a sensitivity of −64 dBm at 20 Mb s^{-1} and −57 dBm at 50 Mb s^{-1} [18]. The necessary dead time to keep afterpulsing low prohibited higher data rates of these 1-SPAD and 4-SPAD receivers. A 64 × 64 SPAD receiver in 0.13 μm CMOS achieved a sensitivity of −49.9 dBm at 400 Mb s^{-1} and 450 nm [19]. A SiPM (Silicon PhotoMultiplier) receiver with −58.4 dBm sensitivity at 400 Mb s^{-1} and −49 dBm at 1 Gb s^{-1} (with 405 nm light, BER = 10^{-3}) was suggested [20]. Concluding, mainly afterpulsing and crosstalk of SPADs limit the possible data rate and the BER [21, 22] and/or require receivers with a high complexity, large chip area and high power consumption [14, 19]. Also, the photon detection probability (PDP) of reported SPAD receivers was only up to about 40% [14, 18, 19]. The energy per bit of a 400 Mb s^{-1} SPAD receiver (incl. ADC functionality for 4-PAM) was 230 pJ/bit [19], whereas an USPAR receiver is thought to achieve less than 50 pJ/bit in RZ on-off keying where no ADC is needed. For PAM, 31 pJ/bit with an ADC in 90 nm CMOS [23] or 3.8 pJ/bit with an ADC in 28 nm CMOS [24], would have to be added for an USPAR receiver. These properties show the weaknesses of SPAD receivers. Despite all this effort, SPAD receivers still leave a gap to the quantum limit of 12.2 dB to 21.5 dB (see figure 1.2).

The sensitivity values of the green circles and squares in figure 1.2 belonging to simple ultra-sensitive PIN and avalanche photodiode receivers (USPAR) will be derived in chapters 6 and 7. The green area is the target region for more advanced USPAR receivers.

1.3 Quantum image sensors

There are photodiode active pixel sensors (APS) and photogate APS. APS are image sensors, which integrate photogenerated charge carriers on the photodetector within each pixel. They contain a transfer gate (transfer transistor) to switch the integrated charge to a storage node, a reset transistor, a source follower as 'pixel output buffer' and a select transistor to switch the signal from the storage node via the source follower to a readout line. This arrangement is called a 4-transistor APS pixel. There is also an arrangement without transfer transistor and without storage node, where the photogenerated charges are integrated directly on the capacitance of the photodetector; this arrangement is called 3-transistor APS pixel. For more details the interested reader may be referred to [6].

Image sensors are called Quantum Image Sensors (QIS), if their sensitivity is so high that they can resolve single photons, i.e. count photons. The photo-sensitivity of image sensors was considerably improved in the last decades. Avalanche multiplication was suggested for low light level imaging in charge-coupled devices (CCDs) using repeated transfer of charges through high-field regions [26, 27]. Nowadays these devices are called EMCCDs, which can resolve single photons when being cooled [28]. The repeated charge transfer limits the frame rate of these image sensors. An APD was switched between linear and Geiger mode to extend the dynamic range of a 1280 × 720 image sensor in 110 nm CMOS with a pixel pitch of

3.8 μm [29]. SPADs were used in a 65 nm CMOS imager with 160 000 6 μm-pixels at a fill factor of 70% to achieve single-photon resolution [30].

The small photodiodes and storing nodes together with avoidance of the reset transistor (using reach-through or bootstrapping reset [31, 32]) allow for a low integration and storage capacitance and in turn for a signal voltage exceeding the read noise level without avalanche gain. The read noise was $0.44e^-_{rms}$ and $0.27e^-_{rms}$ in [32] and [31], respectively [31], used a 0.11 μm CIS (CMOS Image Sensor) process. To achieve single-photon resolution, process modifications were done. The readout worked at 1 kHz frame rate. $1/f$-noise is still preventing the accurate determination of the number of detected photons. Due to shrinking, the pixel sizes went down to 1.1 μm in square in a very advanced and expensive, stacked, backside-illuminated 45 nm CMOS sensor chip on top of a 65 nm CMOS readout chip [33]. In [34], a read noise of $0.22e^-_{rms}$ at room temperature was reported.

Concluding: the fill factor of image sensors without micro lenses is (much) below 100%, the charge transfer from the pinned photodiode to the storage node needs a long time (100 ns, although already process modifications were done [35]) and the readout of many pixels makes image sensors very slow, i.e. not appropriate for data rates of 100 Mb s^{-1} to 1 Gb s^{-1}.

The basic concept of QIS is to exploit very small photodiodes with a very small capacitance and to reduce the node capacitances in the pixels as much as possible. Using minimum-size transistors was not enough and reach-through or bootstrapping reset was used to obtain smaller parasitic capacitances than with a reset transistor. With low capacitances, single photons can be resolved and a few photons can be counted (see figure 1.3). The lower the read noise is, the lower the error rate is. A read noise of $0.12e^-_{rms}$ would be highly desirable (see figure 1.3(c)), but was not experimentally verified yet.

1.4 The principle of ultra-sensitive PIN and avalanche photodiode receivers

The principle of ultra-sensitive optical receivers is to integrate the photogenerated charges on a small capacitance and to read out the voltage change on the capacitance by a low-noise amplifier like in image sensors or QIS, but at much higher data rates. The (parasitic) capacitance of the photodiode itself can be used for charge integration. If the capacitance is very small, the obtained voltage change can be larger than the equivalent input noise voltage of the low-noise amplifier—even if ideally only one or a few photons were absorbed in the photodiode. Because the ultra-low capacitances needed would require a very advanced and expensive nanometer-scale CMOS image sensor technology with photodiode areas below 1 μm^2, PIN-photodiodes and—if APD gain is necessary—APDs based on PIN-photodiode CMOS technology will be used in much cheaper sub-μm CMOS technology[2]. These PIN photodiodes and PIN-APDs also achieve low capacitances, but at larger light-sensitive areas than photodiodes in nanometer CMOS image sensors. Furthermore,

[2] In fact, also 0.18 μm USPAR ICs will be investigated, which belong to the so-called deep-sub-μm range.

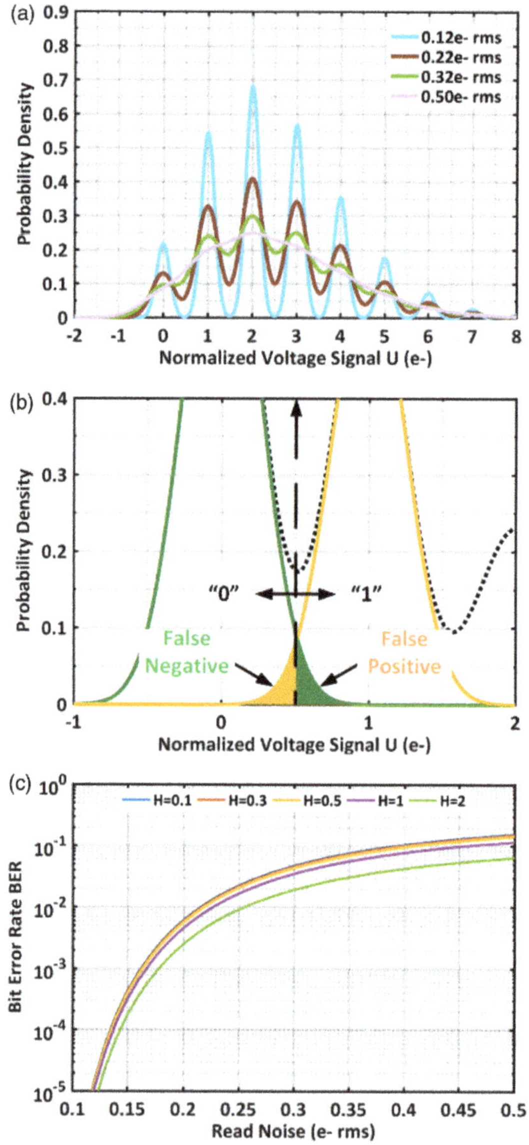

Figure 1.3. Modelled photoelectron counting of quantum image sensor. (a) Influence of the read noise on the probability distribution of the photoelectron voltage signal. (b) Illustration of error probability in single-bit QIS. (c) BER in single-bit QIS in dependence on read noise and quanta exposure H (from $H = 0.1$ to $H = 2$). Reproduced from [33] CC BY 4.0.

finger and dot photodiodes will be investigated to reduce the photodiode capacitance even further and/or to increase the light-sensitive area. The photogenerated charges can also be integrated on a small capacitor with a low-noise amplifier in integrator configuration. Combination/summation of the signals of several low-capacitance photodiodes and low-noise preamplifiers are also possible to obtain large light-sensitive areas. In the following we call such receivers Ultra-Sensitive PIN and Avalanche photodiode Receivers (USPARs).

The advantages of USPARs compared to SPAD receivers and image sensors are:

A. The Geiger efficiency, i.e. the probability of triggering a Geiger avalanche event, and the optical fill factor of SPADs are not limiting, although the finger and dot photodiodes allow a similar detector area.
B. Afterpulsing and optical crosstalk of SPADs is avoided and there is no dead time.
C. In contrast to SPAD receivers with BER = 2×10^{-3}, a BER of 10^{-9} is possible and error correction is not necessary.
D. Photodiode reverse bias, chip area and power consumption will be smaller compared to SPAD receivers.
E. Compared to pixels in nanometer CMOS image sensors, larger light-sensitive areas are achieved and a much higher data rate is possible.

The PIN photodiodes and APDs in PIN-photodiode CMOS technology with a thick absorption zone can be integrated together with MOSFETs [14, 15] and bipolar NPN transistors [36, 37] on the same chip in order to avoid bondpad capacitances and to obtain a low capacitance. In fact, for USPARs, bondpads between photodiode and amplifier must be absent. The thick epitaxial layer of PIN photodiodes and PIN-APDs allows for a high responsivity for red and near infra-red (NIR) light.

For the approach of Razavi [38], a chance of success in USPAR may exist, because the noisy biasing network for the input transistor will not be needed here due to the reset transistor RS (figure 1.4) for recharging the photodiode at the end of each bit.

Figure 1.4 shows a possible circuit for the input stage of a highly sensitive optical receiver. This circuit actually corresponds to a 3-transistor image sensor pixel without the select transistor. The source follower SF has a low input capacitance (gate–drain overlap capacitance C_{GD} plus a fraction of C_{GS}) [39] which will be advantageous for low photodiode capacitances C_{PD}. The photodiode's cathode is charged to VDD by the reset transistor RS and then left floating. The photo-generated electrons drift to the positively charged cathode and are integrated there

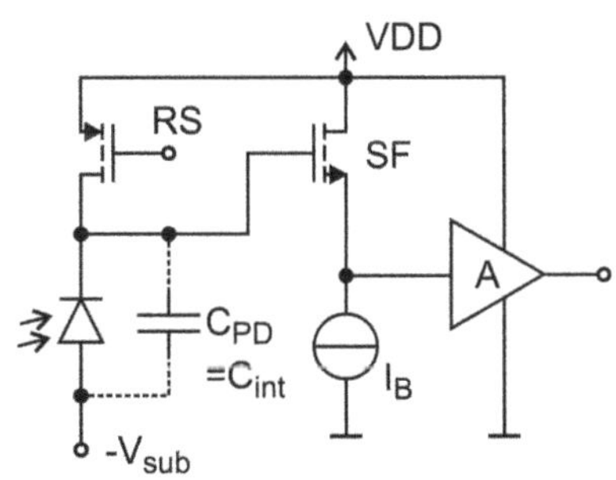

Figure 1.4. Read-out front-end with source follower (SF); RS = reset.

during the bit time. According to the electrons accumulated below the cathode, its potential decreases by the signal voltage U_{Signal}, which is read out after the integration period via SF and amplifier A. The charge Q accumulated during the integration time is given by the number of incident photons N_{ph}, the quantum efficiency η_{d} of the photodiode, the electron charge q and the multiplication gain M, when an APD is used. In figure 1.4, the photodiode capacitance C_{PD} itself acts as integration capacitance C_{int}.

$$U_{\text{Signal}} = \frac{Q}{C_{\text{int}}} = M\eta_{\text{d}}N_{\text{ph}}\frac{q}{C_{\text{int}}} \tag{1.4}$$

The input noise voltage of the amplifier $v_{\text{n}}^{\text{in,amp}}$ results from thermal noise due to the channel resistance expressed by the transconductance g_m of the input transistor (SF in figure 1.4) and the empirical factor $\Gamma = 2/3$ for MOSFETs operated in the saturation region (k = Boltzmann constant, T = absolute temperature, Δf = bandwidth).

$$(v_{\text{n}}^{\text{in,amp}})^2 = 2/3 \times 4kT\frac{\Delta f}{g_m} \tag{1.5}$$

The signal-to-noise ratio SNR at the cathode node is ($(v_{\text{n}}^{\text{in,total}})^2 = (v_{\text{n}}^{\text{in,amp}})^2 + (v_{\text{n}}^{\text{RS}})^2 + (v_{\text{n}}^{\text{ph}})^2$, v_{n}^{RS} from reset (kT/C) noise and v_{n}^{ph} from photon shot noise proportional to $\sqrt{N_{\text{ph}}}$ [40] plus excess noise in case of an APD):

$$\text{SNR} = \frac{U_{\text{Signal}}}{v_{\text{n}}^{\text{in,total}}} \tag{1.6}$$

From the noise theory of optical receivers [41] it is known that the SNR has to be about 12 (Q-factor = 6) for BER = 10^{-9} and 6 (Q-factor=3) for BER = 2×10^{-3}. For the source follower approach a short study gave $v_{\text{n}}^{\text{in,amp}} = 0.13\ \text{mV}_{\text{rms}}$ ($\Delta f = 400$ MHz), $v_{\text{n}}^{\text{RS}} = \sqrt{kT/C_{\text{int}}} = 0.455\ \text{mV}_{\text{rms}}$ for an input node capacitance of 20 fF and $v_{\text{n}}^{\text{ph}} = 0.18\ \text{mV}_{\text{rms}}$ for $M = 1$ and BER = 2×10^{-3}. In comparison to image sensors which need bandwidths below 1 kHz, the bandwidth is much larger in USPAR and at a first glance single-photon resolution will therefore be out of reach due to thermal and reset noise.

Source followers can be used to achieve a low amplifier input capacitance. As next stage a high-gain low-noise stage with MOSFETs or NPN transistors (if a BiCMOS process is used, which allows for higher data rates) can be implemented. Reach-through reset [31] may be interesting when photodiode capacitances in the few fF range are achieved to eliminate kT/C noise and the drain-bulk capacitance of the reset transistor.

The photo-charge can also be integrated on a separate capacitor C_{int} (see figure 1.5). This circuit allows a higher conversion efficiency (signal voltage) for larger photodiode capacitances as long as the inverter gain A is large enough. The integration efficiency is given by

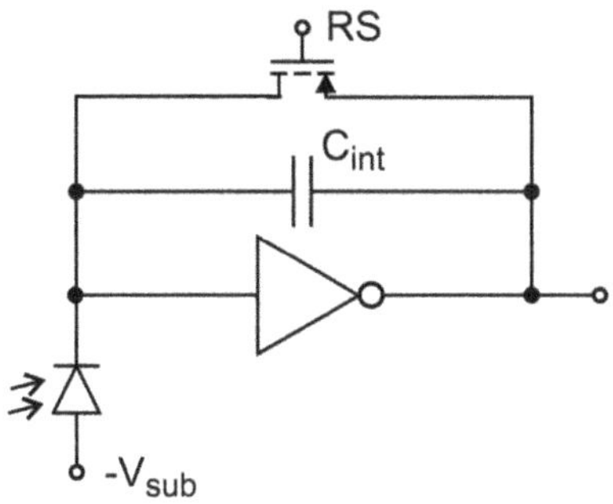

Figure 1.5. Read-out front-end with inverter in integrator configuration; RS = reset.

$$\eta_{int} = \frac{AC_{int}}{(A+1)C_{int} + C_{PD}}, \tag{1.7}$$

where A is the gain[3] of the amplifier and parasitic input capacitances have to be added to C_{PD}. In both circuits the RS transistor's leakage current is negligible since its drain–source voltage U_{DS} and its drain-bulk voltage U_{DB} are close to zero (the signal will be in the mV range; note that the well (bulk B) will be connected to source). For the low photo-charges aimed at, a minimum-size RS transistor is sufficient, which reduces the parasitic capacitance.

Inverters combine high gain and low-noise performance [42] and can be used as input stage of the low-noise amplifier. Furthermore, they allow a very low integration capacitance, which is the sum of the gate–drain overlap capacitances of the NMOS and PMOS transistors in the inverter (about 3 fF in the 0.35 μm CMOS example below and 2 fF in 0.18 μm CMOS). This works well also in worst-case simulations. Thereby also parasitic capacitances of metal connections to polysilicon–polysilicon or metal–metal capacitors are eliminated. A first study using (pre-layout) circuit simulation inclusive all transistor parasitics but without any optimization showed that for C_{PD} = 20 fF and an inverter in 0.35 μm CMOS a data rate of 400 Mb s^{-1} is possible at a sensitivity of—47.3 dBm with 635nm light (BER = 2×10^{-3}) and infinite extinction ratio; at 100 Mb s^{-1} a sensitivity of −54.1 dBm and at 50 Mb s^{-1} a sensitivity of −57.3 dBm is possible. Photon shot noise was converted to the corresponding error in the signal voltage and considered in the calculation of these sensitivities. These three results (green circles) are included in figure 1.2 as start points in the target region of USPAR. For BER = 2×10^{-3} at 400 Mb s^{-1} 270 photons are necessary in a '1' (i.e. 135 photons in average over '0's and '1's, which corresponds to 270 photons in a '1'). In contrast to SPAD receivers in the same technology, the USPAR receiver can achieve a BER of 10^{-9} at 50 Mb s^{-1} with −52.3 dBm sensitivity. Reduction of reset noise has the potential of a 5 dB sensitivity improvement by using correlated double sampling [40] in USPAR in return-to-zero data transmission mode. Inverters in 0.18 μm CMOS add a 2 dB sensitivity

[3] The amplifier in the integrator has a negative gain, since it inverts the polarity of the signal and we take the magnitude of this negative gain.

improvement and the target region for 0.35 μm CMOS in figure 1.2 will reach down by these 5 + 2 dB. With 0.18 μm CMOS in addition higher data rates can be achieved.

Using an APD instead of a PIN photodiode will improve the sensitivity by about 2 dB (because the gain M for low enough excess noise is rather limited [27]). At larger noise levels, i.e. for larger photodiode areas and higher data rates the sensitivity improvement of an APD over a PIN receiver increases. For instance, at 750 Mb s^{-1} with C_{PD} = 100 fF, a 0.35 μm USPAR APD CMOS receiver achieves according to simulation a sensitivity of −43.3 dBm (BER = 2×10^{-3}), which is a considerable improvement compared to −35.8 dBm at 500 Mb s^{-1} (BER = 10^{-9}) with an RF TIA BiCMOS receiver [36] (see figure 1.2). With the multi-dot APD (see figure 5.59) the photodiode diameter of 200 μm [36] should even be surpassable. The green square symbol in figure 1.2 compares the USPAR CMOS APD receiver at 750 Mb s^{-1} to the BiCMOS results at 500 Mb s^{-1} and 1 Gb s^{-1} (brown squares) of [36].

Overall, according to above described estimations the USPAR receivers can outperform state-of-the-art SiPM and SPAD receivers (see figure 1.2) and several measures described in this book promise additional improvements. The way towards USPAR receivers needing even less than 80 photons in the logical '1' will be described in the follow chapters and many surprising problems on the way to the experimental realization of USPAR receivers will be discovered and solved.

References

[1] Woodward T K and Krishnamoorthy A V 1998 1 Gbit s^{-1} CMOS photoreceiver with integrated detector operating at 850 nm *Electron. Lett.* **34** 1252–3

[2] Schaub J D, Li R, Csutak S M and Campbell J C 2001 High-speed monolithic silicon photoreceivers on high resistivity and SOI substrates *J. Light. Technol* **19** 272–8

[3] Csutak S M, Schaub J D, Wu W E and Campbell J C 2002 High-speed monolithically integrated silicon optical receiver fabricated in 130 nm CMOS technology *IEEE Photon. Technol. Lett.* **14** 516–8

[4] Zimmermann H 2010 *Integrated Silicon Optoelectronics* (Heidelberg, Germany: Springer)

[5] Zimmermann H and Heide T 2001 A monolithically integrated 1 Gb s^{-1} optical receiver in 1 μm CMOS technology *IEEE Photon. Technol. Lett.* **13** 711–3

[6] Zimmermann H 2018 *Silicon Optoelectronic Integrated Circuits* (Switzerland: Springer Nature)

[7] Seidl C, Knorr J and Zimmermann H 2004 Single-stage 378 MHz 178 kΩ transimpedance amplifier with capacitive-coupled voltage dividers *Proc. IEEE Int. Solid-State Circuits Conf.* pp 470–1

[8] Seidl C, Knorr J and Zimmermann H 2003 Simple feedback network for bandwidth enhancement of transimpedance amplifiers *Electron. Lett.* **39** 1849–51

[9] Milovancev D, Brandl P, Jukic T, Steindl B, Vokic N and Zimmermann H 2019 Optical wireless APD receivers integrated in 0.35 μm HV CMOS technology with large detection area *Opt. Express* **27** 11930–45

[10] Milovancev D, Jukic T, Vokic N, Brandl P, Steindl B and Zimmermann H 2021 VLC using 800 μm diameter APD receiver integrated in standard 0.35 μm BiCMOS technology *IEEE Photonics J.* **13** 7900513

[11] Haigh F A 1967 *Handbook of the Poisson Distribution* (New York: Wiley)
[12] Hofbauer M, Schneider-Hornstein K and Zimmermann H 2021 *Single-Photon Detection for Data Communication and Quantum Systems* (Bristol, U.K.: IOP Publishing)
[13] Fisher E, Underwood I and Henderson R 2013 A reconfigurable single-photon counting integrating receiver for optical communications *IEEE J. Solid-State Circuits* **48** 1638–50
[14] Zimmermann H, Steindl B, Hofbauer M and Enne R 2017 Integrated fiber optical receiver reducing the gap to the quantum limit *Sci. Rep.* **7** 2652
[15] Steindl B, Hofbauer M, Schneider-Hornstein K, Brandl P and Zimmermann H 2018 Single-photon avalanche photodiode based fiber optical receiver up to 200 Mb s^{-1} *J. Sel. Top. Quantum Electron.* **24** 3801308
[16] Sklar B 2001 *Digital Communications: Fundamentals and Applications* (Englewood Cliffs, NJ: Prentice Hall)
[17] ITU-T 2004 G.975.1: Forward error correction for high bit-rate dwdm submarine systems *Telecommunication Standardization Sector*
[18] Goll B, Hofbauer M, Steindl B and Zimmermann H 2018 A fully integrated SPAD-based CMOS data-receiver with a sensitivity of −64 dBm at 20 Mb s^{-1} *IEEE Solid-State Circuits Lett.* **1** 2–5
[19] Kosman J, Almer O, Abbas T A, Dutton N, Walker R, Videv S, Moore K, Haas H and Henderson R 2019 A 500 Mb s^{-1} −46.1 dBm CMOS SPAD receiver for laser diode visible-light communication *IEEE Int. Solid-State Circuits Conf.* 468–9
[20] Ahmed Z, Singh R, Ali W, Faukner G, O'Brien D and Collins S 2020 A SiPM-based VLC receiver for Gigabit communication using OOK modulation *IEEE Photon. Technol. Lett.* **32** 317–20
[21] Mahmoudi H, Hofbauer M, Steindl B, Schneider-Hornstein K and Zimmermann H 2018 Modeling and analysis of BER performance in a SPAD-based integrated fiber optical receiver *IEEE Photonics J.* **10** 7908411
[22] Mahmoudi H, Hofbauer M, Steindl B, Schneider-Hornstein K and Zimmermann H 2019 Statistical study of intrinsic parasitics in an SPAD-based integrated fiber optical receiver *IEEE Trans. Electron Devices* **66** 497–504
[23] Kuo K-C 2019 A 10-bit 250 MS s^{-1} binary search and two channel SAR ADC by a two-bit per conversion with error tolerance ability *Proc. ISOCC* 1–2
[24] Fan Q and Chen J 2017 A 10-bit 400 MS s^{-1} asynchronous SAR ADC using dual-DAC architecture for speed enhancement *Proc. IEEE MWSCAS* 583–6
[25] Griffith A D, Herrnsdorf J, Almer O, Henderson R K, Strain M J and Dawson M D 2019 High-sensitivity free space optical communications using low size, weight and power hardware arXiv:1902.00495 https://arxiv.org/pdf/1902.00495
[26] Hynecek J 2001 Impactron–a new solid state image intensifier *IEEE Trans. Electron Devices* **48** 2238–41
[27] Hynecek J and Nishiwaki T 2003 Excess noise and other important characteristics of low light level imaging using charge multiplying CCDs *IEEE Trans. Electron Devices* **50** 239–45
[28] https://www.oxinst.com/search-results?search=EMCCD&businesses=andor accessed 12 Sept. 2023
[29] Mori M *et al* 2016 A 1280 × 720 single-photon-detecting image sensor with 100 dB dynamic range using a sensitivity-boosting technique *Proc. IEEE ISSCC* 120–1
[30] Hirose Y *et al* 2019 A 400 × 400-pixel 6 μm pitch vertical avalanche photodiode CMOS image sensor based on 150 ps-fast capacitive relaxation quenching in Geiger mode for synthesis of arbitrary gain images *Proc. IEEE ISSCC* 104–5

[31] Seo M-W, Kawahito S, Kagawa K and Yasutomi K 2015 A 0.27e-rms read noise 220 μV/e-conversion gain reset-gate-less CMOS image sensor with 0.11 μm CIS process *IEEE Electron Device Lett.* **36** 1344–7

[32] Seo M-W *et al* 2017 A 0.44e rms read-noise 32 fps 0.5 Mpixel high-sensitivity RG-less-pixel CMOS image sensor using bootstrapping reset *Proc. IEEE ISSCC* 80–1

[33] Ma J, Masoodian S, Starkey D A and Fossum E R 2017 Photon-number-resolving megapixel image sensor at room temperature without avalanche gain *Optica* **4** 1474–81

[34] Starkey D, Deng W, Ma J, Masoodian S and Fossum E R 2018 Quanta imaging sensors: achieving single-photon counting without avalanche gain *Micro- and Nanotechnology Sensors, Systems, and Applications* ed T George, A K Dutta and M S Islam (Bellingham, WA: International Society for Optics and Photonics, SPIE) vol 10639 pp 271–8

[35] Crooks J *et al* 2013 Kirana: a solid-state megapixel CMOS image sensor for ultra-high speed imaging *Proc. SPIE—IS&T Electronic Imaging* vol 8659 865903

[36] Jukic T, Steindl B, Enne R and Zimmermann H 2016 200 μm APD OEIC in 0.35 μm BiCMOS *Electron. Lett.* **52** 128–30

[37] Jukic T, Steindl B, Enne R and Zimmermann H 2016 400 μm diameter APD OEIC in 0.35 μm BiCMOS *IEEE Photon. Technol. Lett.* **28** 2004–7

[38] Razavi B 2000 A 622 Mb s^{-1} 4.5 pA/$\sqrt{\mathrm{Hz}}$ CMOS transimpedance amplifier *IEEE ISSCC* **9** TA 9.7

[39] Razavi B 2001 *Design of Analog CMOS Integrated Circuits* (Boston, MA: McGraw-Hill)

[40] Theuwissen A J P 2008 CMOS image sensors: state of the art *Solid-State Electron.* **52** 1401–8

[41] Ebeling K J 1993 *Integrated Optoelectronics* (Berlin, Heidelberg: Springer)

[42] Schneider K and Zimmermann H 2006 *Highly Sensitive Optical Receivers* (Berlin, Heidelberg: Springer)

IOP Publishing

Horst Zimmermann

Chapter 2

Basics of photodiodes

In this chapter the basics of photodiodes, optical absorption, photogeneration, quantum efficiency, responsivity, drift and diffusion carrier transport mechanisms, width of the space-charge region and capacitance are introduced and described. The semiconductor equations are presented. The basics of avalanche photodiodes (APDs), impact ionization and breakdown voltage are included.

2.1 Optical absorption and photogeneration

If the photon energy is larger than the bandgap energy of silicon (about 1.12 eV), the photon can be absorbed and an electron–hole pair can be generated in silicon. The optical power P decays in matter exponentially following Lambert–Beer's law[1]:

$$P(\lambda, y) = P_0 e^{-\alpha(\lambda)y}, \tag{2.1}$$

where, P_0 is the optical power at the silicon surface ($y = 0$), $\alpha(\lambda)$ is the optical absorption coefficient depending on the wavelength λ and y is the depth (light incidence perpenticular to the surface assumed). $\alpha(\lambda)$ is the most important optical constant for photodetectors. $\alpha(\lambda)$ determines the 1/e penetration depth of light inside semiconductors. Table 2.1 lists the optical absorption coefficient of silicon and the 1/e penetration depth for some important wavelengths [2, 3] and figure 2.1 depicts the corresponding decay of the optical power in silicon.

The photogeneration rate per volume G in dependence on the depth y in silicon can be obtained by

$$G(y) = \frac{P(y) - P(y + \triangle y)}{\triangle y} \frac{1}{Ah\nu}, \tag{2.2}$$

[1] This dependence holds for light incidence from an infinitely thick medium into the semiconductor. For light incidence through a thin cover layer into the silicon, the optical power deviates from the exponential decay as shown in [1].

doi:10.1088/978-0-7503-5437-0ch2

Table 2.1. Absorption coefficients α of silicon and 1/e penetration depth for several important wavelengths.

Wavelength (nm)	α (μm^{-1})	1/e penetration depth (μm)
850	0.06	16.67
780	0.12	8.33
680	0.24	4.16
635	0.38	2.63
565	0.73	1.37
465	3.6	0.278

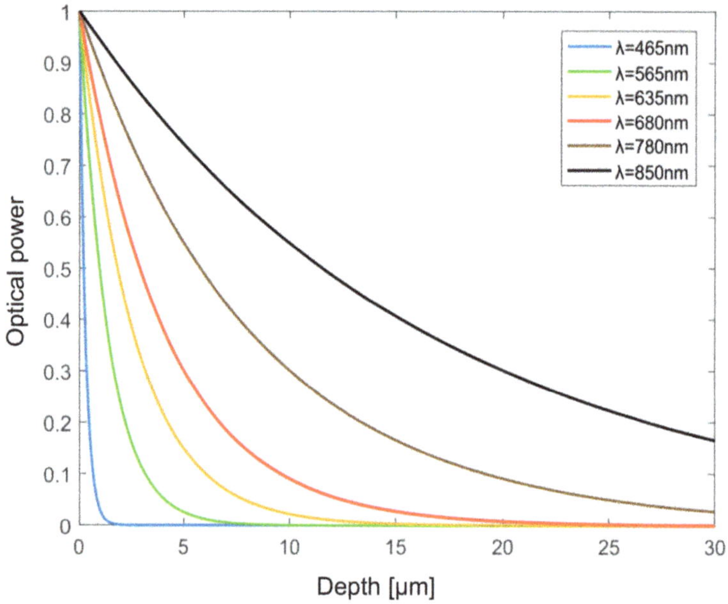

Figure 2.1. Decay of optical power in silicon with depth (normalized to the optical power at the silicon surface). Reproduced with permission from IOPP from [12], Copyright (2021).

where A is the cross section area for light incidence and $h\nu$ is the photon energy $E_{\text{ph}} = \frac{hc}{\lambda}$ of the photon (h = Planck's constant, c = light velocity[2]. For $\triangle y \to 0$, we obtain $G(y) = -\mathrm{d}P(y)/\mathrm{d}y\frac{1}{Ah\nu}$. After differentiating equation (2.1), the photogeneration results:

$$G(y) = \frac{\alpha P_0}{Ah\nu}e^{(-\alpha y)}. \tag{2.3}$$

[2] Be aware that c and λ depend on the medium, where the light is travelling. Only the photon frequency ν is constant.

The photogeneration has to be considered in the semiconductor equations [4, 5] to describe the photocurrent in photodetectors. The interested reader can also refer to [6] for studying these equations and modelling of optoelectronic semiconductor devices. The Poisson equation (2.4), the transport equations (2.6), (2.7) and the continuity equations (2.9), (2.10) belong to the semiconductor equations:

$$\triangle\Psi = -\frac{\rho}{\varepsilon}. \tag{2.4}$$

where Ψ is the potential, ρ represents the charge density and ε is the product of the relative and absolute dielectric constants: $\varepsilon = \varepsilon_r\varepsilon_0$. The charge density is proportional to the elementary charge (electron charge) q and the sum of the hole concentration p (positively charged), of the electron concentration n (negatively charged), of the donor concentration N_D (positively charged), and of the acceptor concentration N_A (negatively charged):

$$\rho = q(p - n + N_D - N_A). \tag{2.5}$$

The transport equations, which define the current densities for electrons and holes, are the sum of the drift and diffusion current densities:

$$\vec{j}_n = qn\mu_n\vec{E} + qD_n\,\mathbf{grad}\,n, \tag{2.6}$$

$$\vec{j}_p = qp\mu_p\vec{E} - qD_p\,\mathbf{grad}\,p. \tag{2.7}$$

The total current density is the sum of the electron and hole current densities:

$$\vec{j} = \vec{j}_n + \vec{j}_p. \tag{2.8}$$

The continuity equations with the inclusion of thermal generation/recombination U_{th} and of photogeneration G[3] due to the penetration of light into the semiconductor can be written as:

$$\frac{\partial n}{\partial t} = \frac{\mathrm{div}\,\vec{j}_n}{q} + U_{th} + G, \tag{2.9}$$

$$\frac{\partial p}{\partial t} = -\frac{\mathrm{div}\,\vec{j}_p}{q} + U_{th} + G. \tag{2.10}$$

There are device simulators, which solve the semiconductor equations, like e.g. ATLAS [7], MEDICI [8] or SENTAURUS [9]. Photogeneration is also implemented in these tools.

[3] G depends on the location in the semiconductor, especially on y as depth in the semiconductor for perpendicular light incidence.

2.2 Drift and diffusion

There are two terms in the transport equations. The first term contains the electric field strength and the second term is determined by carrier diffusion. The electric field follows from (2.4) and obeys:

$$\vec{E} = -\mathbf{grad}\,\Psi. \tag{2.11}$$

The electric field is important for the calculation of the drift velocity $\vec{v}$ of photogenerated carriers in photodiodes:

$$\vec{v} = \mu\vec{E}. \tag{2.12}$$

The mobilities of electrons and holes differ strongly and μ_n or μ_p has to be used for μ in order to calculate the electron drift velocity or the hole drift velocity, respectively. The mobilities μ_n and μ_p are constant only for low electric field strengths. The drift velocities of electrons and holes in silicon saturate for large values of the electric field at 10^7 cm s^{-1}, both. The carrier mobilities also depend on the impurity concentration, i.e. especially on the doping concentration. The carrier mobilities decrease with increasing doping concentration. These are important aspects for photodetector development (see e.g [6]). In device simulators, different models for the carrier mobilities, for generation/recombination and other quantities can be chosen [5, 7–9].

The diffusion of minority carriers is also important for the response speed of photodetectors. Carrier diffusion occurs in semiconductor regions without electric field. The carrier diffusion coefficients D_n and D_p for electrons and holes are determined by the carrier mobilities via the Einstein relation:

$$D_{n/p} = \mu_{n/p}\frac{k_B T}{q} = \mu_{n/p} U_T. \tag{2.13}$$

Therefore, carrier diffusion is usually much slower than carrier drift, because the thermal voltage $U_T = \frac{k_B T}{q}$ (k_B = Boltzmann constant and T = absolute temperature) has a low value (26 mV) around room temperature and the electric field strength at P/N junctions and in the depleted intrinsic region of PIN photodiodes is several thousand V cm^{-1} [6]. Therefore, if light photo-generates carriers only in depleted drift regions, photodiodes show a high bandwidth and short rise- and fall-times of the photocurrent. If carriers are also photogenerated below the depleted space-charge region in the substrate and in highly doped regions, the photocurrent shows a slow diffusion tail [6]. Figure 2.2 explains the difference.

Another important aspect is recombination of photogenerated carriers in regions without electric field. Because minority carriers diffuse the slower, the higher the doping level is, in $N+$ and $P+$ regions as well as in highly doped substrates (e.g. in the $P+$ part of $P-/P+$ epitaxial wafers) many photogenerated carriers recombine before they can reach the drift region and therefore do not contribute to the photocurrent of P/N-junction photodiodes, of PIN photodiodes or of APDs.

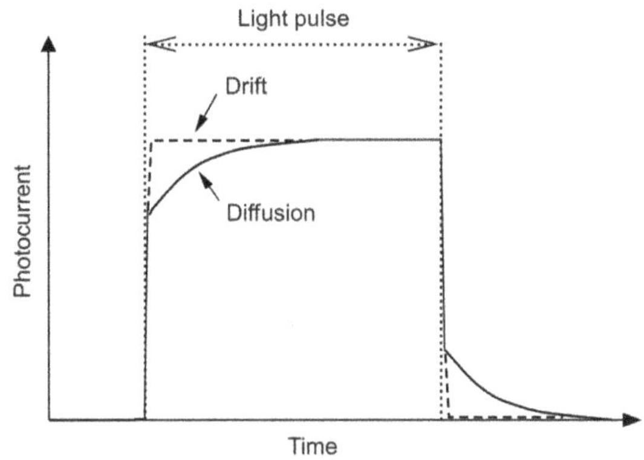

Figure 2.2. Transient response of the photocurrent for carrier drift and diffusion.

2.3 Width of space-charge region and capacitance

For the time response of photodetectors and single-photon avalanche diodes it is not only carrier drift and diffusion that are important. The width of the space-charge region together with the electric field strength in this region determines the drift time and the rise/fall time of the photocurrent of P/N junction and PIN photodiodes. In SPADs, the Geiger mode event can be delayed after photon absorption by drift through the absorption region. Because photon absorption can occur in different depths inside the absorption region, a jitter arises in SPADs. In PIN photodiodes and APDs operated in the Geiger mode, such a jitter may only be visible when only a few photons are absorbed.

The width of the space-charge region W (and the dependence of the electric field on the location inside the space-charge region) can be obtained from the Poisson equation in the so-called depletion approximation [6]. For an abrupt P/N-junction, an analytical solution is obtained:

$$W = \sqrt{\frac{2\varepsilon_r\varepsilon_0}{q}\frac{N_A + N_D}{N_A N_D}(U_D - U)} \tag{2.14}$$

with

$$U_D = \frac{k_B T}{q}\ln\frac{N_A N_D}{n_i^2}, \tag{2.15}$$

where U_D is the built-in voltage of the P/N-junction and n_i is the intrinsic carrier concentration. Because photodiodes work in reverse direction, a negative value has to be inserted for U in equation (2.14).

In PIN photodiodes the absorption zone (I zone) must be so low doped that the I layer is completely depleted at the operating voltage ($W > d_I$, d_I = thickness of I zone). Then, only carrier drift determines the transit time of electrons and holes through the I zone (if the incident light is completely absorbed within the I zone).

It should be mentioned that in a PIN diode, the width of the space-charge region can only be calculated with N_I being the low doping concentration in the I zone:

$$W = \sqrt{\frac{2\varepsilon_r\varepsilon_0}{q}\frac{1}{N_I}(U_D - U)}. \tag{2.16}$$

The total quantum efficiency η (number of electron–hole pairs contributing to the photocurrent divided by the number of photons (total number of photons) hitting the photodiode) is the product of optical quantum efficiency η_o (number of photons transmitting the isolation and passivation layers of the photodiode divided by the total number of photons) and the internal quantum efficiency η_i (number of electron–hole pairs generated and causing the photocurrent divided by the number of transmitted photons):

$$\eta = \eta_o\eta_i. \tag{2.17}$$

The internal quantum efficiency η_i of a PIN photodiode is [10]

$$\eta_i = 1 - \exp(-\alpha d_I). \tag{2.18}$$

In this equation it is assumed that electron–hole pairs photogenerated below the I zone do not contribute to the photocurrent (due to recombination in the highly P^+ doped substrate, considering the PIN photodiodes used in USPAR). So, η_i represents the so-called dynamic quantum efficiency[4].

Photodiodes are often characterized by a responsivity R_o, which can be determined quite easily by measuring the photocurrent and dividing it by the measured incident light power[5]. Its unit is A W^{-1}. Quantum efficiency η and R are strictly related:

$$R_o = \frac{I_{ph}}{P_{opt}} = \frac{q\lambda}{hc}\eta. \tag{2.19}$$

But the space-charge region width is not only important for the transient response and bandwidth as well as the quantum efficiency of a PIN photodiode. It also determines the capacitance of the photodiode, which has to be considered together with input resistance of an amplifier circuit for calculation of rise/fall times and bandwidth of optical sensors or receivers. For SPADs, the capacitance determines the avalanche charge and how fast a quenching or gating circuit can discharge or quench the SPAD. The capacitance of an abrupt P/N-junction, of which one side of the P/N-junction is several orders of magnitude lower doped than the other, is (the lower value of N_A or N_D has to be inserted for $N_{A/D}$):

[4] The static quantum efficiency considers slow carriers diffusing into the drift zone and contributing to the photocurrent.

[5] The responsivity determined with optical power meters usually is a static quantity. CW light, i.e. unmodulated light, has to be used.

$$C_{\mathrm{D}} = A\sqrt{\frac{q\varepsilon_{\mathrm{r}}\varepsilon_0 N_{\mathrm{A/D}}}{2}}\frac{1}{\sqrt{U_{\mathrm{D}} - U}}, \tag{2.20}$$

Real P/N-junctions are not abrupt and an analytical calculation of W and C is not possible. Fortunately, there are device simulators to solve the semiconductor equations, such as for instance ATLAS [7], MEDICI [8] or SENTAURUS [9].

The capacitance of a PIN photodiode, whose I zone is fully depleted is simply that of a plate capacitor with the distance d_{I} between its plates and a dielectric with $\varepsilon_r = 11.9$ for a Si PIN photodiode:

$$C_{\mathrm{PIN}} = \frac{A\varepsilon_{\mathrm{r}}\varepsilon_0}{d_{\mathrm{I}}}. \tag{2.21}$$

A low capacitance of the photodiode allows a high bandwidth of an optical receiver with a transimpedance amplifier or of an optical sensor. The sensitivity of an optical receiver can be improved, i.e. the equivalent input noise current decreases with lower photodiode capacitance (see section 3.7 and chapter 7).

2.4 Impact ionisation

In high-electric field regions within semiconductors, electrons and holes are accelerated strongly and gain high energies, which are sufficient to generate electron–hole pairs by impact ionization. These electron–hole pairs are separated and also accelerated by the high electric field. They can also generate electron–hole pairs and so on. An avalanche is caused and the current grows.

The generation rate G of electron–hole pairs caused by impact ionization is

$$G = \alpha_{\mathrm{n}} n v_{\mathrm{n}} + \alpha_{\mathrm{p}} p v_{\mathrm{p}}, \tag{2.22}$$

where α_{n} and α_{p} are the impact ionization coefficients of electrons and holes. n and p are the electron and hole concentrations. v_{n} and v_{p} are the velocities of electrons and holes, respectively. The ionization coefficients have the unit 1 cm and $1/\alpha_{\mathrm{n,p}}$ can be interpreted as the mean travelled distance between two impact ionization events. G has the unit cm^{-3} s^{-1}. The carrier velocities depend on the electric field [6]. But here the dependence of the ionization coefficients on the electric field is even more important [4]:

$$\alpha_{\mathrm{n,p}}(E) = \frac{qE}{E_{\mathrm{I,n,p}}} e^{(-\Xi_{\mathrm{I,n,p}}/E)}, \tag{2.23}$$

where E is the electric field strength (in V cm^{-1}) and $E_{\mathrm{I,n,p}}$ denotes the high-field, effective ionization threshold energy for electrons, respectively, holes [4]. For electrons in Si, $E_{\mathrm{I,n}}$ is 3.6eV and for holes in Si $E_{\mathrm{I,p}}$ is 5.0 eV due to ionization scattering events. Ξ_{I} is the threshold electric field for carriers to overcome the decelerating effect of ionization scattering [4]. This critical field strength is about 2×10^5V cm^{-1} for silicon.

When we assume that a hole current I_{p0} flows from the left side into the high-field region (which starts at $x = 0$) with the width W, the hole current I_p increases with x and at $x = W$ the hole current M_pI_{p0} leaves the high-field region [4]. The electron current flows into the other direction and increases from W to $x = 0$. At steady state, the sum I of I_n and I_p (the total current) is constant.

Within an interval dx, the hole current changes (increases) proportional to the hole current I_p, proportional to the hole ionization coefficient α_p and proportional to the width of the interval dx. Since high-energy electrons can also produce holes (each carrier generates an electron–hole pair), we have to add a corresponding term for the electrons [4]:

$$dI_p = I_p\alpha_p dx + I_n\alpha_n dx, \tag{2.24}$$

which can be rearranged using $I_n = I - I_p$:

$$\frac{dI_p}{dx} = I_p(\alpha_p - \alpha_n) + \alpha_n I. \tag{2.25}$$

The solution of this equation considering the boundary condition $I = I_p(W) = M_pI_{p0}$ is [4]:

$$I_p(dx) = \frac{I\left[\frac{1}{M_p} + \int_0^x \alpha_n e^{\left[-\int_0^x (\alpha_p-\alpha_n)dx'\right]}dx\right]}{e^{\left[-\int_0^x (\alpha_p-\alpha_n)dx'\right]}}. \tag{2.26}$$

with the hole multiplication factor

$$M_p = \frac{I_p(W)}{I_p(0)}. \tag{2.27}$$

To obtain an expression which has to be fulfilled for avalanche breakdown, equation (2.26) can be written as

$$1 - \frac{1}{M_p} = \int_0^W \alpha_p e^{\left[-\int_0^x (\alpha_p-\alpha_n)dx'\right]}dx. \tag{2.28}$$

2.5 Breakdown voltage

At the avalanche breakdown voltage the hole multiplication factor approaches infinity. The condition for avalanche breakdown therefore is [4]

$$\int_0^W \alpha_p e^{\left[-\int_0^x (\alpha_p-\alpha_n)dx'\right]}dx = 1. \tag{2.29}$$

when holes are injected into the multiplication zone. When electrons initiate the avalanche process, the ionization integral is:

$$\int_0^W \alpha_n e^{\left[-\int_x^W (\alpha_n-\alpha_p)dx'\right]} dx = 1. \tag{2.30}$$

These equations (2.29) and (2.30) are equivalent and valid also for injection of both, electrons and holes. Either equations (2.29) or (2.30) represent the breakdown condition. So, the breakdown voltage determines the electric field and the electric field determines the ionization coefficients in such a way that these breakdown conditions are fulfilled.

These ionization integrals simplify a lot for $\alpha_n = \alpha_p$. However, semiconductor physics and material properties of Si, GaAs, InGaAs and other semiconductor materials do not obey this equality.

To give a complete picture it has to be mentioned that the impact ionisation mechanism occurs for Si junction breakdown voltages above $6E_g/q$ (E_g is the bandgap energy and q the electron charge, E_g of Si is about 1.1 eV at room temperature, $6E_g/q$ is then about 6.6 V). For breakdown voltages below $4E_g/q$ the tunneling effect (band-to-band tunneling) causes the reverse current. Between $4E_g/q$ and $6E_g/q$, the breakdown is due to a mixture of impact ionisation and tunneling [4]. The breakdown voltage depends on the doping at the P/N-junction. For highly doped junctions the tunneling mechanism dominates and for lower doped junctions the avalanche mechanism prevails. The temperature coefficient of the breakdown voltage is negative for the tunneling mechanism and positive for the avalanche effect. This knowledge can be used to conclude which breakdown mechanism contributes for APDs and SPADs.

In USPAR we have to use APDs fabricated in PIN photodiode CMOS technology to obtain a thick absorption region and a low capacitance. These APDs are so-called reach-through APDs [11] and have breakdown voltages above 24 V (see section 5.3). For silicon, the impact ionization coefficients of electrons α and holes β are strongly different, $\alpha \gg \beta$ or $k = \beta/\alpha \ll 1$. Therefore, for APDs the P-type substrate, i.e. the P^-/P^+ epitaxial structure with the thick P^- absorption layer is actually good to achieve a high responsivity (see section 5.3). The APDs in USPAR will be operated below the breakdown voltage in the so-called linear mode. In fact, to keep the excess noise (see section 3.10) in acceptable limits, the APDs will be operated at rather low gains M.

References

[1] Mahmoudi H, Poushi S S K, Steindl B, Hofbauer M and Zimmermann H 2021 Optical and electrical characterization and modeling of photon detection probability in CMOS single-photon avalanche diodes *IEEE Sens. J.* **21** 7572–80

[2] Palik E D 1985 *Handbook of Optical Constants of Solids* (Orlando, FL: Academic Press, Inc.)

[3] Aspnes D E and Studna A A 1983 *Phys. Rev.* **B 27** 985–1009

[4] Sze S M 1981 *Physics of Semiconductor Devices* (New York: Wiley)

[5] Selberherr S 1984 *Analysis and Simulation of Semiconductor Devices* (Wien: Springer)

[6] Zimmermann H 2010 *Integrated Silicon Optoelectronics* (Berlin, Heidelberg: Springer)

[7] Silvaco 2023 Atlas user's manual available: https://www.silvaco.com accessed: 23 March 2023
[8] Synopsys 2023 Taurus-medici http://www.synopsys.com/silicon/tcad/device-simulation/taurus-medici.html accessed: 23 March 2023
[9] Synopsys 2023 Sentaurus http://www.synopsys.com/silicon/tcad/device-simulation/sentaurus-device.html accessed: 23 March 2023
[10] Ebeling K J 1993 *Integrated Optoelectronics* (Berlin, Heidelberg: Springer)
[11] Steindl B, Gaberl W, Enne R, Schidl S, Schneider-Hornstein K and Zimmermann H 2014 Linear mode avalanche photodiode with 1 GHz bandwidth fabricated in 0.35 μm CMOS *IEEE Photon. Technol. Lett.* **26** 1511–4
[12] Hofbauer M, Schneider-Hornstein K and Zimmerman H 2021 *Single-Photon Detection for Data Communication and Quantum Systems* (Bristol: IOP Publishing)

Chapter 3

Basics of noise and sensitivity

The basics of electronic noise, the receiver noise of resistive-feedback (RF) transimpedance amplifiers (TIAs), the noise of integrators as well as the excess noise of avalanche photodiodes (APDs) and photon noise are collected and described. The term sensitivity is introduced. Reset noise also is covered since it is important in some integrating receivers. The advantages of integrating receivers over resistive-feedback TIAs with respect to noise and sensitivity are described and presented as motivation for ultra-sensitive PIN and APD receivers.

3.1 Thermal noise

All atoms and molecules inside matter, which are at the absolute temperature vibrate similarly to the Brownian movement of gas molecules. These vibrations manifest themselves in the temperature, respectively, in the thermal energy of matter. Boltzmann calculated with the constant (now called Boltzmann's constant) $k_{\mathrm{B}} = 1.38 \times 10^{-23}$ J K^{-1} (1 J = 1 Ws = 1 VAs) the thermal energy equivalent. In a resistor being at the temperature T, the thermal vibration energy of the atoms is transferred to the electrons. Thereby, the 'warm' resistor forms a generator for a fluctuating electrical signal, the thermal resistor noise. With a sensitive voltmeter, an alternating voltage across the resistor can be measured. Due to this alternating voltage, the average value of a long enough time period is zero. Therefore, noise has to be characterized by a power (respectively, by squared voltage or squared current) and not by a simple voltage or current. The time-averaged squared open-circuit noise voltage of a resistor R in a frequency bandwidth Δf is

$$\overline{v_{\mathrm{n}}^{2}} = 4k_{\mathrm{B}}TR\Delta f \tag{3.1}$$

with the unit V^2 [1]. The time-averaged squared short-circuit noise current of a resistor R in the frequency bandwidth Δf is

doi:10.1088/978-0-7503-5437-0ch3

$$\overline{i_n^2} = \frac{4k_B T}{R}\Delta f \tag{3.2}$$

with the unit A^2. From these expressions we obtain the effective (root mean square, rms) open-circuit noise voltage

$$v_{n,rms} = \sqrt{\overline{v_n^2}} = 2\sqrt{k_B T R \Delta f} \tag{3.3}$$

and the effective (rms) short-circuit noise current

$$i_{n,rms} = \sqrt{\overline{i_n^2}} = 2\sqrt{\frac{k_B T}{R}\Delta f}. \tag{3.4}$$

The frequency spectrum of this thermal noise is white. Its spectral density is constant (independent of frequency). The thermal noise occurs in passive, linear resistors. Current flow is not necessary. The amplitude distribution of thermal noise is Gaussian.

Figure 3.1 shows the equivalent circuits of a resistor including noise generators. The voltage noise generator is connected in series to the resistor and the current noise generator lies in parallel to the resistor. It depends on the circuit, in which the resistor is used, whether the noise voltage or the noise current generator is applied.

At room temperature, we have $4k_B T = 1.66 \times 10^{-20}$ VAs. The spectral noise voltage density of a 1 kΩ resistor at 300 K is $v_n^2/\Delta f = 16 \times 10^{-18}$ V^2 Hz^{-1} leading to the rms noise voltage density $v_{n,rms} = 4$ nV $Hz^{-1/2}$. So, in a bandwidth of 1 MHz, the rms noise voltage of the 1 kΩ resistor at 300 K is 4 μV. In the same bandwidth, the rms noise current of the 1 kΩ resistor at 300 K is 4 nA.

3.2 Shot noise

The elementary charge q is finite. Therefore, the structure of charge is granular, which leads to the name *shot noise*. Shot noise occurs for current flow over P/N-junctions and in turn it is present in diodes and (bipolar) transistors. The electrons and holes need enough energy to overcome the potential barrier at the P/N-junction of the diode in forward direction, making the crossing of the potential barrier a random event for the electrons and holes. The external current I, which seems to be stationary when the current is large enough, is composed of many random and independent current pulses (single elementary charges). The momentary current deviates statistically from the (measured) mean value I_D. This fluctuation in I is called shot noise. The mean square noise current value is [1]

$$\overline{i_n^2} = \overline{(I - I_D)^2} = \lim_{T\to\infty} \frac{1}{T}\int_0^T (I - I_D)^2 dt. \tag{3.5}$$

For independent current pulses it was shown that

$$\overline{i_n^2} = 2qI_D\Delta f. \tag{3.6}$$

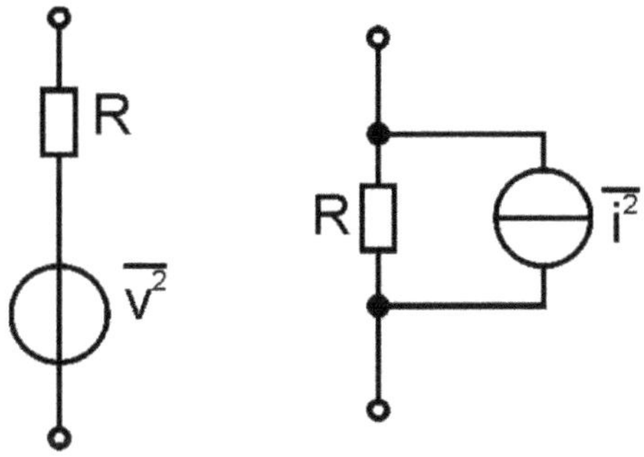

Figure 3.1. Equivalent circuits of a resistor including noise generators.

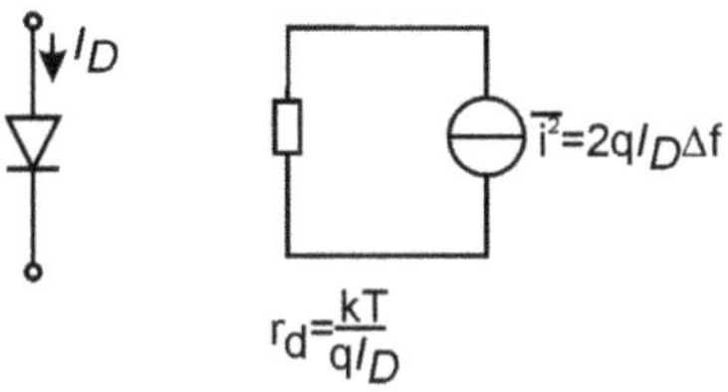

Figure 3.2. Equivalent circuit of a diode including noise generator.

The spectral noise current density $i_n^2/\Delta f$ is constant with respect to frequency and shot noise is white noise like thermal noise. For a current I_D of 1 mA through the diode and a bandwidth of 1 MHz, the rms shot noise current is 18 nA. The amplitude distribution of the shot noise current is Gaussian-like for thermal noise. Shot noise is present in forward and reverse direction of diodes and P/N-junctions.

Figure 3.2 shows the equivalent circuit of a diode including shot noise generator. The current noise generator is connected in parallel to the differential resistor r_d of the diode. This resistor r_d is not a physical resistor, which means that it does not generate thermal noise. r_d is a resistor just for modelling the resistance of a diode, which will be needed to calculate bandwidth and amplification factors of circuits, which contain a diode. Real diodes may have a series resistance, which generates thermal noise. The series resistance is not included in figure 3.2.

3.3 $1/f$ noise

Due to fluctuating charge reversal of defects and surface or interface states in semiconductor devices another kind of noise exists, which is not of white nature and increases at low frequencies. Its increase towards $f = 0$ is approximately proportional to $1/f$. In bipolar transistors $1/f$ noise originates from impurities deep inside the bandgap and crystal defects in the emitter-base depletion zone. In junction field-effect transistors (JFETs), deep impurity states in the channel and at the boundary of the space-charge region are responsible for $1/f$ noise. In MOSFETs, Si/SiO_2-interface states in the channel below the gate cause $1/f$ noise. In general, $1/f$ noise is stronger in MOSFETs than in JFETs and in bipolar transistors. $1/f$ noise is always linked to current flow. The $1/f$ noise dominates at low frequencies over thermal noise and shot noise. Since, $1/f$ noise can be described by [1]

$$\overline{i_n^2} = K_1 \frac{I^a}{f^b} \Delta f, \tag{3.7}$$

it depends on the current I flowing in the device and can be observed in MOSFETs sometimes up into the MHz range. K_1 is a constant depending on the manufacturing process, a is in the range 0.5 to 2 and b is a constant close to 1 (leading to the name *1/f noise*).

3.4 Noise in field-effect transistors

The channel of field-effect transistors (FETs) acts as a resistor and therefore generates thermal noise. The gate leakage current causes shot noise, which is often negligible due to the low gate leakage current (nA in JFETs and pA in MOSFETs). The thermal noise of the channel resistance is the dominating noise source of FETs. The channel resistor's noise is considered in the drain noise current generator $\overline{i_d^2}$ and the gate shot noise current is represented by the noise generator $\overline{i_g^2}$ (see figure 3.3).

The $1/f$ noise can be incorporated in the drain noise current generator $\overline{i_{d,n}^2}$ [1]:

$$\overline{i_{n,D}^2} = 4k_B T\left(\frac{2}{3}g_m\right)\Delta f + K_1 \frac{I^a}{f^b}\Delta f. \tag{3.8}$$

The transconductance g_m of FETs is used instead of 1/(channel resistance) to consider the resistance, which causes the thermal channel noise. The empirical factor 2/3 holds in the saturation region, which is the constant current region, of FETs. In the ohmic region, instead the factor 1 has to be used.

3.5 Noise of PIN-FET receiver

A simple optical receiver can be composed of a PIN photodiode, its load resistor R_L and a FET as amplifier (see figure 3.4). There, a JFET, which is a depletion transistor, is shown [2]. Although the JFET cannot simply be replaced by an enhancement-MOSFET as being implemented in CMOS technology because it will not work with zero gate–source voltage, the noise behaviour will be valid for JFETs and MOSFETs.

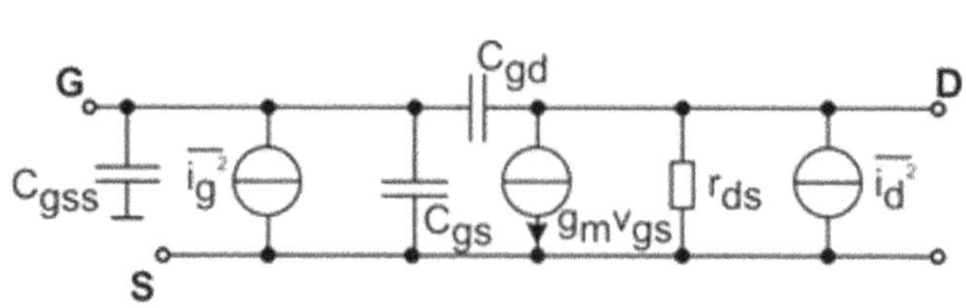

Figure 3.3. Small-signal equivalent circuit of a FET including noise generators.

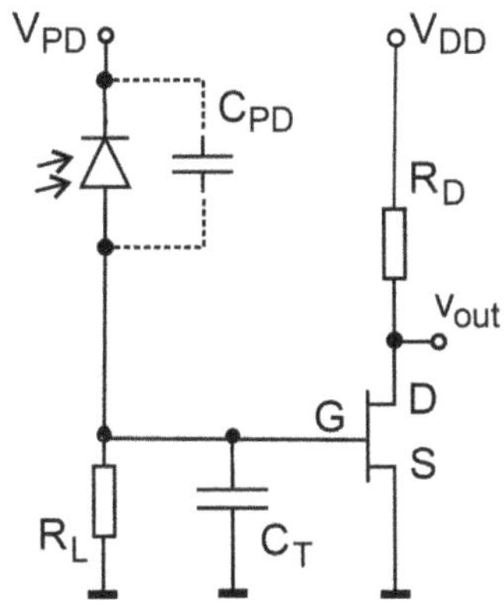

Figure 3.4. PIN-FET receiver.

Above the thermal noise of resistors, the shot noise of the photodiodes and the noise in field-effect transistors was described. For the optical receiver, however, we want to compare the total equivalent input current noise of the receiver to the signal photocurrent. So, we have to transform the drain current noise of the FET to the input. For this purpose we can use the definition of the transconductance g_m of the FET:

$$g_m = \frac{\delta I_\mathrm{D}}{\delta U_\mathrm{GS}}\Big|_{\mathrm{o.p.}}, \tag{3.9}$$

where at the operating point (o.p.) U_DS is constant. We take the drain current noise for δI_D and the transconductance of the FET as known and solve for δU_GS, which we can regard as the input noise noise voltage, which causes the drain current noise:

$$v_\mathrm{n,in,FET} = \frac{i_\mathrm{n,D}}{g_m}. \tag{3.10}$$

after squaring and inserting the thermal noise from equation (3.8), it is obtained (neglecting the $1/f$-noise):

$$v^2_\mathrm{n,\,in,\,FET} = \frac{4k_\mathrm{B}T\left(\frac{2}{3}g_m\right)\Delta f}{g_m^2} = \frac{8k_\mathrm{B}T}{3g_m}\Delta f. \tag{3.11}$$

This noise voltage leads to a noise current through the parallel connection of R_L and C_T.

$$i^2_\mathrm{n,in,FET} = v^2_\mathrm{n,in,FET}\left(\frac{1}{R_\mathrm{L}^2} + (2\pi f C_\mathrm{T})^2\right). \tag{3.12}$$

The thermal noise current of R_L has to be considered in addition and the total input noise current square is

$$i_{\text{n, in}}^2 = \left(\frac{1}{R_\text{L}^2} + (2\pi f C_\text{T})^2\right)\left(\frac{8k_\text{B}T}{3g_m}\Delta f\right) + \frac{4k_\text{B}T}{R_\text{L}}\Delta f. \tag{3.13}$$

The receiver usually is designed with large $g_m R_\text{L}$ and the expression simplifies to

$$i_{\text{n, in}}^2 = \left(4k_\text{B}T\left(\frac{1}{R_\text{L}} + \frac{8}{3}\pi^2 f^2 \frac{C_\text{T}^2}{g_m}\right)\right)\Delta f. \tag{3.14}$$

According to this equation, the FET noise dominates at high frequencies. A small input-node capacitance C_T and a large transconductance of the FET (input transistor of the amplifier) reduce the equivalent input noise current. Equation (3.14) can be used as a good approximation for the resistive-feedback TIA (see section 4.1) for $R_\text{F} = R_\text{L}$ [3]. Section 3.7 contains also a corresponding derivation.

3.6 Sensitivity

The simplest signals in optical data transmission are binary signals with two photocurrents $\langle i_0 \rangle$ and $\langle i_1 \rangle$ (figure 3.5). $\langle i_0 \rangle$ represents the mean value of the logical-zero photocurrent and $\langle i_1 \rangle$ is the mean value of the photocurrent for the logical one. We assume that $\langle i_1 \rangle$ is larger than $\langle i_0 \rangle$. The photocurrents have a certain distribution around the mean values due to the input current noise of the receiver introduced above [2]. The time slot T is the duration of a bit and defines the bit rate B. The bandwidth of the receiver is assumed to be $B/2$ [2]. The derivation shown in the following, however, can easily be done also for bandwidths of $2B/3$ or $3B/4$ as often used in practice.

The variance of the total input noise current is obtained by integration of the spectral noise density $i_{\text{n, in}}^2$ and considering also shot noise due to the leakage current i_r of the photodiode [2]:

$$i_{\text{tot}}^2 = \int_0^{B/2} i_{\text{n, in}}^2 df = \frac{2k_\text{B}TB}{R_\text{L}} + qi_\text{r}B + \frac{4k_\text{B}T\pi^2 C_\text{T}^2 B^3}{9g_m}. \tag{3.15}$$

The instant current values are assumed to have a Gaussian distribution with the mean values $\langle i_0 \rangle$ and $\langle i_1 \rangle$. The probability density for the logical zero, then is

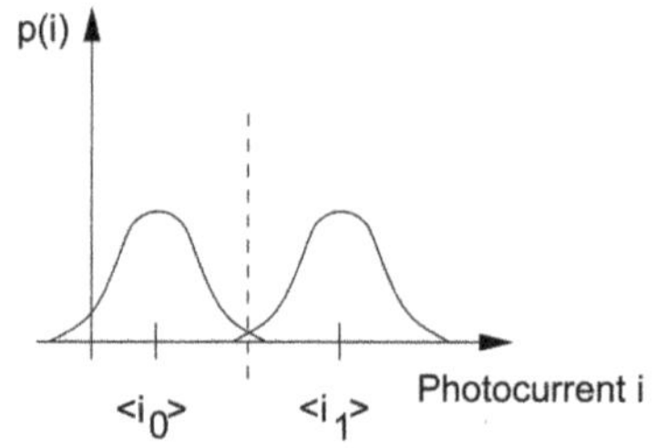

Figure 3.5. Distribution $p(i)$ of the instant i values around the mean logic levels $\langle i_0 \rangle$ and $\langle i_1 \rangle$.

$$p_0(i) = \frac{1}{\sqrt{2\pi\langle\delta i_{\text{tot}}^2\rangle}} \exp\left(-\frac{(i - \langle i_0\rangle)^2}{2\langle\delta i_{\text{tot}}^2\rangle}\right). \tag{3.16}$$

For the logical one we have correspondingly

$$p_1(i) = \frac{1}{\sqrt{2\pi\langle\delta i_{\text{tot}}^2\rangle}} \exp\left(-\frac{(i - \langle i_1\rangle)^2}{2\langle\delta i_{\text{tot}}^2\rangle}\right). \tag{3.17}$$

For equally frequent logical zeros and ones, the mean value of the photocurrent over a long time period is:

$$D_t = (\langle i_1\rangle + \langle i_0\rangle)/2. \tag{3.18}$$

This value can be used as decision threshold. For a current $i \leqslant D_t$, the detected bit is considered as logical zero. The probability of a wrong decision is given by the bit error ratio (BER) for the logical zero

$$\text{BER}_0 = \int_{D_t}^{\infty} p_0(i)di = \frac{1}{\sqrt{2\pi}} \int_Q^{\infty} \exp(-u^2/2)du, \tag{3.19}$$

with $u = (i - \langle i_0\rangle)/\sqrt{\langle\delta i_{\text{tot}}^2\rangle}$ and where the noise distance

$$Q = \frac{\langle i_1\rangle - \langle i_0\rangle}{2\sqrt{\langle\delta i_{\text{tot}}^2\rangle}} \tag{3.20}$$

has been adopted. For the assumptions made, it can be shown that the BER for the logical one is equal to that for the logical zero

$$\text{BER}_1 = \int_{-\infty}^{D_t} p_1(i)di = \text{BER}_0 = \text{BER}. \tag{3.21}$$

Therefore the BER in general is

$$\text{BER} = \frac{1}{\sqrt{2\pi}} \int_Q^{\infty} \exp(-u^2/2) \approx \frac{1}{\sqrt{2\pi}} \exp\left(-\frac{Q^2}{2}\right)\left(1 - \frac{1}{Q^2}\right). \tag{3.22}$$

The error of the approximation is less than 1% for $Q \geqslant 2$. The relation between BER and the noise distance Q is shown in figure 3.6. As examples, BER = 10^{-9} holds at $Q = 6$ and BER = 1.3×10^{12} for $Q = 7$. When error correction [4] is used, the BER can reach up to 2×10^{-3} and $Q = 3$ is sufficient.

The decision threshold D_t is in the middle between the logical '1' and '0' mean photocurrents. For bit sequences with equally frequent logical '1's and '0's it, therefore, can be expressed by the average optical power $\eta\langle P\rangle$, which is converted into photocurrent:

$$D_t = \frac{\langle i_1\rangle + \langle i_0\rangle}{2} = \eta\frac{q}{\hbar\omega}\langle P\rangle. \tag{3.23}$$

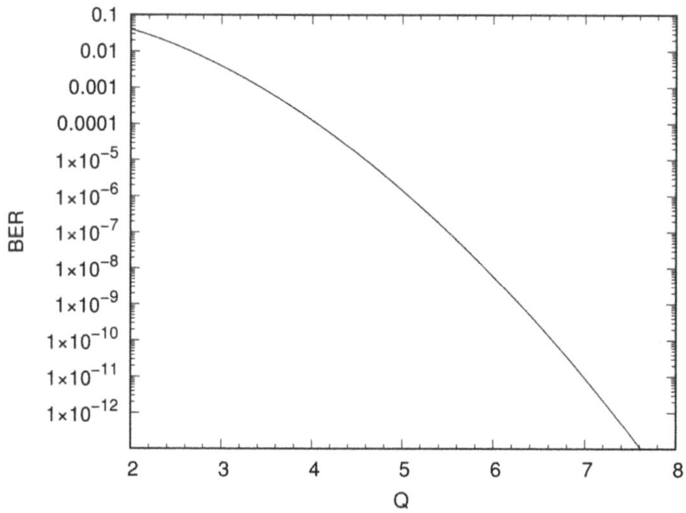

Figure 3.6. BER in dependence on the noise distance Q.

The ratio r of the two photocurrent levels was defined in [2]:

$$r = \langle i_0 \rangle / \langle i_1 \rangle. \tag{3.24}$$

With this definition, the optical power of a binary signal, which is necessary to achieve a certain required BER (BER is represented by the corresponding Q-factor) results in:

$$\langle P \rangle = \frac{1 + r}{1 - r} \frac{h\nu}{\eta q} Q \sqrt{i_{\text{tot}}^2}. \tag{3.25}$$

This quantity is called sensitivity of the optical receiver and it is usually expressed in dBm ($10 \times \log(\langle P \rangle / 1\ \text{mW})$). When a sensitivity of an optical receiver is reported, the wavelength used and the BER have to be mentioned. The sensitivity gets lowest for $r = 0$, i.e. $\langle i_0 \rangle = 0$ or a vanishing '0' optical power. In this case it is

$$\langle P \rangle = \frac{\hbar\omega}{\eta q} Q \sqrt{i_{\text{tot}}^2}. \tag{3.26}$$

After inserting (3.15) into (3.26), we obtain the lowest possible average optical power for a PIN-FET input stage

$$\langle P \rangle = \frac{\hbar\omega}{\eta q} Q \sqrt{\frac{2k_{\text{B}}TB}{R_{\text{L}}} + q i_{\text{L}} B + \frac{4k_{\text{B}}T\pi^2 C_{\text{T}}^2 B^3}{9g_{\text{m}}}}. \tag{3.27}$$

$\langle P \rangle$ can be measured with optical power meters. Usually the number of logical ones and zeros are equal and $\langle P \rangle$ is half of the peak power in a logical one (if the extinction ratio is infinite). $\langle P \rangle$ is called the optical sensitivity for a certain BER (note that Q corresponds to a BER). The wavelength also has to be mentioned when a sensitivity is given. In real experiments the BER and therefore the sensitivity also depends on the length of the used pseudo-random bit sequence (PRBS). The sensitivity S is often expressed in terms of dBm:

$$S = \log_{10}\left(\frac{\langle P\rangle}{1\ \mathrm{mW}}\right)\ \mathrm{dBm}. \tag{3.28}$$

In reality, especially in the gigabit range, laser diodes have to be biased above the threshold to allow fast data rates. The extinction ratio (=$1/r$) is therefore less than infinity in practice and a somewhat larger $\langle P\rangle$ is necessary than the minimum value for $r = 0$. For ultra-sensitive optical receivers aiming close to the quantum limit, however, r close to zero or an extinction ratio of the light source being very large are required.

Let us investigate the influence of C_T and R_L on $\langle P\rangle$ and the sensitivity (neglecting the shot noise of the detectors leakage current i_L). For an $R_L = 3.2\ \mathrm{k}\Omega$, $\eta = 0.5$, $g_m = 1$ mS, and $Q = 6$ (BER $= 10^{-9}$) at 1 Gbit s^{-1}, we obtain the results shown in figure 3.7. For $C_T = 0$, the optical power of about 1.2 μW is still necessary due to the thermal noise contribution of R_L. With raising input-node capacitance C_T (i.e. with growing photodiode capacitance), there is a pronounced increase of the optical power needed. In fact, we should interprete the results in figure 3.7 coming from large C_T to smaller ones: if we take 1 pF as the capacitance of a P/N-junction photodiode and 0.1 pF as the capacitance of a PIN photodiode, we see the clear advantage of the PIN photodiode and its smaller capacitance. The optical power needed reduces from 3.3 μW to about 1.2 μW. Figure 3.8 plots the corresponding

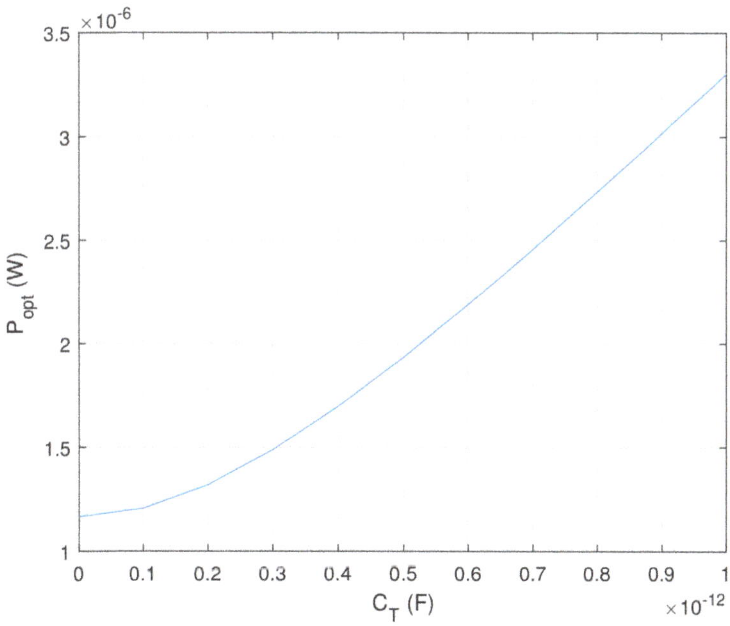

Figure 3.7. Average optical power needed for $Q = 6$ and $B = 1$ Gbit s^{-1} with a wavelength of 650 nm.

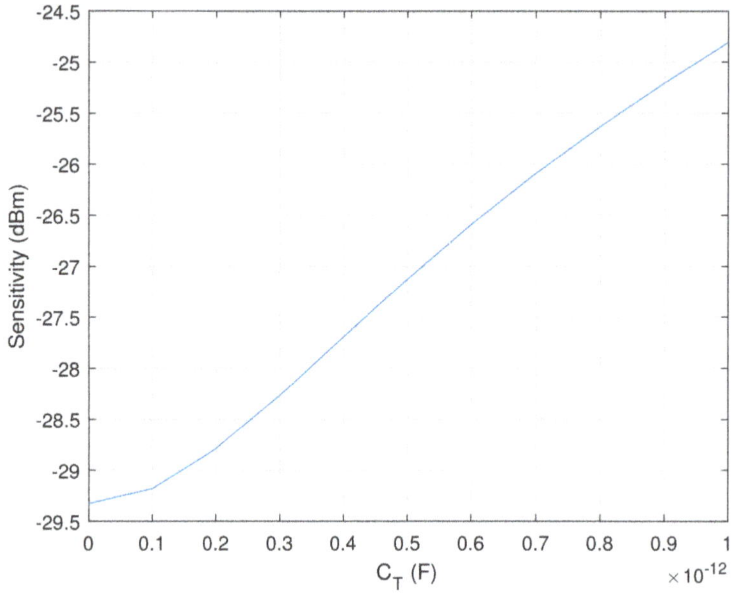

Figure 3.8. Obtained sensitivity for $Q = 6$ and $B = 1$ Gbit s^{-1} with a wavelength of 650 nm.

sensitivity. If the input node capacitance could be made to zero, the sensitivity due to the thermal noise contribution of R_L could be about −29.3 dBm. For $C_T = 0.5$ pF, the sensitivity of about −27.1 dBm is obtained at 1 Gb s^{-1}. Considering again the P/N-junction photodiode with 1 pF capacitance and the PIN photodiode with 0.1 pF, the sensitivity improves from about −24.8 dBm to about −29.2 dBm. The advantage of PIN photodiodes is quite obvious.

One of the purposes of the research towards USPAR ICs is the further reduction of the capacitance of PIN and APDs. These PIN and APs with ultra small capacitance will not only be useful for integrating USPAR chips but they will also enable the sensitivity improvement of resistive-feedback TIAs.

3.7 TIA noise

The transimpedance amplifier with RF is classified as parallel–parallel feedback circuit [5]. Figure 3.9 shows the RF-TIA with its noise sources. The amplifier possesses the equivalent input noise voltage and noise current sources $\overline{v_{ia}^2}$ and $\overline{i_{ia}^2}$, respectively. The thermal noise of the feedback resistor R_F is represented by $\overline{i_f^2}$ (see figure 3.9(b)).

To obtain the total input noise generators $\overline{v_i^2}$ and $\overline{i_i^2}$ (see figure 3.9(c)), we assume the input as open and shorted. For an open input (or a very high ohmic source like a photodiode), v_{ia} cannot generate an output signal and we obtain

$$i_i = i_{ia} + \frac{v_{ia}}{R_F} + i_f. \tag{3.29}$$

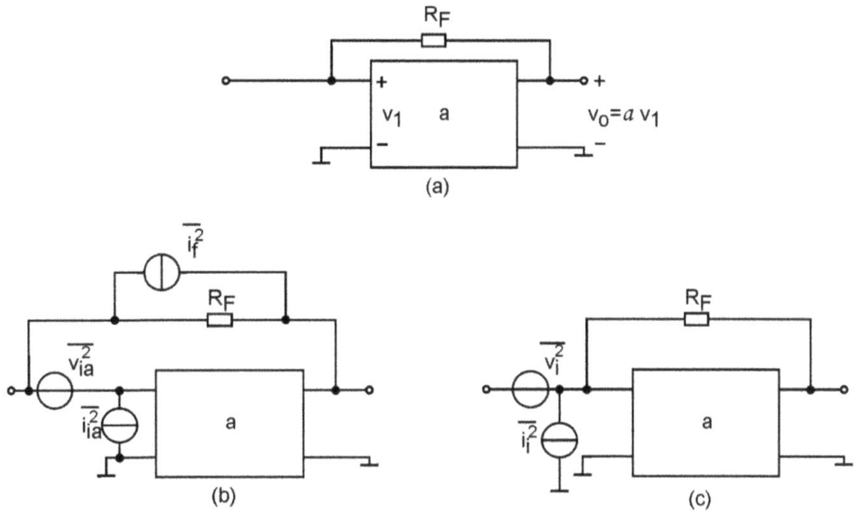

Figure 3.9. (a) Amplifier with parallel–parallel feedback (TIA). (b) TIA with noise generators. (c) Representation with two equivalent input noise generators.

The noise sources are independent from each other and it follows:

$$\overline{i_{\mathrm{i}}^2} = \overline{i_{\mathrm{ia}}^2} + \frac{\overline{v_{\mathrm{ia}}^2}}{R_{\mathrm{F}}^2} + 4k_{\mathrm{B}}T\frac{1}{R_{\mathrm{F}}}\Delta f. \tag{3.30}$$

The second term is often negligible. The thermal noise current of the feedback resistor adds to the input noise current of the amplifier.

For a shorted input, the noise currents are shorted and cannot generate an output signal. The two input noise generators have to generate the same output signal in both models (figures 3.9(b) and (c)) and therefore [5]

$$\overline{v_{\mathrm{i}}^2} = \overline{v_{\mathrm{ia}}^2}. \tag{3.31}$$

The PIN-FET noise model is a good approximation for the noise of TIAs with feedback resistors $R_{\mathrm{F}} = R_{\mathrm{L}}$. For ultra-sensitive optical receivers, the input noise current of the feedback resistor R_{F}[1] (or of R_{L}) is a severe issue. When we omit the resistor R_{L} or R_{F} in equation (3.27) (neglecting the shot noise from the detector's leakage current), the change of $\langle P \rangle$ and the sensitivity (see figures 3.10 and 3.11) does not look to be large (about 1.5 μW instead of 1.9 μW for $C_{\mathrm{T}} = 0.5$ pF and about 3.1 μW instead of 3.3 μW for $C_{\mathrm{T}} = 1$ pF as well as $S = -28$ dBm instead of -27.2 dBm for $C_{\mathrm{T}} = 0.5$ pF). This is because of the high data rate $B = 1$ Gb s^{-1}, for which the noise of the transistor dominates. If we reduce the data rate to 100 Mb s^{-1}, $\langle P \rangle$ and the sensitivity improve much more for a receiver without feedback resistor (see figures 3.12 and 3.13).

[1] See equation (3.30).

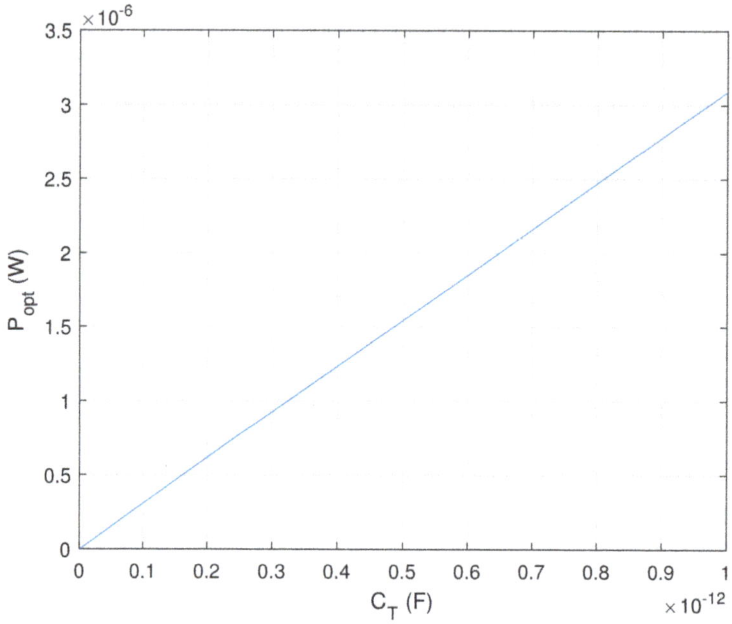

Figure 3.10. Average optical power needed by a resistorless receiver for $Q = 6$ and $B = 1$ Gb s^{-1} with a wavelength of 650 nm.

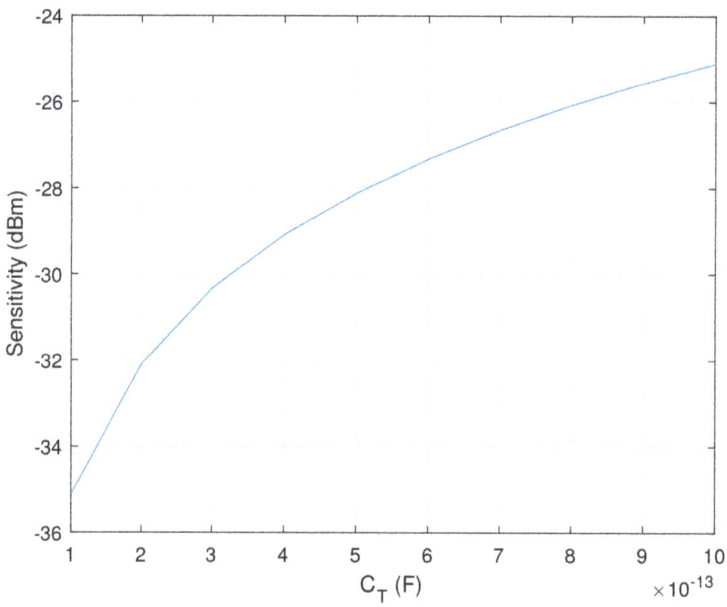

Figure 3.11. Obtained sensitivity of a resistorless receiver for $Q = 6$ and $B = 1$ Gb s^{-1} with a wavelength of 650 nm.

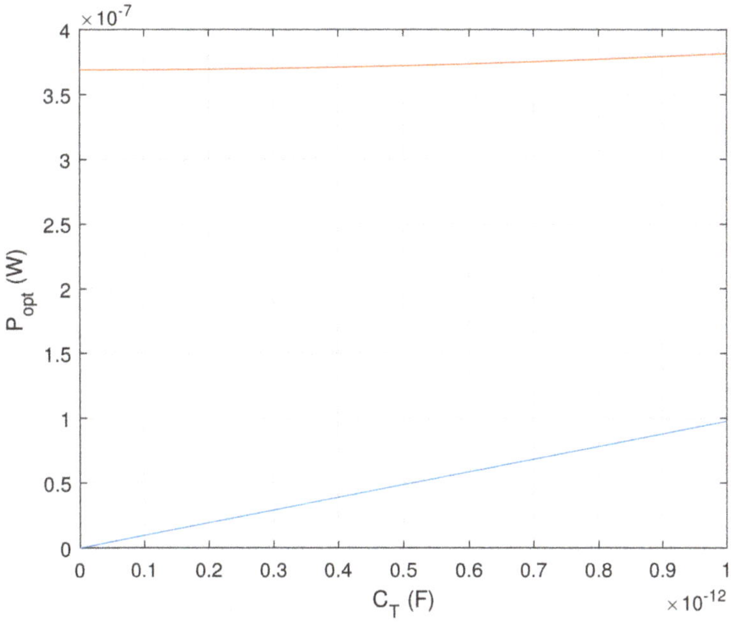

Figure 3.12. Comparison of the average optical power needed by a receiver with 3.2 kΩ feedback resistor (red curve) and by a resistorless receiver (blue curve) for $Q = 6$ and $B = 100$ Mb s^{-1} with a wavelength of 650 nm.

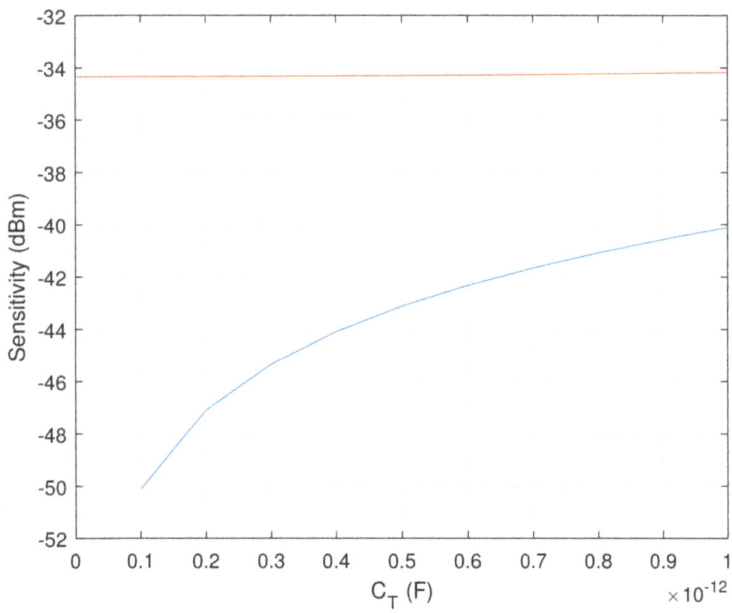

Figure 3.13. Comparison od the obtained sensitivity of a receiver with 3.2 kΩ feedback resistor (red curve) and of a resistorless receiver (blue curve) for $Q = 6$ and $B = 100$ Mb s^{-1} with a wavelength of 650 nm.

For $C_T = 0.5$ pF, $\langle P\rangle$ is 0.05 μW without resistor compared to about 0.32 μW with 3.2 kΩ resistor (see figure 3.12). And what is most interesting, for $C_T \to 0$, $\langle P\rangle \to 0$. In fact, we can see this as motivation for USPAR ICs.

Looking at figure 3.13, the sensitivity for $C_T = 0.5$ pF improves from about −34.2 dBm to—43.2 dBm, if the 3.2 kΩ resistor is omitted. Reducing C_T to 0.1 pF even leads to a sensitivity of −50 dBm at 100 Mb s^{-1} without resistor.

3.8 Noise bandwidth

To calculate the noise at the output of an amplifier, it is necessary to integrate the product of input spectral noise density and squared frequency dependent gain over the frequency. If the spectral noise density is constant as for thermal noise and shot noise as well as for simple amplifier frequency responses, the concept of noise bandwidth simplifies the determination of the output noise [5].

We assume the equivalent input noise voltage $\overline{v_i^2}$ and that the spectral density $\overline{v_i^2}/\Delta f = S_i(f) = S_{i0}$ is constant, i.e. independent of the frequency. The output noise voltage spectral density $S_o(f) = \overline{v_o^2}/\Delta f$ is then the product of the input noise voltage spectral density and the square of the frequency dependent voltage gain $A(f)$ of the amplifier.

$$\overline{v_o^2} = \int_0^\infty S_o(f)df = \int_0^\infty |A(if)|^2 S_{i0}df = S_{i0}\int_0^\infty |A(if)|^2 df \tag{3.32}$$

Let us assume an amplifier with a simple low-pass behavior of first order (a single-pole frequency response), with the low-frequency gain A_0 and with the −3 dB bandwidth f_g

$$A(jf) = \frac{A_0}{1 + j\dfrac{f}{f_g}}. \tag{3.33}$$

The squared magnitude of the gain $A(f)$ is plotted in figure 3.14 for a bandwidth f_g = 223 MHz.

The goal is now to replace the integral by the product of A_0^2 times the noise bandwidth f_n, which results in the same output noise as the integral.

$$\overline{v_o^2} = S_{i0}A_0^2 f_n \tag{3.34}$$

$$f_n = \frac{1}{A_0^2}\int_0^\infty |A(jf)|^2 df \tag{3.35}$$

Inserting equation (3.33) leads to

$$f_n = \int_0^\infty \frac{df}{1 + \left(\dfrac{f}{f_g}\right)^2} = \frac{\pi}{2}f_g = 1.57 f_g. \tag{3.36}$$

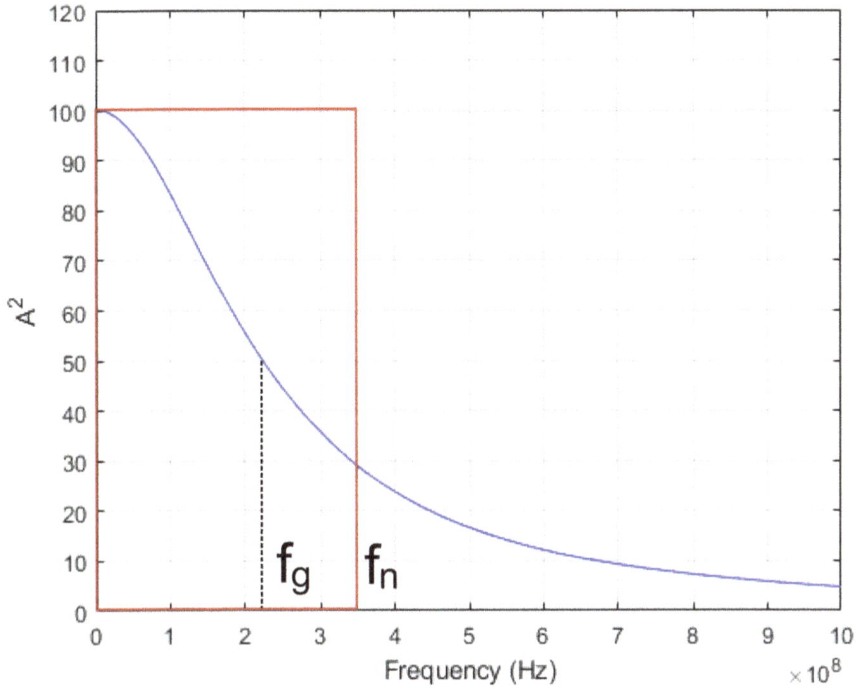

Figure 3.14. Squared magnitude of the voltage gain for a single-pole frequency response (blue) and rectangle $A_0^2 \times f_n$ (red).

This means that the noise bandwidth of any single-pole amplifier is 1.57 times larger than its −3 dB bandwidth. An amplifier with the frequency response of equation (3.33) generates an output noise as if it had an abrupt band edge at $1.57\,f_g$. A two-pole frequency response with complex poles at 45° to the negative real axis shows a noise bandwidth of $1.11\,f_g$ [5].

3.9 Noise of integrator

For RF-TIAs the noise increases for larger photodiode capacitances [6]. This circumstance should serve as motivation to investigate the dependence of the output noise of the integrating receiver, i.e. of the capacitive-feedback TIA, on the capacitance of the photodiode C_{PD}.[2] Figure 3.15 shows the principle of such a TIA. The amplifier is characterized by its transconductance g_m, its input capacitance C_{in} and its output resistance r_o. There is a feedback via the capacitive voltage divider consisting of C_F and C_T ($C_T = C_{PD} + C_{in}$). A load capacitance C_L is added to consider the input capacitance of a next amplifier stage. The input noise current of a next amplifier stage can be considered in the noise current source i_n, which is incorporated in the small-signal equivalent circuit depicted in figure 3.16.

[2] I would like to thank Simon Laube from my research group for unhesitatingly deriving the equations presented in this section.

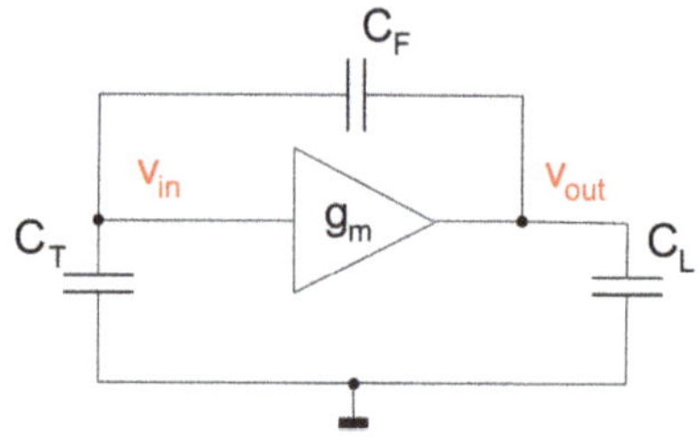

Figure 3.15. Capacitive-feedback TIA.

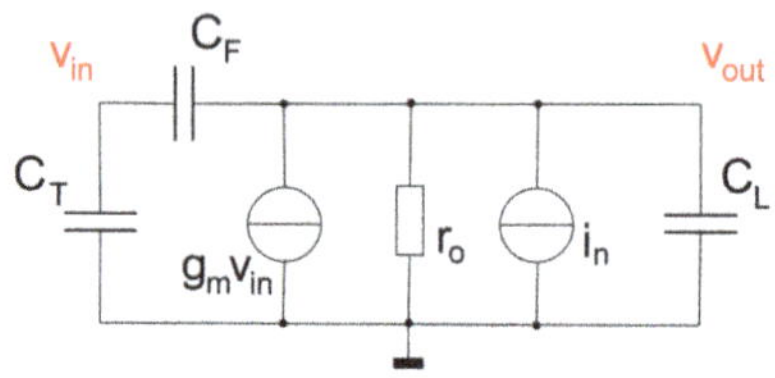

Figure 3.16. Small-signal equivalent circuit of capacitive-feedback TIA ($C_T = C_{PD} + C_{in}$).

The (signal) output current of the amplifier is determined by $g_m v_{in}$ and the output noise current source i_n of the amplifier is, e.g. the drain current noise of MOSFETs (NMOS common-source amplifier and PMOS load or NMOS and PMOS transistors of an inverter). If a load resistor is used, it can be considered in r_o.

First, we want to know how the voltage divider ratio v_{in}/v_{out} depends on C_F and C_T. For its derivation, we use Ohm's law $V = Z \cdot I$ with the AC-resistance (impedance) Z instead of R. For Z, we have $Z_F = 1/(j\omega C_F)$ and $Z_T = 1/(j\omega C_T)$ and the output voltage v_{out} is

$$v_{out} = v_T + v_F = i\left(\frac{1}{j\omega C_T} + \frac{1}{j\omega C_F}\right), \tag{3.37}$$

where the current i flows through the series connection of C_T and C_F. The input voltage is simply

$$v_{in} = i\frac{1}{j\omega C_T} \tag{3.38}$$

and we obtain the voltage divider ratio

$$\frac{v_{in}}{v_{out}} = \frac{\frac{1}{j\omega C_T}}{\frac{1}{j\omega C_T} + \frac{1}{j\omega C_F}} = \frac{\frac{1}{C_T}}{\frac{1}{C_T} + \frac{1}{C_F}}, \tag{3.39}$$

$$\frac{v_{in}}{v_{out}} = \frac{1}{C_T}\frac{C_F C_T}{C_T + C_F} = \frac{C_F}{C_T + C_F} = \frac{C_F}{C_{PD} + C_{in} + C_F}. \tag{3.40}$$

Using the method of Ochoa [7], the output voltage can be calculated dividing the short-circuit output current (meaning $v_{out} = 0$) by the admittance for a short at the input (meaning $v_{in} = 0$ or C_T is shorted):

$$v_{out} = \frac{-i_n + v_{in}(j\omega C_F - g_m)}{j\omega(C_F + C_L) + \dfrac{1}{r_o}}. \tag{3.41}$$

Inserting v_{in} from equation (3.40) results in:

$$v_{out}\left[\frac{1}{r_o} + j\omega(C_F + C_L) - \frac{C_F}{C_T + C_F}(j\omega - g_m)\right] = -i_n \tag{3.42}$$

and

$$\begin{aligned}\frac{v_{out}}{i_n} &= -\frac{1}{\dfrac{1}{r_o} + j\omega(C_F + C_L) - \dfrac{C_F}{C_T + C_F}(j\omega - g_m)}\\ &= -\frac{r_o}{1 + g_m r_o\dfrac{C_F}{C_T + C_F} + j\omega r_o\left(C_L + \dfrac{C_F C_T}{C_T + C_F}\right)}.\end{aligned} \tag{3.43}$$

The term $g_m r_o \frac{C_F}{C_T + C_F}$ demonstrates the importance of the capacitive feedback with C_F. The reader may feel uncomfortable about the used Ochoa method for derivation of equation (3.43). Therefore, it shall be shown in the following that the derivation by using Kirchoff's laws leads to the same result. The first equation is from the Kirchoff current rule:

$$i_F + g_m v_{in} + i_{ro} + i_{CL} + i_n = 0. \tag{3.44}$$

The current through C_T and C_F is equal:

$$i_F = i_T \tag{3.45}$$

and the voltage across C_F is the difference of output and input voltages:

$$v_F = v_{out} - v_{in} \tag{3.46}$$

Inserting v_{in} from equation (3.40) into equation (3.44) and using $i_F = v_F \cdot j\omega C_F = (v_{out} - v_{in}) \cdot j\omega C_F = v_{out}(1 - C_F/(C_T + C_F)) \cdot j\omega C_F$ results in

$$v_{out} j\omega C_F\left(1 - \frac{C_F}{C_T + C_F}\right) + g_m\frac{C_F}{C_T + C_F}v_{out} + \frac{v_{out}}{r_o} + v_{out} j\omega C_L = -i_n. \tag{3.47}$$

$$\frac{v_{out}}{i_n} = -\frac{1}{\dfrac{1}{r_o}\left[j\omega r_o C_F\left(1 - \dfrac{C_F}{C_T + C_F}\right) + g_m r_o\dfrac{C_F}{C_T + C_F} + 1 + j\omega r_o C_L\right]}. \tag{3.48}$$

$$\frac{v_{\text{out}}}{i_{\text{n}}} = -\frac{r_{\text{o}}}{j\omega r_{\text{o}} C_{\text{F}}\left(1 - \frac{C_{\text{F}}}{C_{\text{T}} + C_{\text{F}}}\right) + j\omega r_{\text{o}} C_{\text{L}} + g_m r_{\text{o}} \frac{C_{\text{F}}}{C_{\text{T}} + C_{\text{F}}} + 1}. \tag{3.49}$$

$$\frac{v_{\text{out}}}{i_{\text{n}}} = -\frac{r_{\text{o}}}{j\omega r_{\text{o}}\left(C_{\text{F}} - \frac{C_{\text{F}}^2}{C_{\text{T}} + C_{\text{F}}} + C_{\text{L}}\right) + g_m r_{\text{o}} \frac{C_{\text{F}}}{C_{\text{T}} + C_{\text{F}}} + 1}. \tag{3.50}$$

$$\frac{v_{\text{out}}}{i_{\text{n}}} = -\frac{r_{\text{o}}}{j\omega r_{\text{o}}\left(\frac{C_{\text{F}}(C_{\text{T}} + C_{\text{F}}) - C_{\text{F}}^2}{C_{\text{T}} + C_{\text{F}}} + C_{\text{L}}\right) + g_m r_{\text{o}} \frac{C_{\text{F}}}{C_{\text{T}} + C_{\text{F}}} + 1}. \tag{3.51}$$

$$\frac{v_{\text{out}}}{i_{\text{n}}} = -\frac{r_{\text{o}}}{j\omega r_{\text{o}}\left(\frac{C_{\text{F}} C_{\text{T}}}{C_{\text{T}} + C_{\text{F}}} + C_{\text{L}}\right) + g_m r_{\text{o}} \frac{C_{\text{F}}}{C_{\text{T}} + C_{\text{F}}} + 1}. \tag{3.52}$$

This is the same result as obtained with the Ochoa method in equation (3.43). In the limiting case $C_{\text{F}} \to 0$, we obtain

$$\frac{v_{\text{out}}}{i_{\text{n}}} = -\frac{r_{\text{o}}}{1 + j\omega r_{\text{o}} C_{\text{L}}} \tag{3.53}$$

and for an open output (no load, $C_{\text{L}} = 0$)

$$\frac{v_{\text{out}}}{i_{\text{n}}} = -r_{\text{o}}. \tag{3.54}$$

Then the output noise voltage is simply given by the voltage, which is caused by the drain noise current across the output resistance. This is expected and we can be quite sure that the model for the dependence of the output noise voltage of the integrating receiver (capacitive-feedback TIA) is correct.

Next, the spectral noise power densities have to be applied. We know $S_{\text{i}}(f) = 8/3 k_{\text{B}} T g_m$ from equation (3.8) and we omit the contribution of the $1/f$-noise. Since we have a noise voltage at the output, $S_v(f)$ has to be determined. This is easily possible by applying equation (3.52) or (3.43):

$$S_v(f) = S_{\text{i}}(f)\left|\frac{r_{\text{o}}}{1 + g_m r_{\text{o}} \frac{C_{\text{F}}}{C_{\text{T}} + C_{\text{F}}} + j\omega r_{\text{o}}\left(C_{\text{L}} + \frac{C_{\text{F}} C_{\text{T}}}{C_{\text{T}} + C_{\text{F}}}\right)}\right|^2. \tag{3.55}$$

The total output noise voltage is then obtained via

$$\overline{v_{\text{out, n}}^2} = \int_0^{\infty} S_v(f) df \tag{3.56}$$

and the rms output noise voltage is

$$v_{\text{out,n}} = \sqrt{v_{\text{out,n}}^2}\,. \qquad (3.57)$$

The numerical integration was done from 1 Hz to 10 GHz and the result, the dependence of the rms output noise voltage on the photodiode capacitance, is shown in figure 3.17 for the parameters of a CMOS inverter in 0.35 μm CMOS ($W_n = 1.0\ \mu$m, $W_p = 2.5\ \mu$m).

We should be aware that the load capacitance has a strong impact on the rms output noise voltage. This is not astonishing since the load capacitance determines the bandwidth. $1/f$-noise was neglected in calculating the curve in figure 3.17. Depending on the CMOS technology used, the $1/f$-noise may increase the noise more or less.

Figure 3.17 shows that the rms output noise voltage increases by only about 16% for an increase of the photodiode capacitance from 3.5 fF to 13.5 fF. Since the rms output noise voltage does not strongly depend on the capacitance of the photodiode, the influence of this model on the Q-factor will be neglected in chapter 6. The influence of the C_{PD}-model on the noise voltage can be considered in the choice of the bandwidth, when the noise is determined from equation (3.8) for the use in the Q-factor.

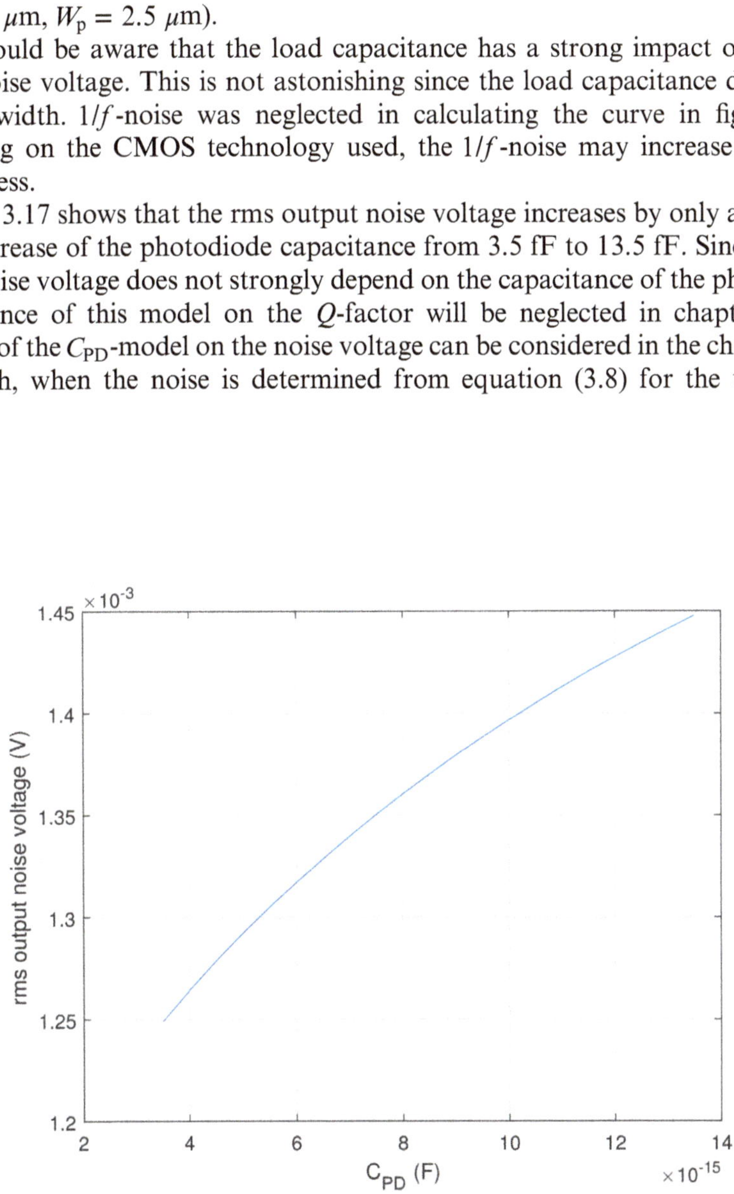

Figure 3.17. Rms output noise voltage for $g_m = 0.4$ mS, $r_o = 41.7$ kΩ, $C_F = 0.7$ fF, $C_{in} = 6.5$ fF and $C_L = 20$ fF considering only thermal noise.

3.10 Excess noise of APDs

The avalanche effect is based on the impact ionization in the crystal lattice by charge carriers, which possess enough energy to generate an electron–hole pair [8]. The generated carriers are also accelerated in the high electric field region and can generate more electron–hole pairs. An avalanche builds up and compared to a PIN photodiode with the same (primary) quantum efficiency, M times more photocurrent is generated by the APD:

$$M = \frac{I_{\mathrm{APD}}}{I_{\mathrm{PIN}}}. \tag{3.58}$$

This means that the shot noise of the PIN photodiode is also amplified with M. In addition the avalanche gain varies statistically (because carriers with enough energy for impact ionization can loose their energy to the crystal lattice without generating an electron–hole pair), which increases the noise of the APD additionally. This increased noise is called *excess noise* and the excess noise factor F describes it. F considers the standard deviation of the stochastic impact behavior.

$$F = \frac{\overline{i_{\mathrm{n,\,APD}}^2}}{M^2 \overline{i_{\mathrm{n,\,PIN}}^2}} \tag{3.59}$$

$$\overline{i_{\mathrm{n,\,APD}}^2} = M^2 2qI_{\mathrm{PIN}}F \tag{3.60}$$

The model of McIntyre [9, 10] constitutes a link between impact ionization and excess noise (see equation (3.61)). Thereby, the coefficient k represents the ratio of impact ionization coefficients of holes (β) and electrons (α).

$$F = kM + (1 - k)\left(2 - \frac{1}{M}\right) \tag{3.61}$$

$$k = \frac{\beta}{\alpha} \tag{3.62}$$

For silicon reach-through APDs, $k = 0.02$ is possible and for $M = 150$, the excess noise factor is 4.9 [11]. For silicon epitaxial structures, $k = 0.06$ was reported and for $M = 100$, F was 7.9. The deviation from these values, however, can be significant, because the excess noise of APDs does not only depend on the material and on these fabrication types but also on the doping profiles and dimensions [12]. The engineering of APDs has also to consider the required bandwidth because for high avalanche gains there is a constant gain-bandwidth product [13].

The group of the author characterized linear mode APDs in 0.35 μm PIN-photodiode CMOS [14–16] and 0.35 μm HV CMOS [17]. Both APDs have a thick, fully depleted absorption zone and a separate multiplication zone leading to high responsivities (R_0), high bandwidths and low capacitance. The PIN-photodiode-based APD (PIN-APD) achieved a bandwidth of 1.02 GHz at an avalanche gain of

61.4 with 670 nm light [15]. At a higher reverse bias (higher gain) 505 MHz bandwidth was achieved [16]. With different edge doping, the bandwidth was 1.15 GHz and the responsivity was 20.4 A W^{-1} [14]. The PIN-photodiode CMOS APD showed an excess noise factor $F = 8$ and the HV APD achieved $F = 6$, both at $M = 50$ and with an optical power of about 1 μW. The ratios of the hole-to-electron impact ionization coefficients were 0.0789 and 0.0849, respectively [18, 19]. For a modulation doped APD in PIN-photodiode CMOS technology the impact ionization coefficient was 0.417 [19]. Figure 3.18 compares the excess noise factors for these three ratios of the impact ionization coefficients. The difference in the excess noise factors for $k = 0.0789$ and $k = 0.0849$ is not large. The dark currents were below 10^{-13} A for an APD diameter of 86 μm [17, 20].

The model of McIntyre was derived in one dimension, i.e. for uniform electric field distributions. A newer model is the so-called dead-space model, which was used in many investigations of APDs [21–24]. The dead-space considers the drift distance, which the carriers need to acquire enough energy for impact ionization in the high-field region. As a consequence, the excess noise factors obtained with the dead-space model are lower than those obtained with the McIntyre model [24]. However, the dead-space model is, like the McIntyre model, also not appropriate to describe the excess noise in our dot APDs (see chapter 5), which have three-dimensional electric-field distributions.

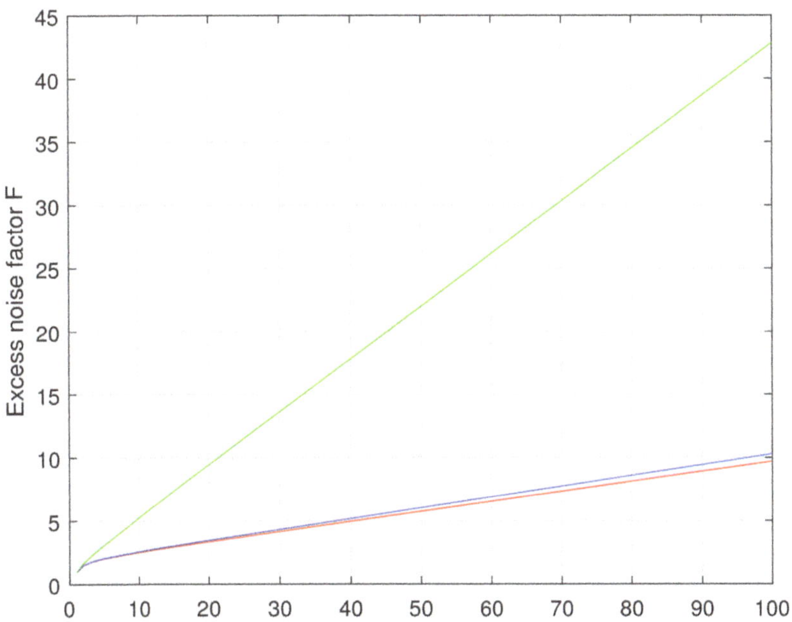

Figure 3.18. Excess noise factors for $k = 0.0789$ (red), $k = 0.0849$ (blue) and $k = 0.417$ (green) in dependence on the avalanche gain M.

3.11 Photon noise

Most light emitters like light emitting diodes (LEDs) and laser diodes are Poissonian light emitters. The number of counted photons with an ideal detector in different time intervals having all the same duration is not constant. This deviation (from the mean value) constitutes photon noise. The Poisson distribution [25]

$$p(x, \mu) = \frac{\mu^x}{x!} e^{-\mu} \tag{3.63}$$

describes the probability $p(x, \mu)$ that x photons are present in a time interval, when the mean value (over a large number of time intervals) is μ. For already quite small numbers $\mu \gtrsim 5$, the Poisson distribution tends towards the Gaussian distribution with the mean value μ and the standard deviation $\sigma = \sqrt{\mu}$ [26]

$$p(x) = \frac{1}{2\pi\sigma^2} e^{-(x-\mu)^2/2\sigma^2}. \tag{3.64}$$

The standard deviation of the Poisson distribution was derived in [27]:

$$\sigma = \sqrt{\mu}. \tag{3.65}$$

So, for the target range of USPAR of down to about 50 photons, we certainly can use the photon noise proportional to the square root of the average photon number.

The photon noise holds e.g. for the signal voltage of CCD image sensor pixels and active pixel sensors, where photogenerated charges are integrated on the photodiode's or photogate's capacitance [28].

It also should be mentioned that the signal-to-noise ratio (SNR) for photon noise is

$$\mathrm{SNR} = \frac{\mu}{\sqrt{\mu}} = \sqrt{\mu}. \tag{3.66}$$

This SNR from photon noise represents the final, physical limit for ultra-sensitive receivers. Of course electronic noise and other non-ideal photodetector and circuit properties will lead to worse SNRs.

For USPAR ICs, we have to apply photon noise and models for the non-ideal behavior in the charge integrators used.

3.12 Reset noise

The reset noise occurs in switched capacitors. The reset transistor is used to discharge a capacitor and this transistor works as a switch. When it is switched on, its channel resistance R_{on} causes thermal noise. At the end of the on-state, there is a noise voltage which cannot be predicted. This momentary noise voltage is present after the on-state on the capacitor. The expression for this noise voltage can be derived for a sample&hold circuit consisting of a switching transistor and the hold capacitor C_{H} (see figure 3.19).

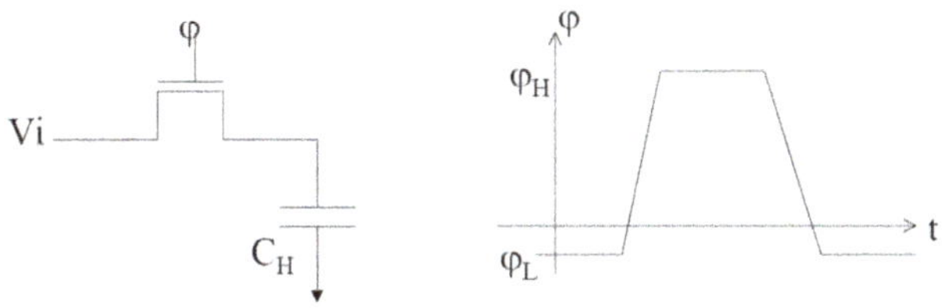

Figure 3.19. Sample&hold circuit.

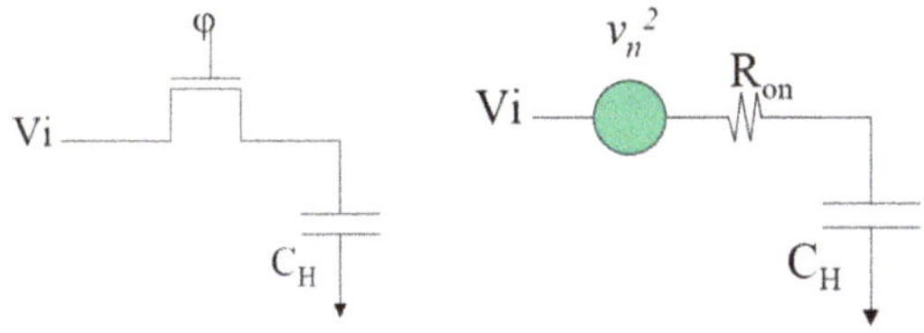

Figure 3.20. Sample&hold circuit (left) and the model for considering its noise (right).

For the gate signal φ being high (φ_H, assuming an n-channel MOSFET), the hold capacitor is charged to the input voltage V_i with the time constant $\tau = R_{on}C_H$. To be exact, the thermal noise voltage $v_n = \sqrt{4k_BTR_{on}\Delta f}$ of the MOSFET's channel resistance R_{on} is overlain to V_i (see figure 3.20).

The transfer function $H(f)$ of this circuit is

$$H(f) = \frac{1}{1 + j2\pi f R_{on} C_H}. \tag{3.67}$$

To obtain the total noise voltage v_{oT} at the output, i.e. between the MOSFET and C_H, the integral

$$\overline{v_{oT,\,rms}^2} = 4k_BTR_{on}\int_0^{\infty} \frac{1}{|1 + j2\pi f R_{on} C_H|^2},\, df \tag{3.68}$$

has to be solved. The result is

$$v_{oT,rms} = \sqrt{\frac{k_BT}{C_H}} \tag{3.69}$$

where the origin (R_{on}) is not present anymore. Due to this result, the reset noise is also called *kT/C*-noise. For a hold capacitor with a value of 1 pF and for room temperature (300 K), the rms noise voltage is 64.3 μV. This is a physical limit, which cannot be avoided.

In the English language, 'noise' means unwanted phenomena, which can be in addition to electronic noise also any disturbance or random fluctuation of signals. Here, for the sample&hold circuit, we have to consider channel charge injection (see figure 3.21) [29]. When the MOSFET is switched on, there are charge carriers in its channel. This channel charge Q_C is [30]

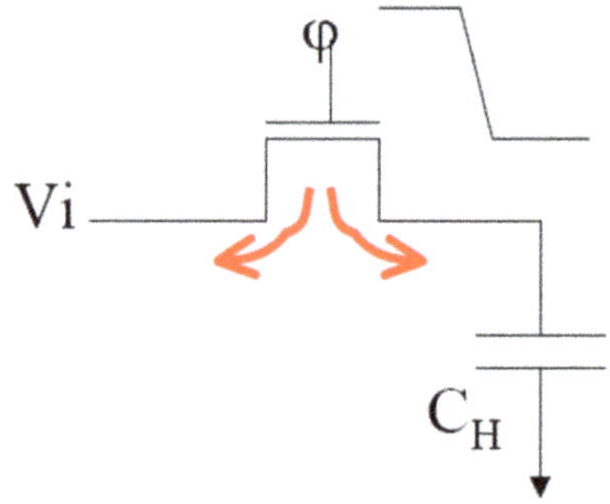

Figure 3.21. Separation of the channel charge during switching off.

$$Q_C = WLC_{OX}(\varphi_H - V_{th} - V_i). \tag{3.70}$$

When the gate voltage φ is made low, the n-channel MOSFET turns off and the channel charge must leave the channel. In the common approximation, half of this charge goes to the input and the other half goes to the output. This causes a change of the voltage ΔV on C_H [31]:

$$\Delta V = -\frac{Q_C}{2C_H} = -\frac{WLC_{OX}(\varphi_H - V_{th} - V_i)}{2C_H} \tag{3.71}$$

where V_{th} is the threshold voltage. This voltage ΔV leads to the voltage

$$V_o = V_i - \frac{Q_C}{2C_H} = -\frac{WLC_{OX}(\varphi_H - V_{th} - V_i)}{2C_H}. \tag{3.72}$$

on C_H instead of V_i, which causes an additional error to the reset noise.

And what is more, there is another error caused by clock feed-through. The MOSFETs possess an overlap capacitance between gate and drain as well as between gate and source (figure 3.22). This capacitance C_{OV} is equal to C_{GS} and C_{GD}. When we look from the gate towards the hold capacitor C_H, we see the series connection of C_{OV} and C_H (see figure 3.23).

The consequence of this is that the voltage error

$$\Delta V_{OV} = \varphi_H \frac{C_{OV}}{C_{OV} + C_H} \tag{3.73}$$

is caused by the voltage step φ_H at the gate. For an older technology with a threshold voltage $V_{th} = 0.7$ V and a transistor with $W = 10\ \mu$m and $L = 2$ µm, $C_{OX} = 1.38$ fF μm^{-2}, $C_{OV} = 3$ fF, $C_H = 1$ pF, $\varphi_H = 5$ V and $\varphi_L = 0$ V, the channel charge error is up to 59 mV (depending on V_i) and the clock feed-through error is 15 mV.

All these errors (due to reset noise, channel charge and clock feed-through) increase for small hold capacitor values. In USPAR this problem is quite severe, because the integration capacitor has to be very small.

Fortunately, there is a suggestion from textbooks [29, 30] to deal with these errors. Figure 3.24 shows the sample&hold circuit with an additional transistor M2

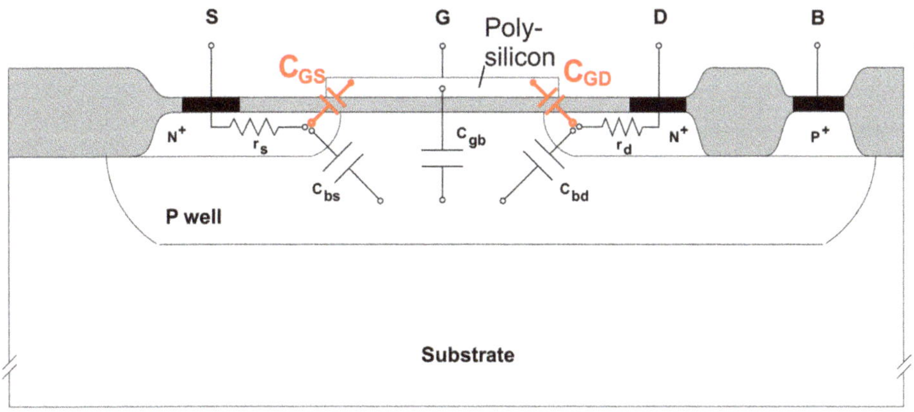

Figure 3.22. Cross-section of MOSFET with parasitic components. The gate–drain C_{GD} and gate–source C_{GS} overlap capacitances are drawn in red color.

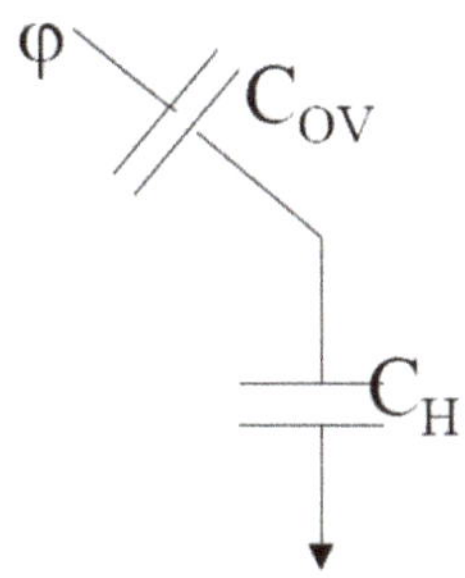

Figure 3.23. Series connection of gate–drain overlap capacitance and hold capacitor forming a capacitive voltage divider.

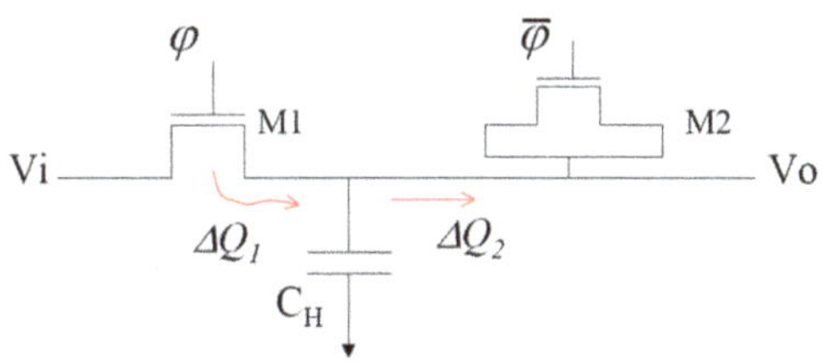

Figure 3.24. Sample&hold circuit with compensation transistor.

for compensation purposes. M2 is connected as MOS-capacitor (source and drain are connected, the gate-to-channel capacitance is exploited as capacitor; the gate–drain and gate–source overlap capacitances are also present of course) and must have the half width of M1: $W_2 = W_1/2$ (It is understood that M1 and M2 have equal gate lengths.). For $W_2 = W_1/2$, M2 includes an overlap capacitance, which equals the gate–drain overlap capacitance of M1. Because $L_1 = L_2$ in addition, the channel area of M2 is half that of M1 and the channel charge error caused by M1 is also compensated. The clocks of both transistors switch in opposite directions at the same time (They need equal amplitudes and slew rates of course.) Therefore, the charge which is injected by M1 is subtracted by M2. In practice, it is hard to get it running perfectly, but the errors caused by the channel charge and by clock feed-through are reduced a lot.

References

[1] Gray P R and Meyer R G 1993 *Analysis and Design of Analog Integrated Circuits* (New York: Wiley)

[2] Ebeling K J 1993 *Integrated Optoelectronics* (Berlin, Heidelberg: Springer)

[3] Zimmermann H 2004 *Silicon Optoelectronic Integrated Circuits* (Berlin, Heidelberg, Germany: Springer)

[4] Sklar B 2001 *Digital Communications: Fundamentals and Applications* (Englewood Cliffs, NJ: Prentice Hall)

[5] Gray P R, Hurst P J, Lewis S H and Meyer R G 2001 *Analysis and Design of Analog Integrated Circuits* (New York: Wiley)

[6] Zimmermann H 2018 *Silicon Optoelectronic Integrated Circuits* (Switzerland: Springer)

[7] Ochoa A 2016 *Feedback in Analog Circuits* (Switzerland: Springer International Publishing)

[8] Sze S M and Ng K K 2007 *Physics of Semiconductor Devices* (Hoboken, New Jersey: Wiley)

[9] McIntyre R J 1966 Multiplication noise in uniform avalanche diodes *IEEE Trans. Electron Devices* **13** 164–8

[10] Sze S M 1981 *Physics of Semiconductor Devices* (New York: Wiley)

[11] Elmer P 2016 Avalanche photodiodes user guide http://www.perkinelmer.com/CMSResources/Images/44-6538APP_AvalanchePhotodiodesUsersGuide.pdf

[12] Yuan P, Wang S, Sun X, Zheng X G, Holmes A and Campbell J C 2000 Avalanche photodiodes with an impact-ionization-engineered multiplication region *IEEE Photon. Technol. Lett.* **12** 1370–2

[13] Bin L, Xiaohong Y, Weihong Y, Qianqian L, Rong C and Qin H 2014 A high-speed avalanche photodiode *J. Semicond.* **35** 074009

[14] Gaberl W, Steindl B, Schneider-Hornstein K, Enne R and Zimmermann H 2014 0.35 μm CMOS avalanche photodiode with high responsivity and responsivity-bandwidth product *Opt. Lett.* **39** 586–9

[15] Steindl B, Gaberl W, Enne R, Schidl S, Schneider-Hornstein K and Zimmermann H 2014 Linear mode avalanche photodiode with 1 GHz bandwidth fabricated in 0.35 μm CMOS *IEEE Photon. Technol. Lett.* **26** 1511–4

[16] Gaberl W, Schneider-Hornstein K, Enne R, Steindl B and Zimmermann H 2014 Avalanche photodiode with high responsivity in 0.35 μm CMOS *SPIE Optical Engineering* **53** 043105-1–4

[17] Steindl B, Enne R, Schidl S and Zimmermann H 2014 Linear mode APD with high responsivity integrated in high-voltage CMOS *IEEE Electron Device Lett.* **35** 897–9

[18] Jukic T 2017 Empfänger mit integrierter Lawinenfotodiode *PhD Thesis* TU Wien

[19] Jukic T, Brandl P and Zimmermann H 2018 Determination of the excess noise of avalanche photodiodes intergated in 0.35-μm CMOS technologies *SPIE Opt. Eng.* **57** 044101-1–5

[20] Enne R, Steindl B, Schneider-Hornstein K and Zimmermann H 2014 pn photodiode in 0.35 μm high-voltage CMOS with 1.2 GHz bandwidth *SPIE Opt. Eng.* **53** 116114

[21] Hossain M M, Ray S, Cheong J S, Qiao L, Baharuddin A N A P, Hella M M, David J P R and Hayat M M 2017 Low-noise speed-optimized large area CMOS avalanche photodetector for visible light communication *J. Light. Technol.* **35** 2315–24
[22] Hayat M, Sargeant W and Saleh B 1992 Effect of dead space on gain and noise in si and gaas avalanche photodiodes *IEEE J. Quantum Electron.* **28** 1360–5
[23] Hayat M, Saleh B and Teich M 1992 Effect of dead space on gain and noise of double-carrier-multiplication avalanche photodiodes *IEEE Trans. Electron Devices* **39** 546–52
[24] Pauchard A, Besse P-A and Popovic R 2000 Dead space effect on the wavelength dependence of gain and noise in avalanche photodiodes *IEEE Trans. Electron Devices* **47** 1685–93
[25] Haigh F A 1967 *Handbook of the Poisson Distribution* (New York: Wiley)
[26] http://slittlefair.staff.shef.ac.uk/teaching/phy217/lectures/instruments/L14/index.html accessed 08.08.2022
[27] https://www.lpl.arizona.edu/griffith/ptys521/3ErrorAnalysis3.pdf accessed 08.08.2022
[28] Nishida Y, Koike J, Ohtake H, Abe M and Yoshikawa S 1989 Design concept for a low-noise CCD image sensor based on subjective evaluation *IEEE Trans. Electron Devices* **36** 360–6
[29] Baker R J 2005 CMOS Circuit Design *Layout and Simulation* (Hoboken: Wiley Interscience)
[30] Razavi B 2001 *Design of Analog CMOS Integrated Circuits* (New York: McGraw-Hill)
[31] Adusumalli R, Thottathil R and Avalur K K G 2019 A 10-bit 250 MS s^{-1} binary search and two channel SAR ADC by a two-bit per conversion with error tolerance ability *IEEE Int. Conf. on Modelling of Systems, Circuits and Devices* 64–7

Chapter 4

Basics of optical receivers

In this chapter the basics of transimpedance amplifiers (TIAs) with resistive feedback (RF), of current amplifiers and integrating optical receivers are summarized. The most interesting examples of these receiver types from the state-of-the-art are introduced. Their applicability to ultra-sensitive optical receivers is discussed.

4.1 Resistive-feedback transimpedance amplifiers

Transimpedance amplifiers (TIAs) with RF (see figure 4.1) convert the photocurrent delivered by the photodiode to an output voltage. C_{PD} is the capacitance of the photodiode. The amplifier has an inverting gain $A(f)$.

The most important advantage of a transimpedance amplifier is that it achieves a larger bandwidth than a simple load resistor (see figure 4.2) connected in series to the photodiode. The −3 dB bandwidth f_{BW} of the photodiode with the load resistor R_f is:

$$f_{BW} = \frac{1}{2\pi R_f C_{PD}} \tag{4.1}$$

The gain $A(f)$ of the amplifier is exploited to enhance the bandwidth of the optical receiver in TIA configuration [1]:

$$f_{BW}^{TIA} = \frac{A(f)}{2\pi R_f C_{PD}} \tag{4.2}$$

In practice at high data rates, $A(f)$ is often (much) smaller than the low-frequency gain of the amplifier. We also see from this equation that a PIN photodiode leads to a larger bandwidth of the optical receiver because its capacitance is typically an order of magnitude smaller than the capacitance of a P/N-junction photodiode. In a comparison [2], the TIA achieved a bandwidth of 180 MHz compared to 79.6 MHz of the optical receiver consisting of photodiode and load resistor having the same value R_f. Furthermore, the PIN photodiode eliminated gain peaking of the TIA

doi:10.1088/978-0-7503-5437-0ch4

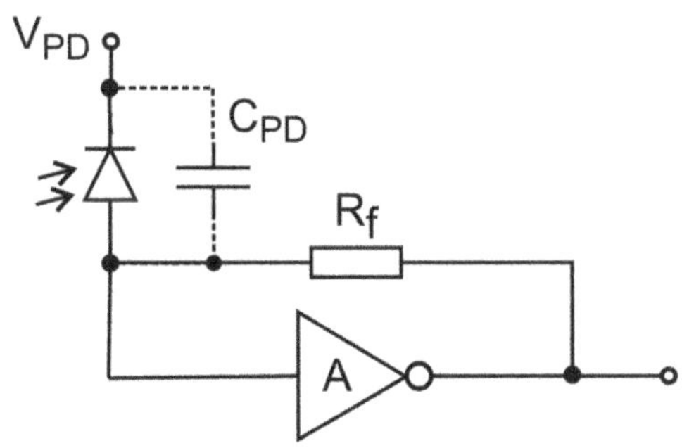

Figure 4.1. Optical receiver with transimpedance amplifier having a feedback resistor R_f.

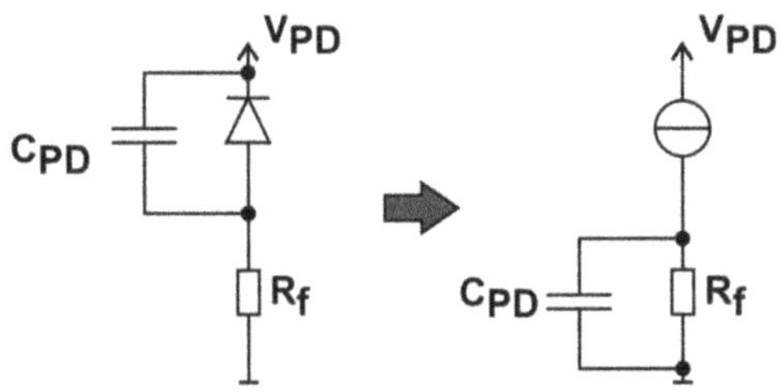

Figure 4.2. Optical receiver with photodiode and load resistor R_f.

compared to the P/N junction photodiode. An additional advantage of the TIA is that the amplifier makes the output low ohmic compared to the feedback resistor value.

TIAs with RF possess an input current noise of 5 nA from the feedback resistor alone (i.e. noise from transistors not considered and this will add of course), when a feedback resistor of 800 kΩ and a bandwidth $B = 900$ MHz is assumed. With 50 photo electrons, the charge of 8×10^{-18} As corresponds to a photocurrent of 8 nA within a 1 ns bit duration (for a data rate of 1 Gb s^{-1} in NRZ). With 10 ns bit duration (data rate = 100 Mb s^{-1}, $B = 90$ MHz) the photocurrent is only 0.8 nA (a factor 10 less, because the amount of charge (of 50 photo electrons) flows within a bit duration of 10 ns), but the resistors noise current is about 1.6 nA (only a factor of $\sqrt{10}$ less). Even larger feedback resistor values would be needed, for which the CCVD [3] and CCDR [4] approaches cannot be applied successfully anymore. This means that RF TIAs cannot be used for the realization of ultra-sensitive optical receivers without SPADs.

An important aspect of monolithically integrated TIAs has to be addressed. Integrated resistors introduce a bandwidth by themselves (contrary to resistors, which are mounted on PCBs). This is because in bipolar silicon technologies, in complementary metal oxide semiconductor (CMOS) technologies and in BiCMOS technologies, the resistors are fabricated on top of the field oxide (FOX) or on top of the shallow trench isolation (STI). In both cases, there is therefore silicon dioxide below the integrated resistors, which consist of (doped) polysilicon (see figure 4.3).

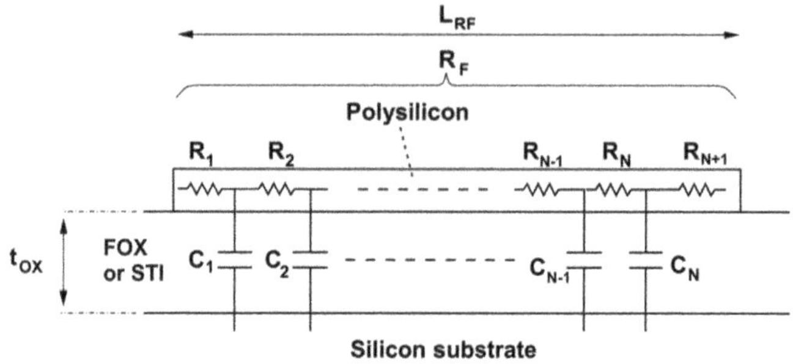

Figure 4.3. Integrated resistor consisting of polysilicon on top of FOX or STI together with R–C chain for modelling its bandwidth.

This means that the polysilicon forms a plate capacitor together with the silicon substrate and the oxide as dielectric material.

Usually, integrated resistors can be modelled in circuit simulation by a chain of RC elements, as shown in figure 4.3. The sum of the resistors in the chain is equal to R_f and the sum of the capacitances is equal to the total capacitance of the polysilicon resistor. Of course, the polysilicon must have a high sheet resistance, i.e. it has to be lightly doped (much lower doped than gate polysilicon), to reduce the length and thereby the area of the polysilicon resistor to lower the capacitance. For a sheet resistance of 100 Ohm per square, the bandwidth of a 40 kΩ polysilicon resistor was 61 MHz [2]. For a sheet resistance of 1 kΩ per square, the bandwidth increased to 610 MHz. The bandwidth of the feedback resistor will limit the TIA bandwidth, especially for small photodiode capacitances, for which high values of R_f are possible for a given TIA bandwidth f_{BW}^{TIA}.

Let us now look at some interesting TIAs described in the state of the art.

RF TIAs are very popular and are integrated together with PN-, PIN photodiodes and APDs. The group of the author worked a lot on RF TIAs' inclusive bandwidth extension to 378 MHz (R_f = 178 kΩ) with capacitive coupled voltage divider feedback (CCVD) [3] and to 615 MHz (R_f = 200 kΩ) capacitive coupled distributed resistors (CCDR) [4] at equivalent input noise current densities of 1.25 pA Hz $^{-1/2}$ and 1.5 pA Hz $^{-1/2}$, both with PIN photodiodes, as well as APD receiver ICs [5–7]. Without these CCVD and CCDR tricks, the bandwidth of the integrated 200 kΩ polysilicon feedback resistor limited the bandwidth to 67.5 MHz.

With integrated 200 μm diameter APD 0.35 μm high-voltage CMOS receivers for optical wireless communication at 675 nm, BER = 10^{-9} and 622 Mb s^{-1}, the sensitivity was −33.6 dBm corresponding to about 4000 photons in a '1' and at 1 Gb s^{-1} it was −35.5 dBm [7]. At 2 Gb s^{-1} [7] the sensitivities were −30.6 dBm (400 μm APD diameter) [6] and −29.7 dBm (600 μm APD diameter) [8, 9] using 0.35 μm PIN photodiode BiCMOS. At 500 Mb s^{-1} and 200 μm photodiode diameter −35.8 dBm were achieved [10]. A receiver optoelectronic integrated circuit with integrated 800 μm APD and 120 kΩ polysilicon feedback resistor in 0.35 μm PIN-photodiode CMOS for optical free-space communication achieved sensitivities of −50.4 dBm, −47.65 dBm and −42.7 dBm at data rates of 2.5 Mb s^{-1}, 5 Mb s^{-1} and 10 Mb s^{-1}, respectively [11]. The integrated simulated equivalent input noise current of this TIA was 774 pA $_{rms}$ at a bandwidth of 4 MHz. In 55 nm BiCMOS a TIA for an off-chip photodiode achieved a simulated input noise current density of 0.7 pA Hz $^{-1/2}$ at a bandwidth of 250 MHz [11].

4.2 Current amplifiers

There were suggestions to use a capacitive feedback network TIA to avoid the thermal noise of the feedback resistor [12]. In fact this circuit was a current amplifier not needing a reset transistor. But, a high-ohmic biasing circuit for the input transistor caused an input noise current of 4.5 pA Hz $^{-1/2}$ at a bandwidth of 550 MHz. The principle of this current amplifier is depicted in figure 4.4(c). This circuit enables the lowest input noise current compared to the common-gate TIA (figure 4.4(a)) and the RF TIA (figure 4.4(b)).

Here it should be mentioned that the common-gate TIA usually has the highest input noise current of these three topologies. This is because the noise currents of M_2 and R_D are directly referred to the input. Since the DC voltage across R_D must be large to reduce the noise current and to obtain a high gain, the headroom for M_2 is small, causing a large noise contribution of M_2 [12].

The TIA with RF is better with respect to noise because the feedback resistor does not carry the drain current of a transistor and its resistor value can be larger than R_D. There are, however, three drawbacks of the RF TIA [12]: first, R_F, the photodiode capacitance C_D and the input capacitance of M_1 cause a pole at the input, which reduces the stability and worsens the pulse response. Second, the gain at low supply voltages is limited since the voltage across R_D is only $V_{DD} - V_{GS2} - V_{GS1}$. Therefore, R_D and M_2 contribute also considerably to the noise. Third, M_1 may contribute as much input noise current as R_F since the input-noise voltage $V_{n,M1}$ of M_1 causes an input noise current of $V_{n,M1}/R_F$ [12].

The TIA shown in figure 4.4(c) achieves a better performance because a capacitive feedback is used, i.e. there is no thermal noise from a feedback resistor. The capacitor C_1 senses the signal voltage across capacitor C_2 and returns a corresponding AC current to the input [12]. For gain $A_1 \gg 1$, the ratio

$$I_{out}/I_{in} = 1 + C_2/C_1 \tag{4.3}$$

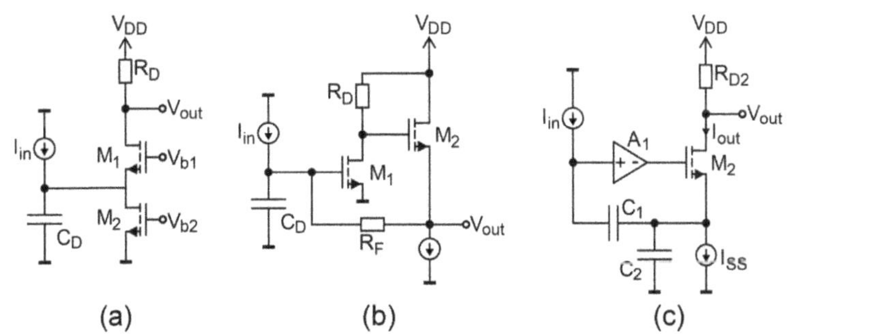

Figure 4.4. Three TIA topologies. (a) Common-gate TIA, (b) TIA with RF, (c) current amplifier [12].

enables the operation as a current amplifier. Together with R_{D2} the transimpedance T_{ca} is obtained [12]:

$$T_{ca} = R_{D2}(1 + C_2/C_1) \tag{4.4}$$

The closed-loop −3 dB bandwith of this current amplifier is

$$f_{ca} = \frac{(1 + A_1)g_{m2}}{2\pi C_2}. \tag{4.5}$$

The current amplifier of figure 4.4(c) convinces over the RF TIA of figure 4.4(b) due to three advantages: first, the network of C_1 and C_2 for definition of the gain does not cause any noise. Second, the capacitance from the input node to ground reduces the DC loop-gain rather than deteriorating the stability. Third, the equivalent input noise voltage of the first stage (A_1) leads to the input noise current [12]

$$I_{n,A1} = \frac{V_{n,A1}C_2 s}{1 + C_2/C_1}. \tag{4.6}$$

Compared to the common-gate TIA of figure 4.4(a), where R_D and M_2 contribute directly their noise, the current amplifier of figure 4.4(c) has the advantage that the current noise of R_{D2} and of the current source I_{SS} is divided by $1 + C_2/C_1$. Another advantage is that the transimpedance gain for the circuit in figure 4.4(c) is $(1 + C_2/C_1)R_{D2}$ instead of R_D for the circuit in figure 4.4(a).

For $C_2 = 0.5$ pF and $R_{D2} = 500\ \Omega$, the input noise current of the current amplifier was lower than that of the RF TIA up to 640 MHz [12]. For a proper design, it was reported that the noise of M_2 is negligible.

The detailed current amplifier circuit is shown in figure 4.5. The voltage drop across R_{D1} must be large (i.e. its resistor value will be large) to reduce its noise contribution [12]. Therefore, capacitive coupling to the second stage with C_s was necessary. M_3 doubled the drain current of M_1 over the current through R_{D1}. V_{GS3} was large to reduce its noise contribution. A gain stage with M_{10} and an output buffer with M_{11} for a 50 Ω load followed the current amplifier core. The integrated inductor increased the bandwidth [12].

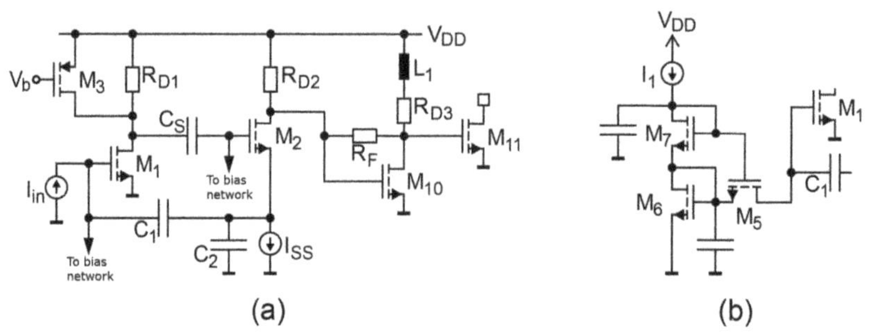

Figure 4.5. (a) Implementation of current amplifier TIA, (b) bias network [12].

For small intersymbol interference (ISI), the lower cut-off frequency of the 620 Mb s^{-1} receiver was chosen as 50 kHz, which required resistors on the order of 2 MΩ at the gates of the transistors M_1 and M_2 (see figure 4.5(a)) [12]. It was also important that these resistors had a small parasitic capacitance. Therefore, the input bias network shown in figure 4.5(b) was used. There, the current I_1 and transistor M_7 determine the channel resistance of M_5. Transistors M_6 and M_1 form a current mirror. I_1 was set to obtain 15 kΩ for g_{m7}^{-1}. The gate width of M_7 was 100 times that of M_5. Therefore, the channel resistance of M_5 was 100 g_{m7}^{-1} [12]. It was mentioned that a large DC component of the photocurrent requires additional measures.

The current amplifier was fabricated in 0.6 μm CMOS and operated with a power supply of 3 V with a power consumption of 30 mW. The die had an area of 400×500 μm^2. The measured bandwidth was 550 MHz at a total transimpedance of 8.7 kΩ and the lower cut-off frequency was 40 kHz. The average input noise current of 4.5 pA $Hz^{-1/2}$ was reported [12], which was determined according to [13].

The principle suggested in [12] was implemented in a 56 MΩ 1.8MHz TIA with an input noise current density of 65 fA $Hz^{-1/2}$ [14]. This capacitive feedback amplifier was fabricated in 0.18 μm CMOS and dissipated 0.436 mW at a supply voltage of 1.8 V. This current amplifier was realized in a differential topology. C_1 was 110 fF and C_2 had a value of 5.5 pF.

For the approach of Razavi [12], a chance of success in USPAR may exist, because the noisy biasing network for the input transistor will not be needed here due to the reset transistor (figure 1.4) for recharging the photodiode at the end of each bit.

4.3 Integrating optical receivers

The integrate&dump receiver introduced in [15] integrated the photogenerated charges on a capacitor and amplified the voltage on the capacitor. In addition, a reset transistor was needed. The principle of this receiver is depicted in figure 4.6. The integration of the photocurrent is done on the input node capacitance C_T, which is composed by the photodiode capacitance C_{PD}, the input capacitance of the amplifier C_{IN} and the capacitance of their interconnection. The reset transistor is not connected to ground in order not to disturb the operating point of the amplifier (see figure 4.6(a)). The on-chip or off-chip electronics has to supply the right voltage to perform the reset to the DC operating voltage of the amplifier input. For this purpose, the replica amplifier, whose input is connected to its output, is implemented to generate the right DC input voltage during the reset when the reset field-effect transistor is conducting.

So, before integration a reset has to be done. Then the voltage on the input node capacitance C_T corresponds to the dark case of the photodiode. If such a receiver is used for non-return-to-zero (NRZ) data transmission, some photogenerated charges are lost (to the output of the replica amplifier). Such a receiver is therefore better appropriate for return-to-zero (RZ) data transmission, where the reset can be done before a light pulse is sent. Readout should be done after the light pulse and before the next reset follows. The amplifier works in open-loop configuration and amplifies

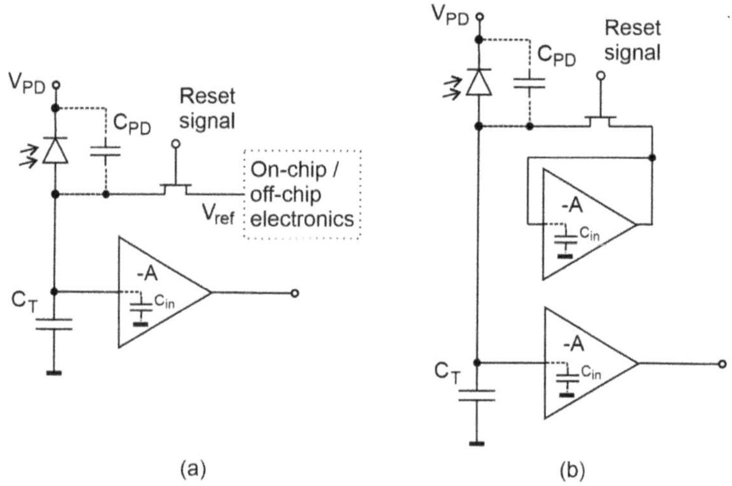

Figure 4.6. (a) Implementation of an integrator with reset to V_{ref}, which is delivered by on-chip or off-chip electronics. (b) Implementation of an integrator where V_{ref} is delivered by a replica (dummy) amplifier [15].

the signal voltage $V(t)$, which occurs during integration of the photogenerated current on the input-node capacitance (assuming NRZ and T_b being the duration of a bit; n is an integer number).

$$V(t) = \frac{1}{C_T}\int_{nT_b}^{t} I(\tau)d\tau, \quad nT_b \leqslant t \leqslant (n+1)T_b \tag{4.7}$$

For the duration of the light pulse T_b and the average photon arrival rate n_r during a logical '1', the signal voltage on C_T is

$$V_1 = n_r T_b \frac{q}{C_T}, \tag{4.8}$$

where q is the electron charge.

The integrate&dump receiver achieves a better signal-to-noise ratio than RF TIAs [15, 16], because there is no thermal noise of a feedback resistor. The thermal noise of the switching transistor (reset transistor), however, causes the so-called reset noise or $k_B T/C$ noise (k_B is the Boltzmann constant and T the abolute temperature). Please see the derivation of the reset noise in chapter 3.

An integrating current sensor in 0.35 μm CMOS with an input current noise density of 4 fA $Hz^{-1/2}$ at 100 kHz was reported [17]. This current sensor can be used for input DC currents below 10 pA and in impedance spectroscopy with the ability to detect capacitance variations in the sub-aF range. In 45 nm CMOS technology, a 2 Gb s^{-1} integrate&dump receiver was realized [16]. A differential implementation with feedback capacitors of 40 fF and integration capacitors of 620 fF was used. The integration period was 300 ps and the reset phase lasted 200 ps. The power consumption was less than 100 mW at a supply voltage of 2.1 V. It was concluded that this receiver is appropriate for 16-level I-Q QAM.

References

[1] Razavi B 2012 *Design of Integrated Circuits for Optical Communications* (Hoboken, NJ: Wiley)

[2] Zimmermann H 2010 *Integrated Silicon Optoelectronics* (Heidelberg, Germany: Springer)

[3] Seidl C, Knorr J and Zimmermann H 2004 Single-stage 378 MHz 178 kΩ transimpedance amplifier with capacitive-coupled voltage dividers *2004 IEEE International Solid-State Circuits Conference* 470–1

[4] Seidl C, Swoboda R and Zimmermann H 2005 Optical receiver IC with capacitive coupled distributed resistors resulting in effectively reduced parasitic capacitance *Electron. Lett.* **41** 1301–2

[5] Brandl P, Jukic T, Enne R, Schneider-Hornstein K and Zimmermann H 2016 Optical wireless APD receiver with high background-light immunity for increased communication distances *IEEE J. Solid-State Circuits* **51** 1663–73

[6] Jukic T, Steindl B, Enne R and Zimmermann H 2016 400 μm diameter APD OEIC in 0.35 μm BiCMOS *IEEE Photon. Technol. Lett.* **28** 2004–7

[7] Milovancev D, Brandl P, Jukic T, Steindl B, Vokic N and Zimmermann H 2019 Optical wireless APD receivers integrated in 0.35 μm HV CMOS technology with large detection area *Opt. Express* **27** 11930–45

[8] Milovancev D 2019 Free-space optical communication with CMOS and BiCMOS receivers *PhD Thesis* Technische Universität Wien

[9] Milovancev D, Jukic T, Vokic N, Brandl P, Steindl B and Zimmermann H 2021 VLC using 800 μm diameter APD receiver integrated in standard 0.35 μm BiCMOS technology *IEEE Photonics J.* **13** 7900513

[10] Jukic T, Steindl B, Enne R and Zimmermann H 2016 200 μm APD OEIC in 0.35 μm BiCMOS *IET Electron. Lett.* **52** 128–30

[11] Gasser C and Zimmermann H 2021 Highly sensitive monolithic optoelectroni receiver with large-area avalanche photodiode for optical wireless communication at low data rates *Opt. Eng.* **60** 075109–1-075109-8

[12] Razavi B 2000 A 622 Mb s^{-1} 4.5 pA/$\sqrt{\text{Hz}}$ CMOS transimpedance amplifier *2000 IEEE International Solid-State Circuits Conference. Digest of Technical Papers* 162–3

[13] Taylor S S and Thomas T P 1994 A 2 pA/$\sqrt{\text{Hz}}$ 622 Mb s^{-1} GaAs MESFET transimpedance amplifier *Proc. of IEEE Int. Solid-State Circuits Conference – ISSCC '94* 254–5

[14] Salvia J, Lajevardi P, Hekmat M and Murmann B 2009 A 56 MΩ CMOS TIA for MEMS applications *2009 IEEE Custom Integrated Circuits Conference* 199–202

[15] Jindal R P 1990 Silicon MOS amplifier operation in the integrate and dump mode for gigahertz band lightwave communication systems *J. Light. Technol.* **8** 1023–6

[16] Gathman J F B T 2013 A 45 nm SOI CMOS integrate-and-dump optical sampling receiver *IEEE Trans. Circuits and Systems* I **60** 469–78

[17] Ferrari G, Gozzini F, Molari A and Sampietro M 2009 Transimpedance amplifier for high sensitivity current measurements on nanodevices *IEEE J. Solid-State Circuits* **44** 1609–16

IOP Publishing

Ultra-Sensitive PIN and Avalanche Photodiode Receivers

Horst Zimmermann

Chapter 5

Low-capacitance photodiodes

Beginning with the well known PIN photodiode and its capacitance, measures for reducing the photodiode capacitance are introduced and explained. Via finger photodiodes we will come to dot photodiodes. Finally solutions for increasing the light-sensitive area with multi-dot photodiodes having reasonably low overall capacitances will be presented. The same approach will be applied to conventional N+/P-well APDs and to APDs exploiting the electric-field-line crowding effect.

5.1 PIN photodiodes—a predictive insight

There are many applications, which benefit from low photodiode capacitances. Optical receivers with resistive-feedback TIAs can achieve a higher bandwidth, i.e. a higher data rate, and/or less noise, i.e. a better optical sensitivity, if the capacitance of the photodiode is reduced [1–3]. In addition, integrated optical sensors achieve a better performance [4] and the output voltages of image sensor pixels increase [5] for lower photodiode capacitances. A special medical application, optical coherence tomography (OCT) with an integrated arrayed waveguide grating (AWG) requiring light-sensitive areas of 500 μm^2 for waveguide-coupling [6], also benefits from a low photodiode capacitance. PIN photodiodes are very popular because of their high bandwidth and low capacitance [7, 8]. Here we will investigate how their capacitance can be further reduced.

PIN photodiodes (see figure 5.1) contain a thick ($d_E \geqslant 10\,\mu m$) low-doped, so-called intrinsic (I) region, which is fully depleted and used as absorption zone for the incident light. Therefore, the rise and fall times of the photocurrent are dominated by the fast carrier drift transport mechanism and—what is also important here—the capacitance of PIN photodiodes $C_{PIN}$is smaller than that of PN-junction photodiodes, which have a thinner space-charge region SCR (for the same applied reverse voltage). The area capacitance of a PIN photodiode is well given by that of a plate capacitor with the dielectric constant of silicon:

doi:10.1088/978-0-7503-5437-0ch5

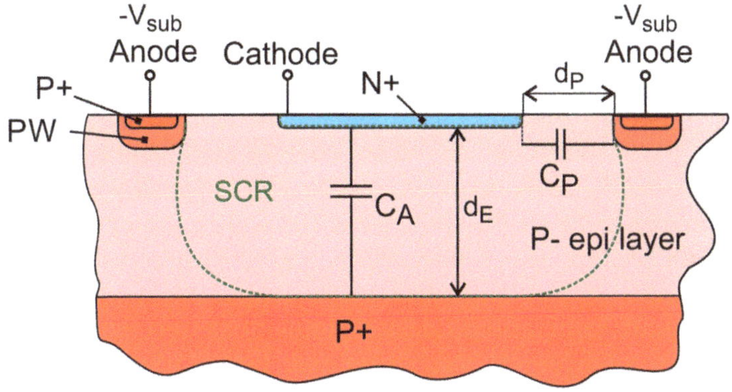

Figure 5.1. Cross-section of PIN photodiode (PW: P-well).

$$C_{\mathrm{PIN}} = \frac{\varepsilon_{\mathrm{r}}\varepsilon_0 A}{d_E}, \tag{5.1}$$

where A is the N+ area, d_E is the thickness of the I-region (low-doped epitaxial layer), ε_{r} is the relative dielectric constant (11.9 for silicon) and ε_0 is the vacuum dielectric constant. In PIN photodiodes the epitaxial layer is doped in the order of 10^{13} cm^{-3} compared to usual IC substrates of 10^{15} cm^{-3}. Therefore, the space-charge region extends 10 times wider in PIN photodiodes (if the I-region is thick enough) than in P/N-junction photodiodes. The operating (reverse) voltage of PIN photodiodes is sufficient to deplete the intrinsic epitaxial layer completely. So roughly, the capacitance of PIN photodiodes is typically 10 times smaller than that of P/N photodiodes with the same junction area (which practically is a little bit larger than the light-sensitive area).

The PIN photodiode integrated in a 0.6 μm BiCMOS process reduced the area capacitance (C_{A}) of an N+/P-substrate photodiode of 0.134 fF μm^{-2} to 0.0105 fF μm^{-2}[9]. The PIN photodiode (see figure 5.1) had a thick, fully depleted epitaxial layer with the thickness $d_E = 15$ μm and the doping concentration of 2×10^{13} cm^{-3}. With an N+ cathode diameter of 20 μm, which is large enough for coupling red light from a single-mode fiber with 4 μm core diameter, the PIN photodiode has a capacitance of 16 fF for the standard epitaxial layer thickness and standard design rules (inclusive perimeter capacitance C_{P}) in 0.35 μm PIN-photodiode complementary metal oxide semiconductor (CMOS) [10].

Let us compare the light-sensitive area and the capacitance for photodiodes of active pixel sensors (APS) and PIN photodiodes. The light-sensitive area of APS pixels is in the order of 1 μm^2 compared to 314 μm^2 for the PIN diode with 20 μm diameter mentioned above. The capacitance of an APS photodiode is on the order of 1 fF, whereas the capacitance of this PIN photodiode is 16 fF. There is a clear advantage for the PIN photodiode with respect to light-sensitive area in relation to capacitance. It is possible, however, to use a thicker epitaxial layer and a wider gap around the cathode to diminish the capacitance of the PIN photodiode.

PIN finger photodiodes (figure 5.2) reduce the area of the P/N-junction for the same light-sensitive area and therefore possess a smaller capacitance than the PIN photodiode with a homogeneous, continuous cathode. The capacitances of PIN finger diodes in 0.6 μm BiCMOS were reduced by factors of 3.2 to 8.5 for finger distances from 10 μm to 30 μm at 3 V reverse bias compared to PIN photodiodes [11]. A similar capacitance reduction is feasible in 0.35 μm PIN photodiode CMOS. Another advantage of finger photodiodes is the higher responsivity for light with short wavelength because less photogenerated carriers can recombine in the smaller N+ regions [12–14].

For an elaborate comparison, device simulations for the finger photodiode consisting of only one N+ cathode finger (see figure 5.3) were performed. The

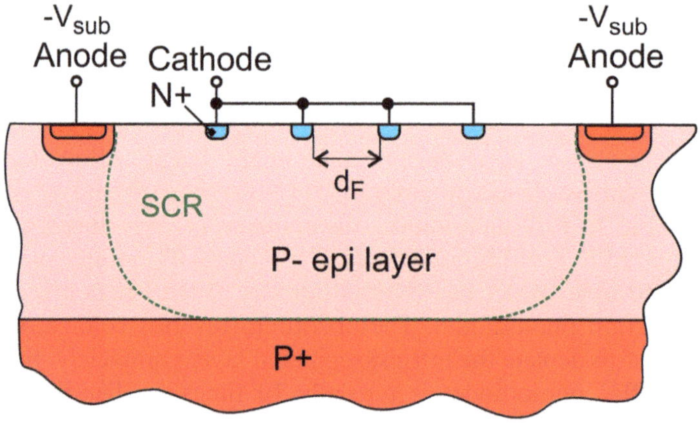

Figure 5.2. Cross-section of PIN finger photodiode.

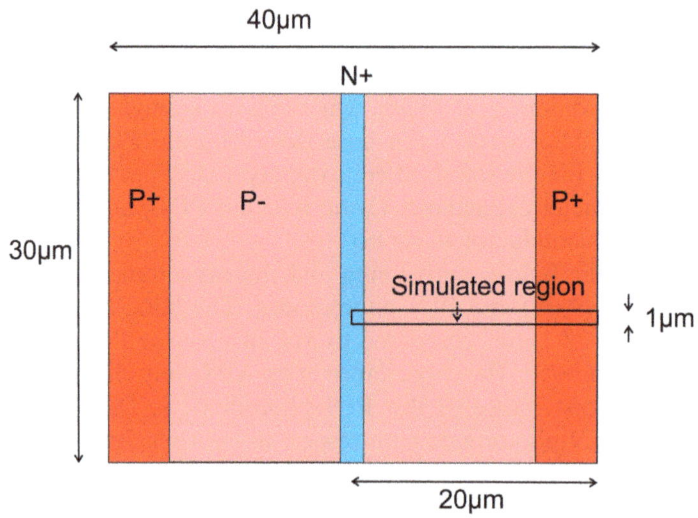

Figure 5.3. Top view of one-finger PIN photodiode.

extension of the N+ region is 4 μm in horizontal direction. The width of the gaps to the P+ areas is 13 μm on both sides of the N+ cathode. The P+ regions have a width of 5 μm each. This is not a typical finger photodiode like that investigated in [11], but leads already to a light-sensitive area of 900 μm^2 (the distance between the two P+ anodes is 30 μm). We can consider this photodiode also as a combination of lateral and vertical PIN photodiode. Due to the symmetry, it is sufficient to simulate only half of the device. Furthermore, it is a good approximation to make a two-dimensional simulation in Cartesian coordinates. In such a simulation, ATLAS [15] assumes the thirrd dimension with an extension of 1 μm. Figure 5.3 depicts the simulated region with a size of 20 μm times 1 μm. In such a way a CPU time and RAM intensive three-dimensional device simulation can be avoided.

The two-dimensional simulation area for the one-finger PIN photodiode shown in figure 5.3 is displayed in figure 5.4. In this simulation area, the N+ cathode has only an extension of 2 μm in x-direction (horizontal direction). There is a 0.4 μm deep shallow trench isolation (STI)[1] with a width of 13 μm between the N+ cathode and the P+ anode. The P+ anode in the right top corner of figure 5.2 and on the left and right edges of figure 5.3 has a width of 5 μm. In addition, there is a bottom anode to contact the P+ substrate. Both anodes are on the same potential (0 V) and the cathode voltage is stepped up to +24 V in the simulations. The P-doping of 1.3×10^{13} cm^{-3} corresponds to a resistivity of 1000 Ωcm.

The simulated capacitance of the one-finger PIN photodiode with this structure is shown in figure 5.5. This result is valid for the simulated region (see figure 5.3) with

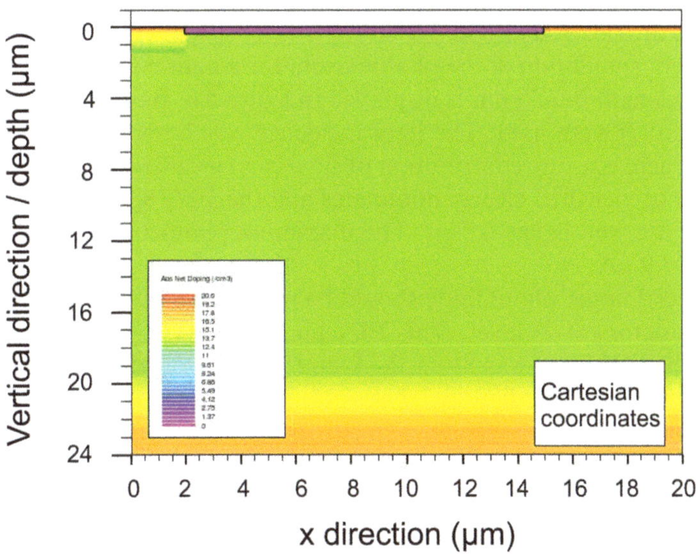

Figure 5.4. Two-dimensional simulation area of one-finger PIN photodiode.

[1] Assuming a 0.18 μm CMOS process.

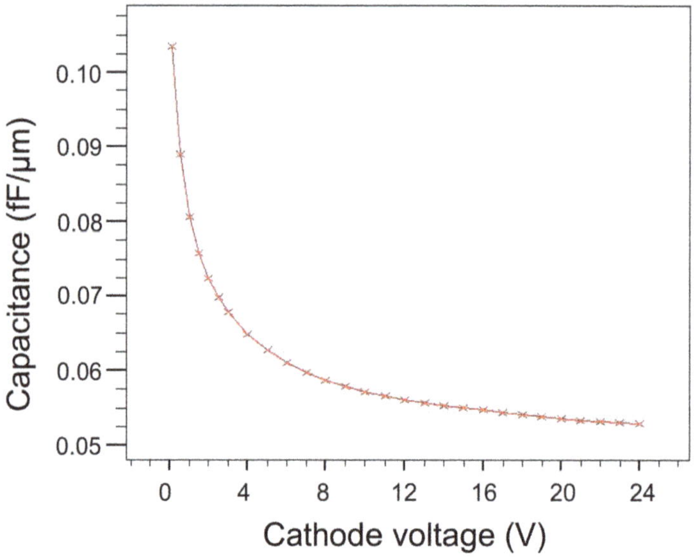

Figure 5.5. Capacitance of one-finger PIN photodiode for a 1 μm thick device.

an area of 20 μm^2. Be aware that the depth (direction perpendicular to the plane of projection) of the simulation region shown in figure 5.4 is 1 μm. To obtain the capacitance of the complete device, we have to multiply the value of about 0.055 fF $\mu m^{-1}$for 24 V cathode potential by 30 and by 2, i.e. by 60, which gives a total capacitance of 3.3 fF. So, the capacitance of the one-finger PIN photodiode with the light-sensitive area of 900 μm^2 is 3.3 fF for a reverse bias of 24 V.

The transient simulation of the photocurrent for a light pulse with a duration of 5 ns and a wavelength of 635 nm is displayed in figure 5.6. Basically the same models as described in [7] were used. The light incidence was homogenous between $x = 0$ and $x = 15$. There is a tiny contribution of slow carrier diffusion visible in figure 5.6. However, the carrier drift clearly dominates and the 10%–90% rise/fall times of the photocurrent are well below 0.5 ns. The maximum photocurrent was about 65 pA times 60, i.e. 3.9 nA.

A very special finger photodiode should be mentioned, which was fabricated in a silicon-on-insulator (SOI) layer. This PIN photodiode was a lateral finger photodiode [16]. The SOI layer was 8 μm thick and phosphorus doped to a resistivity of 700–1300 Ωcm. The interdigitated N+ and P+ fingers had a distance of about 100 μm. The total device area (light-sensitive area) was 3.66 mm^2. The device was covered with a 376 nm thick silicon-dioxide anti-reflection coating with the first and second transmission maxima at 430 nm and 600 nm, respectively. The measured external quantum efficiencies were 88.5% at 430 nm and 75.4% at 638 nm. The mesured rise/fall times were 9.7 ns and 11.2 ns for 400 nm light and 13.3 ns and 12.6 ns for 638 nm, both with 100 V reverse voltage. The corresponding bandwidths estimated with $f_{-3\,\mathrm{dB}} = 2.4/(\pi(t_r + t_f))$are 36.6 MHz and 29.5 MHz, respectively. The dark reverse current was below 0.25 nA below 60 V and increased to 0.66 nA at

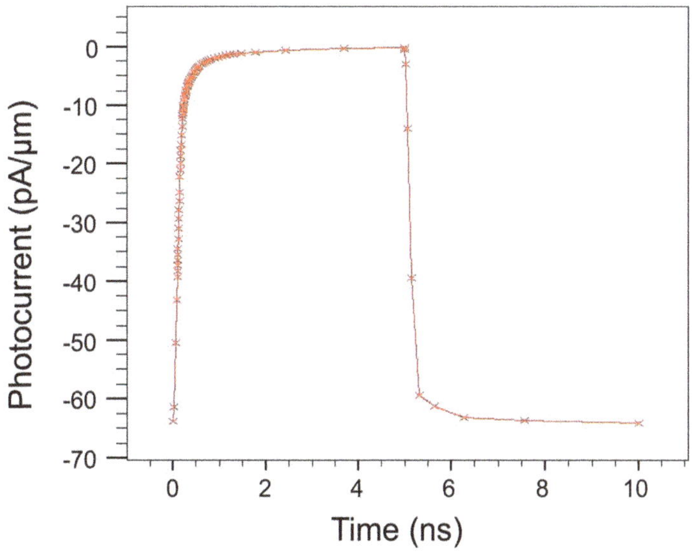

Figure 5.6. Transient response of the photocurrent of the one-finger PIN photodiode for a 1 μm thick device.

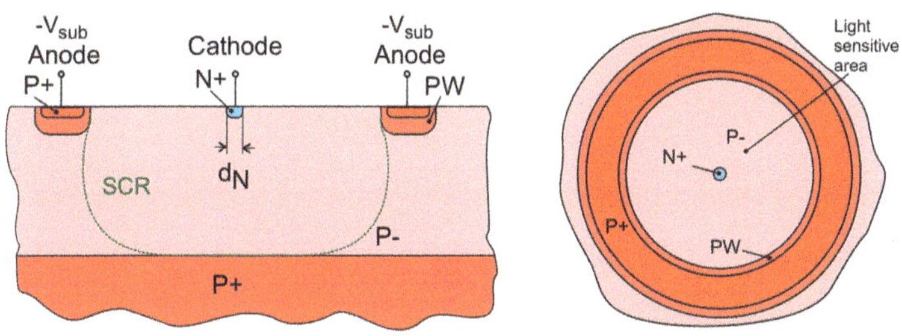

Figure 5.7. Cross-section (left) and top view (right) of dot PIN photodiode.

100 V reverse voltage. But what is most interesting is the capacitance of the device. The measured $C(V)$-curve ended with a value of 4.0 pF at 100 V reverse bias, whereby it has to be mentioned that the capacitance of the bondpad and of the ceramic package of 1.35 pF was included. The capacitance of the device, therefore, was only 2.65 pF at 100 V. At 10 V reverse bias, the capacitance was 5 pF (inclusive of bondpad and package) [16]. The capacitance of 2.65 pF and the light-sensitive area correspond to a value of 0.724 aF μm^{-2}. We will see below in the book that this is the lowest capacitance per area ever reported. But the dot photodiodes introduced below achieve much higher bandwidths at considerably lower operating voltages.

Since the capacitance reduces with the area of the cathode (for PIN-photodiodes on P-/P+ wafers, which actually are often called NIP-photodiodes because the light penetrates through the N+ cathode), a further reduction of the capacitance is possible by implementing an N+ dot as cathode as depicted in figure 5.7. For a 24 μm thick 1000 Ωcm epitaxial layer, device simulations with ATLAS [15] for a device diameter (inner diameter of the P-well substrate contact ring) of 30 μm with an N+ diameter of 4 μm in the center (see figure 5.8) resulted in a capacitance of 1.1 fF at 24 V reverse bias (see figure 5.9). The rise and fall times are about 0.5 ns for red light (635 nm) at 24 V reverse bias (see figure 5.10). For the same dimensions and STI between the cathode and substrate contact ring (see figure 5.11), the capacitance is 0.9 fF (see figure 5.12). This smaller capacitance can be explained by a smaller

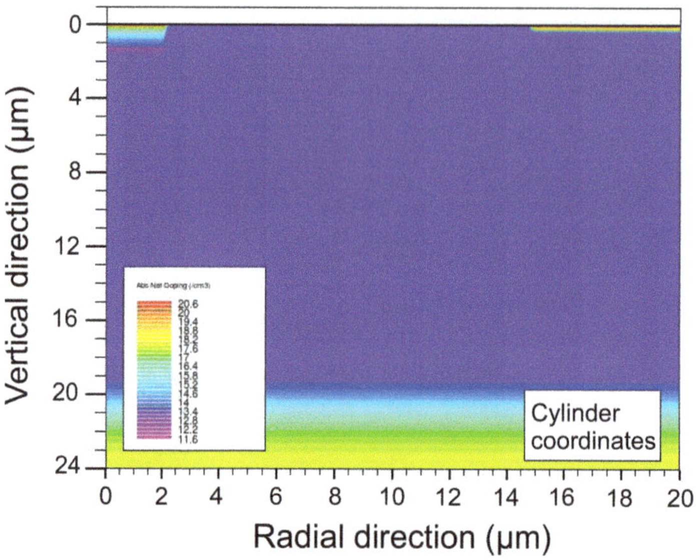

Figure 5.8. Structure and doping used for device simulation of dot PIN photodiode.

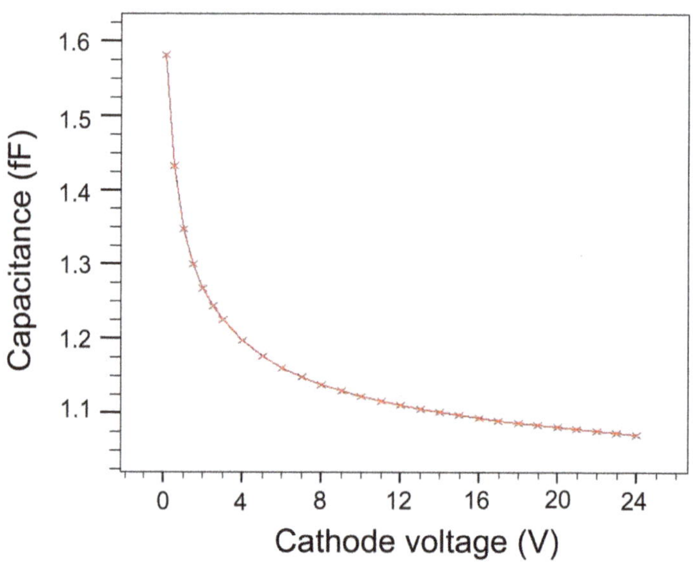

Figure 5.9. Capacitance of a dot-PIN photodiode with 4 μm N+ diameter and 30 μm inner P substrate contact diameter.

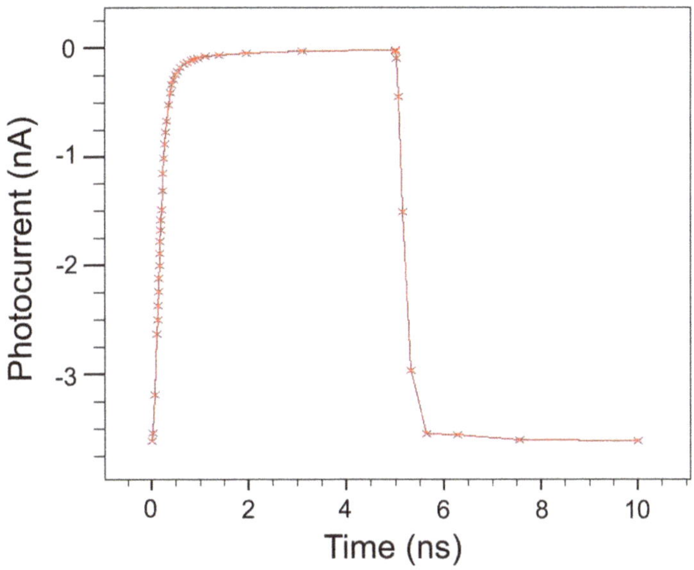

Figure 5.10. Transient response of a dot-PIN photodiode with 4 μm N+ diameter and 30 μm inner P-substrate contact ring diameter for 24 V reverse voltage.

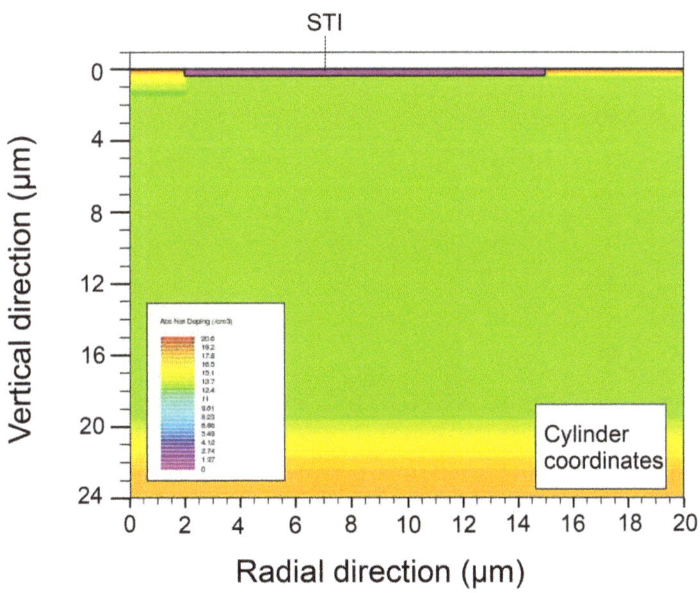

Figure 5.11. Structure and doping used for device simulation of dot PIN photodiode with STI.

N+/P- junction area due to the STI, which has a depth of 0.4 μm in the 0.18 μm CMOS technology used. An N+ dot diameter of 2 μm reduces the capacitance of the dot PIN photodiode further. According to device simulations with ATLAS, the capacitance is then 0.41 fF at 24 V reverse bias (see figure 5.13). But the whole device

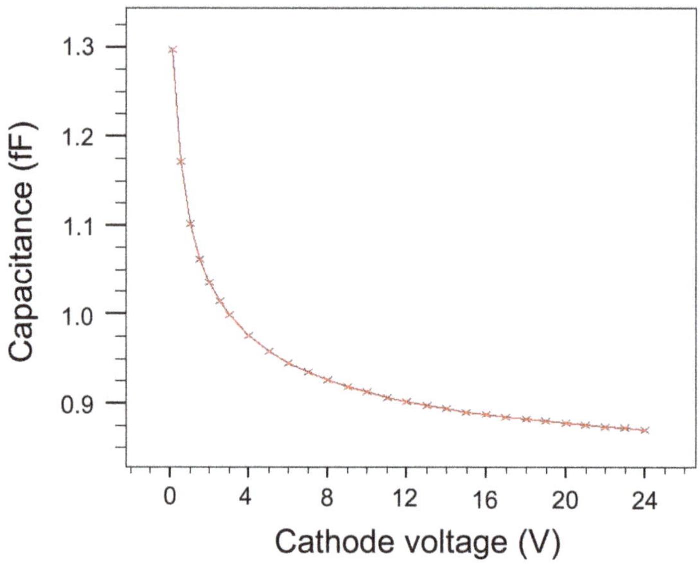

Figure 5.12. Capacitance of a dot-PIN photodiode with 4 μm N+ diameter and 30 μm inner P-substrate contact ring diameter with STI.

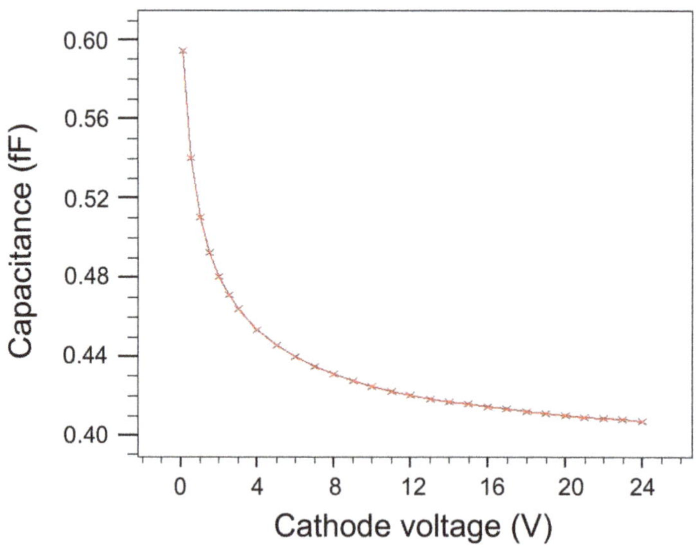

Figure 5.13. Capacitance of a dot-PIN photodiode with 2 μm N+ diameter and 30 μm inner P-substrate contact ring diameter with STI.

radius is not fully depleted anymore because the electric field crowds strongly at the N+ with the low radius of 1 μm. The 10%–90% rise and fall times of the photocurrent increase to about 1.8 ns as a consequence (see figure 5.14).

In the literature, very large photodetectors for x-ray and γ-ray spectroscopy with low capacitance were described [17–19]. They are called silicon drift detectors

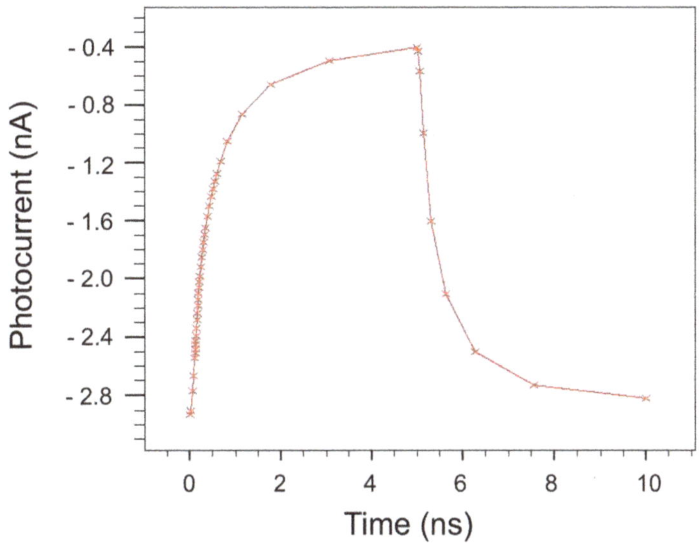

Figure 5.14. Transient response of a dot-PIN photodiode with 2 μm N+ diameter and 30 μm inner P-substrate contact ring diameter with STI for 24 V reverse voltage.

(SDDs). They belong to another class of low-capacitance photodetectors. Several electrodes are exploited to deplete a very low-doped and very large silicon volume. Their capacitance is practically only that of their readout electrode. In [17], a sensitive area of 30 mm^2 was reported with a readout N+ electrode of 8mm times 50 μm and a capacitance of 0.8 pF. In [18], the light-sensitive area was 13 mm^2 at a capacitance of 0.1 pF. However, the drift time from their boundary region was up to 0.7 μs [18] although the electric field strength were in the range from 100 V cm^{-1} to 500 V cm^{-1} [19]. The operating voltages of these SDDs were up between 90 V [17] and 120 V [19]. Another requirement of these SDDs was very low-doped wafers and backside processing, which excludes their integration in mainstream CMOS technology.

The circular dot-PIN photodiode is intended to be integrated together with a low-noise amplifier on the same chip. The circuits have to be isolated from the negative substrate potential, which is the anode voltage of the integrated PIN photodiode. Since we know that the length of the metal line to the amplifier input should be as short as possible in order not to add too much input node capacitance, the deep N-well inside which the circuits have to be placed for insulation reasons should be located close to the anode (substrate contact) ring of the dot-PIN photodiode. The deep N-well, however, has to be connected to the most positive potential of the circuit, i.e. to its positive power supply voltage VDD (3.3 V in 0.35 μm CMOS or 1.8 V in 0.18 μm CMOS), to operate the PN-junctions to P-wells (be aware that the NMOS transistors need a P-well) inside the deep N-well in reverse direction (see section 5.5, where the isolation of circuits is explained in more detail). The amplifier input is about at VDD/2 and because the cathode of the photodiode is connected to the amplifier input, the cathode of the PIN photodiode is also about at VDD/2. The consequence is that the deep N-well is at higher positive potential than the cathode of the PIN photodiode. And what is more, the deep N-well penetrates deeper into the silicon than the P-well substrate contact (and deeper than the cathode). Therefore, the deep N-well may attract photogenerated electrons, which reduces the photocurrent to the amplifier input.

Device simulations using 0.18 μm CMOS doping and VDD = 1.8 V for the transient response with homogenous light incidence inside the inner radius of the P+/P-well substrate contact (surface anode) with a diameter of 30 μm revealed indeed that a noticeable portion of the photocurrent is lost to the deep N-well for red light. Using a Gaussian beam reduced the current loss to the deep N-well because of fewer electrons being photogenerated with increasing radial position towards the anode ring. Figure 5.15 shows the electron path to the deep N-well.

The distance between the deep N-well and the substrate contact ring (anode) of the dot-PIN photodiode, therefore, must not be reduced below about 10 μm. This is a consequence of the thick low-doped epitaxial layer as absorption zone of the dot-PIN photodiode.

An important aspect of integrated photodiodes is also the optical transmission through the isolation and passivation stack on CMOS-integrated circuits. Inside of this stack, several different oxide layers and a silicon nitride or oxynitride layer on top for passivation are present. The thicknesses and the optical refraction index of these layers are not accurately controlled in CMOS fabrication. As a consequence, optical interference inside this stack leads to the formation of standing waves, which means some wavelengths are well entering the silicon below and other wavelengths are reflected in part [7]. In the model for single-photon avalanche diodes (SPADs)

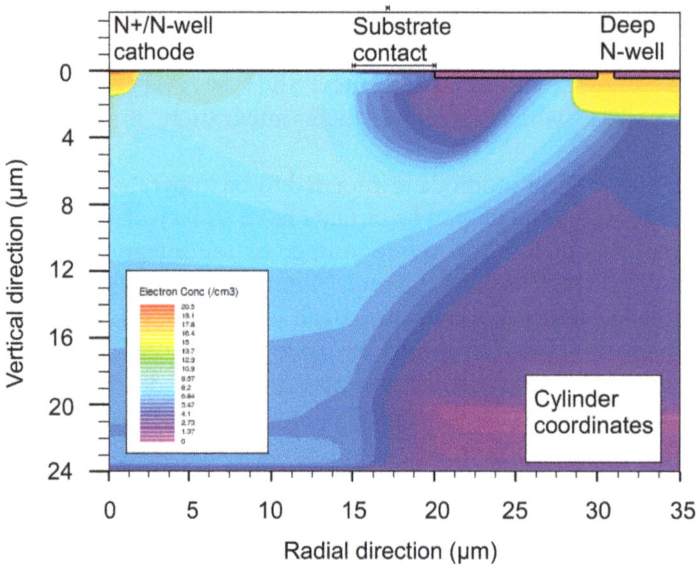

Figure 5.15. Electron distribution in a dot-PIN photodiode with N+/N-well cathode and 30 μm inner P-substrate contact ring diameter with STI.

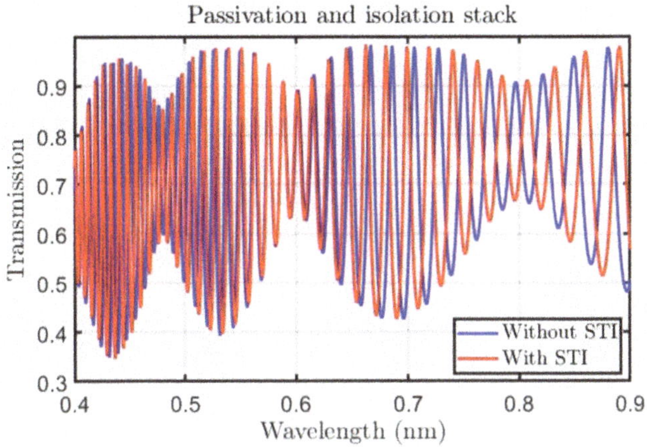

Figure 5.16. Optical transmission through the isolation and passivation stack of the 0.18 μm PIN-photodiode CMOS process.

[20, 21], the optical transmission through the isolation and passivation stack was calculated with an electro-magnetic simulator. This simulator was also used to calculate the optical transmission through the isolation and passivation stack of the 0.18 μm PIN-photodiode CMOS technology. The results are shown in figure 5.16 for the nominal layer thicknesses.

Since the dot-PIN photodiodes in 0.18 μm CMOS can be realized with and without STI between cathode and anode, there are two isolation stacks with two different thicknesses. The STI causes some noticeable differences in the optical transmission, especially at longer wavelengths. For certain wavelengths, the transmission can exceed 95%, however, for certain other wavelengths the transmission can be more than halved.

In the 0.18 μm CMOS process there is no thin anti-reflection coating, but there is a so-called UV window. For this UV window a hole is etched into the silicon nitride passivation layer on top of the chip. The motivation for this UV window is the noticeable absorption of ultraviolet (UV) light in silicon nitride. The remaining isolation stack (oxide) is thinner and the interval between neighboring maxima is larger, but there are still pronounced transmission maxima and minima (see figure 5.17).

There is a tolerance for the remaining oxide thickness in the UV window. The transmission for two oxide thicknesses is compared in figure 5.18 for the case without STI. The oxide thickness tolerances seem to be a smaller problem for wavelengths around 600 nm as the positions of the transmission maxima do not change for different oxide thicknesses.

5.2 PIN photodiodes in 0.18 μm CMOS

Four different PIN photodiodes were designed and fabricated in 0.18 μm high-voltage (HV) CMOS. Each of the four different PIN-photodiode types was realized with and without opto window. The versions with opto window of the first two PIN

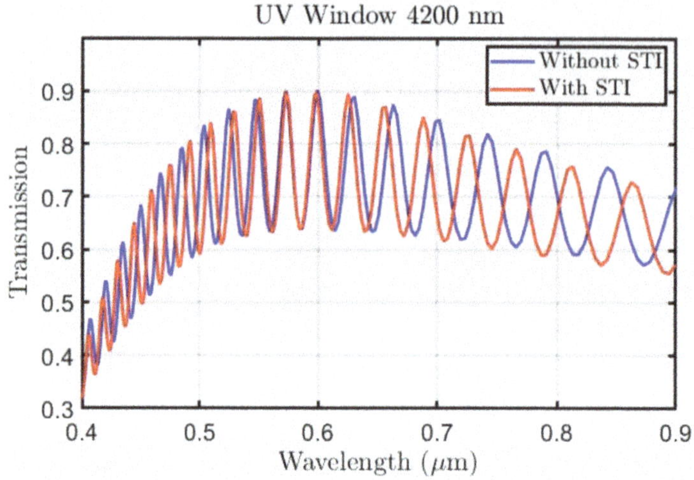

Figure 5.17. Optical transmission for the UV window of the 0.18 μm PIN-photodiode CMOS process with a remaining oxide thickness of 4200 nm.

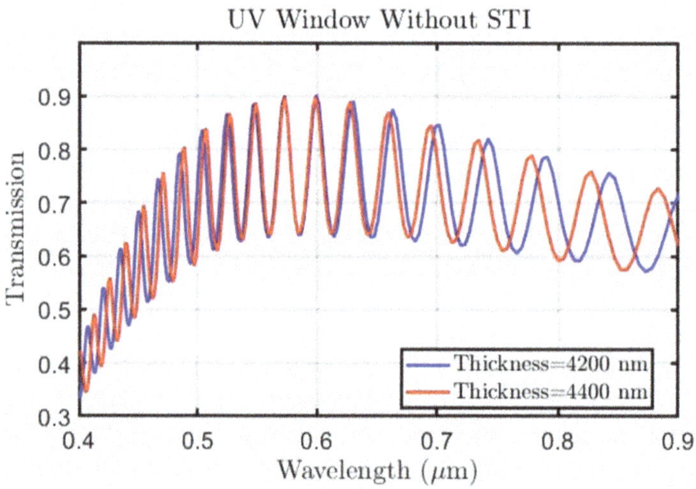

Figure 5.18. Optical transmission with UV window of the 0.18 μm PIN-photodiode CMOS process for remaining oxide thicknesses of 4200 nm and 4400 nm.

photodiodes (PIN1 and PIN2) are illustrated in figure 5.19. Due to the small radius of the cathode dots, consisting of N+ and N-well, of 1.0 μm, these diodes were called dot PIN photodiodes [22]. It should be mentioned that the radius of 1 μm is a design value and that after the process the radius of the N-well is a few 100 nm larger than the radius of the N+ region due to thermal diffusion and the thermal budget of the HV CMOS process.

The motivation for this small cathode radius was the achievable low capacitance. A small conducting sphere has a capacitance of 4 $\pi\varepsilon\varepsilon_0 r$ [23]. The cathode dots

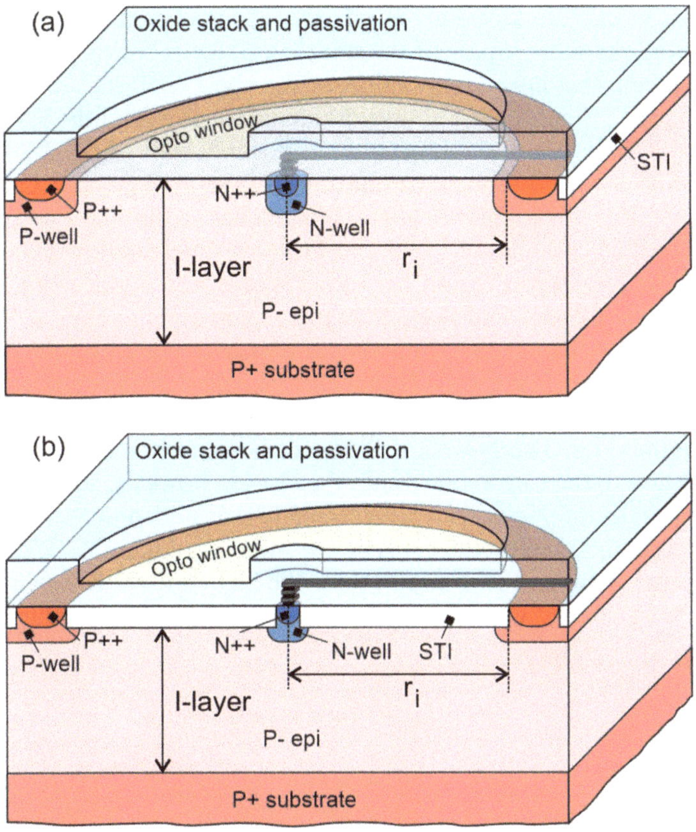

Figure 5.19. Dot PIN-photodiodes with opto window. (a) PIN1. (b) PIN2 with STI. Reproduced from [22] CC BY 4.0.

represent half spheres inside the silicon (inside the low-doped I-layer). Therefore, the capacitance of the dot PIN photodiodes can be considered as

$$C = 2\pi\varepsilon\varepsilon_0 r. \tag{5.2}$$

This equation should represent a good approximation since the inner radius r_i and the I-layer thickness are much larger than the cathode radius (radius of the half sphere). Inserting $\varepsilon = 11.9$ for silicon and $r = 1.0$ μm results in a capacitance of 0.67 fF for PIN1 and PIN2, which looked quite promising for ultra-sensitive PIN-photodiode receivers.

The cathode dots are located in a 24 μm low-doped epitaxial layer with a grown thickness of 24 μm (I-layer in figure 5.19) and a resistivity of 1000 Ωcm. The epitaxial layer was grown on a P+ wafer as substrate.

All devices PIN1, PIN2, PIN3, and PIN4 have two anodes, a surfacc P+/P-well ring anode with an inner radius r_i of 15 μm and a bottom P+ substrate anode. Since the wafers are grinded back to about 300 μm thickness after fabrication, the reverse side of the chips can be conductively glued onto a printed circuit board. Both anodes are connected to the same supply of the reverse voltage. The width of the P+/P-well surface anode ring is 5 μm (design value; again: the P-well diffuses a few 100 nm during the process). Since the N+ source/drain and the P-well of the 1.8 V N-channel MOSFETs as well as the P+ source/drain and the N-well of the 1.8 V P-channel MOSFETs were used, no process modification of the 0.18 μm HV CMOS process were necessary. Just the standard wafers were exchanged by the low-epi wafers.

The radius of the cathodes of PIN3 and PIN4 was increased to 2.0 μm to reduce the width of the low-doped region between cathode and surface ring anode. This should increase the electric field strength in this region and thereby the bandwidth of the devices. The larger cathode radius should reduce the electric field line crowding at the cathode spots and therefore a higher electric field should be spread out into the low-doped region between cathode and surface ring anode. Figure 5.20 shows these spot PIN-photodiodes. The N-well of PIN4 further should reduce the width of the low-doped region to the ring anode. Since the N-well, however, increases the area of the P/N junction to the P-epitaxial layer, the STI was added in PIN4 to reduce the P/N-juction area and thereby the capacitance somewhat again.

The difference between PIN1 and PIN2 is the STI of PIN2 between cathode dot and surface ring anode, which means that in PIN2 the silicon surface between

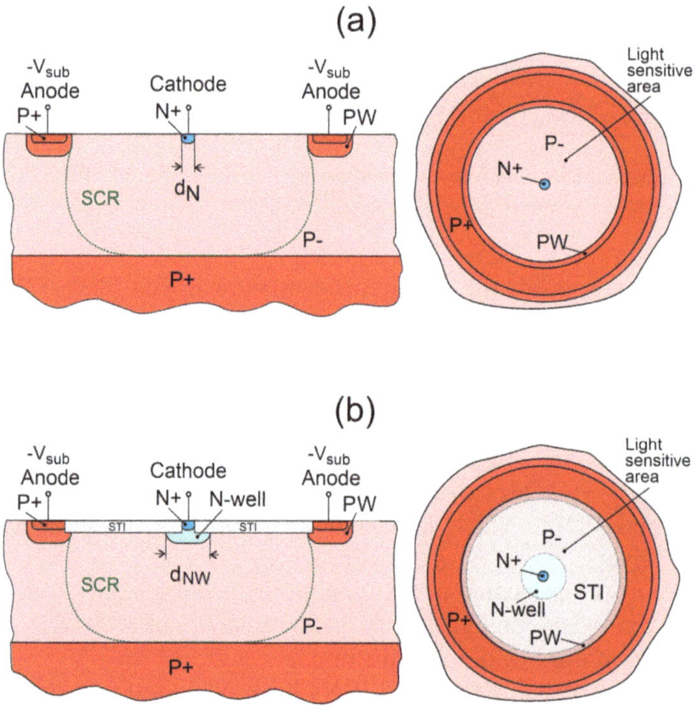

Figure 5.20. Spot PIN-photodiodes (not in scale). (a) PIN3. (b) PIN4 with STI. Reproduced from [24] CC BY 4.0.

N+ and P+ is at $y = 0.4$ μm^2, whereas in PIN1 it is at $y = 0$. In the first version of PIN1 and PIN2, the devices are covered with the complete isolation and passivation stack, whereas a hole is etched through the passivation layer (silicon nitride) and a part of the isolation stack (silicon dioxide) is also removed in the second version of PIN1 and PIN2 within the opto window. These second versions are shown in figure 5.19. The cathode in the center of the devices needs a metal trace over the intrinsic (I) region in the highest metal level to ensure a low capacitance to the silicon substrate. For the second versions, therefore, the opto window must not be drawn to the metal line and has to be kept away for a certain design rule distance in order to exclude etching of the metal.

The breakdown voltage of the dot PIN photodiodes is larger than 60 V [22] and up to about 30 V the reverse current stayed below 1 pA. The responsivities were measured with 635 nm light from a laser diode coupled into the devices with a single-mode fiber having a cropped end. An optical power of less than 1 μW was applied using a calibrated optical power meter. The (photo) current was determined with a Keysight B2987A electrometer. Table 5.1 lists the responsivities of the PIN1 and PIN2 devices.

The responsivity of the devices with the opto window is a little bit larger than without the opto window because of smaller reflections in the remaining oxide. Furthermore, PIN1 has a somewhat larger responsivity than PIN2, which has the STI. But what should be highlighted is that the responsivities do not surge. Therefore, we can exclude impact ionization, which was exploited in [25] for a field-crowding SPAD with a cathode radius of 0.6 μm, up to 30 V reverse bias although the cathode radius of PIN1 and PIN2 is only 1 μm.

For PIN3 and PIN4 the complete dark current characteristics and reverse characteristics with light were measured. Figure 5.21 shows the current–voltage characteristics in reverse direction of PIN3 and PIN4 and compares them to the photocurrent curve for 160 nW of 405 nm light. The dark current is below 1 pA up to about 30 V reverse voltage. The STI does not lead to a larger dark current. The

Table 5.1. Responsivities of dot PIN photodiodes in 0.18 μm HV CMOS for 635 nm. Reproduced from [22] CC BY 4.0.

Reverse voltage (V)	PIN1 No opto window ($A W^{-1}$)	PIN1 Opto window ($A W^{-1}$)	PIN2 No opto window ($A W^{-1}$)	PIN2 Opto window ($A W^{-1}$)
0	0.379	0.387	0.351	0.382
20	0.386	0.405	0.355	0.390
25	0.383	0.401	0.356	0.390
30	0.383	0.405	0.360	0.395

[2] The thickness of the oxide of the STI is 0.4 μm.

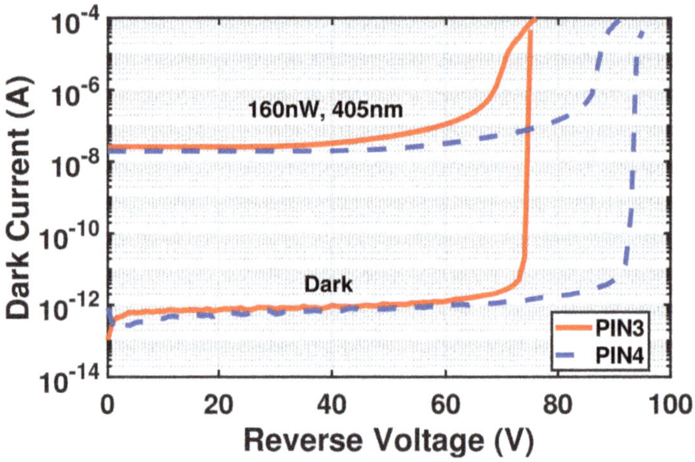

Figure 5.21. Reverse current–voltage characteristics of PIN3 and PIN4 in darkness and with 160 nW of 405 nm light. Reprodcued from. Reproduced from [24] CC BY 4.0.

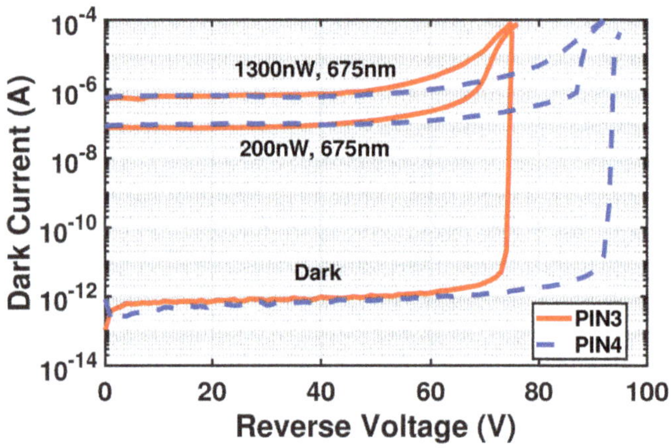

Figure 5.22. Reverse current–voltage characteristics of PIN3 and PIN4 with 675 nm light compared to dark characteristics. Reproduced from. Reproduced from [24] CC BY 4.0.

breakdown voltages of PIN3 and PIN4 are about 75 V and 94 V, respectively. The N-well increases the breakdown voltage of PIN4 due to its much smaller doping compared to that of the N+ of PIN3. The responsivites of PIN3 and PIN4 for 405 nm light are 0.16 A W^{-1} and 0.12 A W^{-1}, respectively, corresponding to 49% and 36.7% quantum efficiencies. There is a more pronounced reflection at this wavelength in the isolation stack on top of PIN4 due to the STI adding 400 nm oxide thickness.

The photocurrent characteristics of PIN3 and PIN4 for 675 nm light are compared to the dark reverse characteristics in figure 5.22. The responsivities at 675 nm for PIN3 and PIN4 are 0.52 A W^{-1} and 0.50 A W^{-1}, respectively,

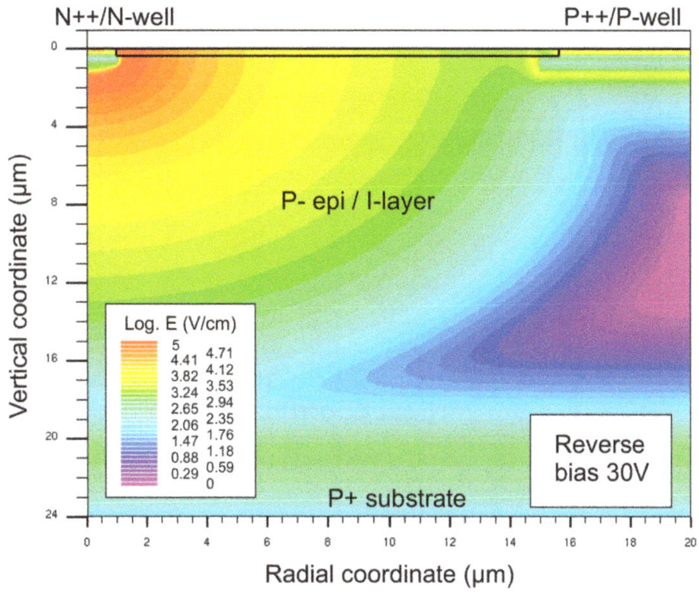

Figure 5.23. Logarithmic electric field inside PIN2 with STI at 30 V reverse voltage. Reproduced from [22] CC BY 4.0.

corresponding to 95.5% and 91.8% quantum efficiencies. The measurement uncertainty of these values was given with 5% [24].

As we will see in the following, the electric field computed with the device simulator ATLAS [15] confirms this finding. The contour plot of the electric field of PIN2 with the STI is presented in figure 5.23 at 30 V reverse bias. The electric-field distribution approximates very well a hemisphere and the intrinsic absorption region inside the surface anode ring is fully depleted. We can conclude that the preconditions for equation (5.2) are fulfilled and that we can expect a very low capacitance. The electric field decreases strongly at about 5 μm depth below the P+/P-well anode ring (in the range $r = 15\ \mu$m to 20 μm). Therefore, we do not take into account the area of the surface ring anode when we determine the light-sensitive area. Then, we obtain the light-sensitive area of 706.9 μm^2 considering the inner radius of 15 μm of the surface anode ring.

To enable a more detailed comprehension, the electric field along a vertical cut line at $r = 0$ (see figure 5.24) and horizontal cut line at $y = 0$ for PIN1 and $y = 0.4\ \mu$m for PIN2 with the STI (see figure 5.25) is compared. The electric field reaches to a depth of about 19 μm. Between 19 μm and 24 μm depth the electric field strength bumps due to a doping gradient from the P+ substrate into the epi layer because of boron diffusion into the 24 μm grown epi layer during epitaxy and the HV CMOS process. What is important for the transient bahavior of the devices is that the electric field strenght in the I-region of PIN1 is a little bit higher than in PIN2 (see figures 5.24 and 5.25). PIN1 will therefore exhibit a larger bandwidth than PIN2.

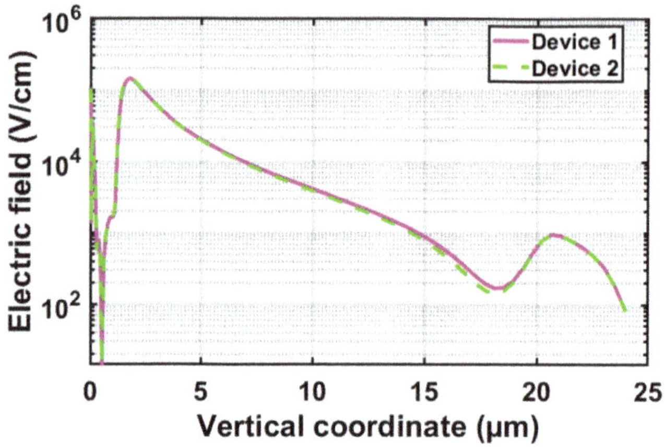

Figure 5.24. Electric field in the center (at $r = 0$) of PIN1 and PIN2 at 30 V reverse voltage. Reproduced from [22] CC BY 4.0.

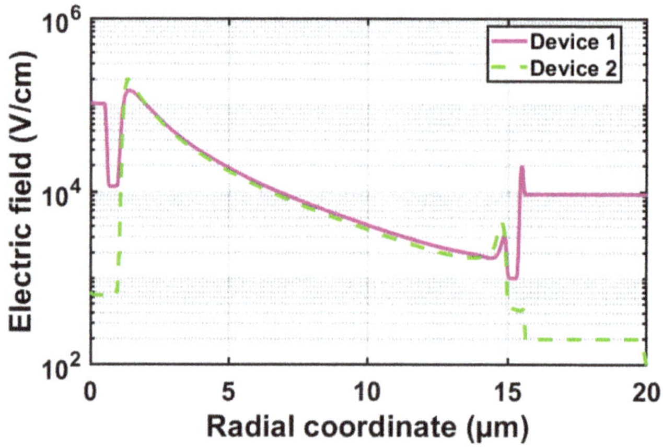

Figure 5.25. Electric field in PIN1 and PIN2 along the silicon surface (PIN1: $y = 0$, PIN2: $y = 0.4\ \mu$m) at 30 V reverse voltage. Reproduced from [22] CC BY 4.0.

The electric field distributions inside PIN3 and PIN4 are shown in figures 5.26 and 5.27. They were calculated at 24 V reverse bias. The complete low-doped region between cathode and anode ring and between cathode and P+ substrate is depleted. Obviously, the depletion is somewhat better in PIN4 especially below the cathode.

To enable a better comparison, the electric field is plotted along a horizontal cut line at the silicon surface (in PIN3 at the depth $y = 0$ and in PIN4 at the depth $y = 0.4\ \mu$m because of the STI) in figure 5.28. The electric field strength is above 1 kV cm^{-1} in PIN3 between the cathode and the surface anode ring, which is between $r = 15\ \mu$m and $r = 20\ \mu$m. The electric field strength in the range $r = 5\ \mu$m to $r = 15\ \mu$m is larger in PIN4 than in PIN3.

Comparing the electric field strength in vertical direction at $r = 0$ (in the center of the cathodes) in PIN3 and PIN4 (figure 5.29), we see that in PIN3 the electric field

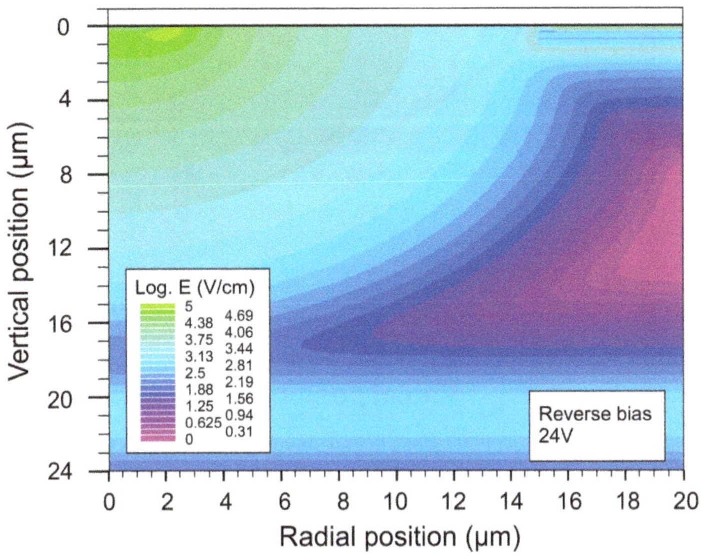

Figure 5.26. Logarithmic electric field inside PIN3 at 24 V reverse voltage. Reproduced from [24] CC BY 4.0.

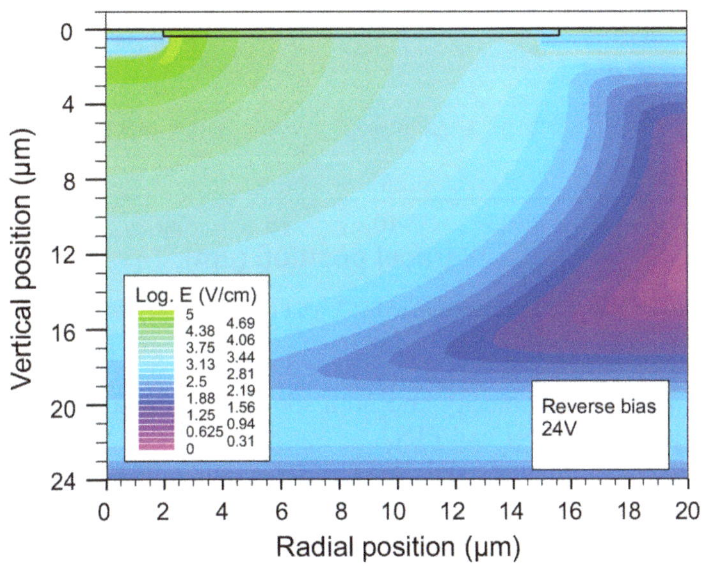

Figure 5.27. Logarithmic electric field inside PIN4 with STI at 24 V reverse voltage. Reproduced from [24] CC BY 4.0.

falls down to about 0.1 kV cm^{-1} at a depth $y = 18\ \mu m$. In PIN4, the electric field strength is larger from $y = 2\ \mu m$ to $y = 20\ \mu m$. The bump at about $y = 21\ \mu m$ is due to the doping gradient of the P+ substrate. Summarizing the findings from the electric-field simulations: PIN4 should possess a larger bandwidth than PIN3

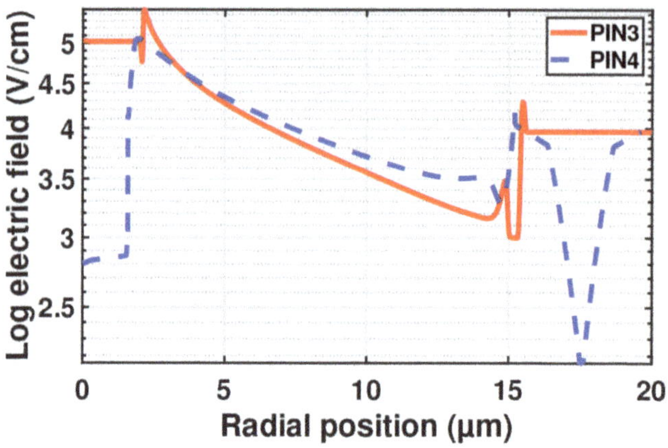

Figure 5.28. Electric field in PIN1 and PIN2 along the silicon surface (PIN3: $y = 0$, PIN4: $y = 0.4\ \mu$m) at 24 V reverse voltage. Reproduced from [24] CC BY 4.0.

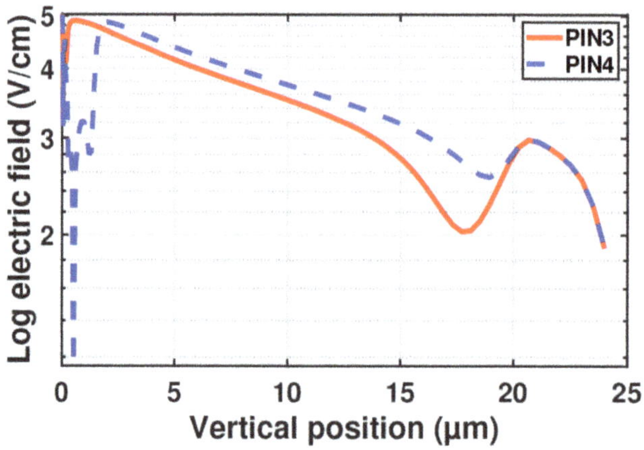

Figure 5.29. Electric field in the center (at $r = 0$) of PIN3 and PIN4 at 24 V reverse voltage. Reproduced from [24] CC BY 4.0.

because of a higher drift velocity and both spot PIN-photodiodes should show a larger bandwidth than PIN1 and PIN2.

The capacitance of the single-dot devices is too small to be measurable accurately even with a precision LCR meter. Therefore, the capacitance of the devices had to be simulated. ATLAS and a small-signal ac analysis at 1 MHz were used to obtain the $C(V)$ curves depicted in figure 5.30. The capcitance of PIN 1 decreases to 0.903 fF at 30 V reverse bias, whereas the final value of PIN2 is 0.806 fF. Taking the light-sensitive area of PIN1 and PIN2 of 706.9 μm^2 into account, we obtain 1.28 aF/μm^2 and 1.14 aF/μm^2 for PIN1 and PIN2, respectively.

The capacitance values of 0.903 fF and 0.806 fF are slightly larger than the expected value from equation (5.2) of 0.67 fF. We should, however, be aware that the N-well diffused a few 100 nm during the HV CMOS process, which increased the cathode radius above the design value of 1.0 μm. The reason for the smaller

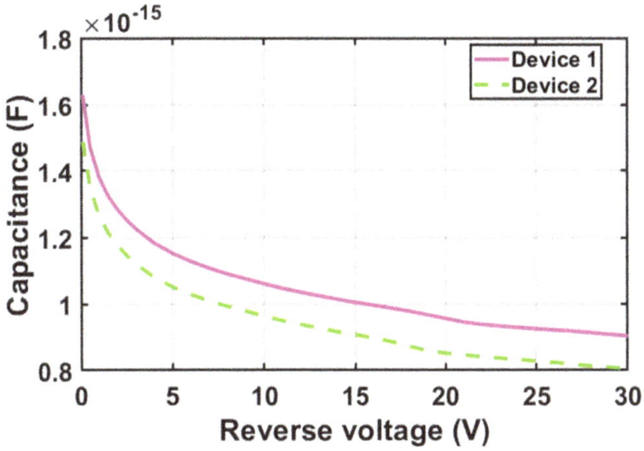

Figure 5.30. Capacitance of PIN1 and PIN2. Reproduced from [22] CC BY 4.0.

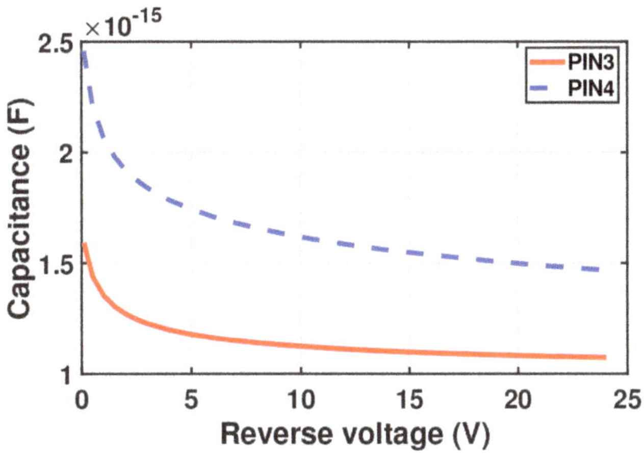

Figure 5.31. Capacitance of PIN3 and PIN4. Reproduced from [24] CC BY 4.0.

capacitance of device 2 is that the STI diminishes its P/N-junction area compared to PIN1.

The capacitance–voltage characteristics of PIN3 and PIN4 are presented in figure 5.31. The capacitance of PIN4 is 1.5 fF at 20 V reverse bias and falls to 1.47 fF at 24 V. PIN3 shows a smaller capacitance of 1.07 fF at 24 V due to the smaller P/N-junction area of the N+ region. Compared to PIN1 and PIN2, however, the capacitances of PIN3 and PIN4 are larger because of the cathode radius of 2 μm instead of 1 μm.

The transient responses of the dot PIN photodiodes were measured with a 675 nm laser directly modulated with a Sympuls BMG 12GIG bit pattern generator. An optical power of 190 μW was coupled via a 50/125 μm multi-mode fiber into the light-sensitive area of the devices. The DC reverse bias was fed via a bias-tee into the photodiodes. A Tektronix TDS 6142C oscilloscope recorded the transient responses.

The obtained step responses of PIN1 and PIN2 (both with the opto window[3]) are compared in figure 5.32. PIN1 shows a (small) advantage as expected from the electric field distributions. Due to the little differences between the step responses of PIN1 and PIN2, only the voltage dependence of the transient behavior of PIN1 was presented in [22]. Figure 5.33 presents the voltage dependence of the transient response of PIN1. Table 5.2 summarizes the measured 10%–90% rise and fall times of PIN1 and PIN2. Because of the thick low-doped epitaxial layer the rise and fall times drop to below about 1.5 ns already for 20 V reverse bias. PIN1 shows somewhat shorter rise and fall times than PIN2 as anticipated above.

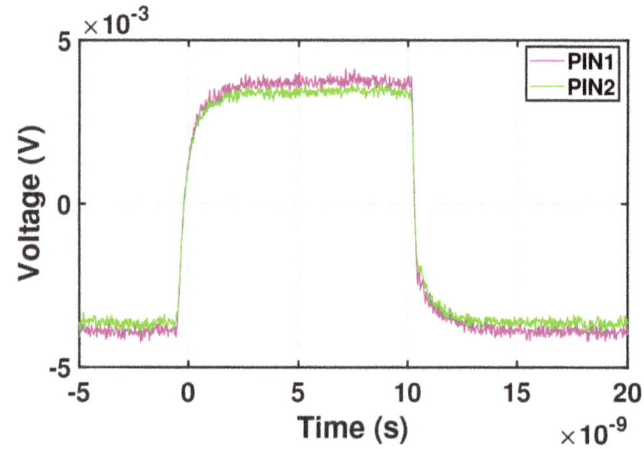

Figure 5.32. Measured step response of PIN1 and PIN2 at 30 V reverse voltage. Reproduced from [22] CC BY 4.0.

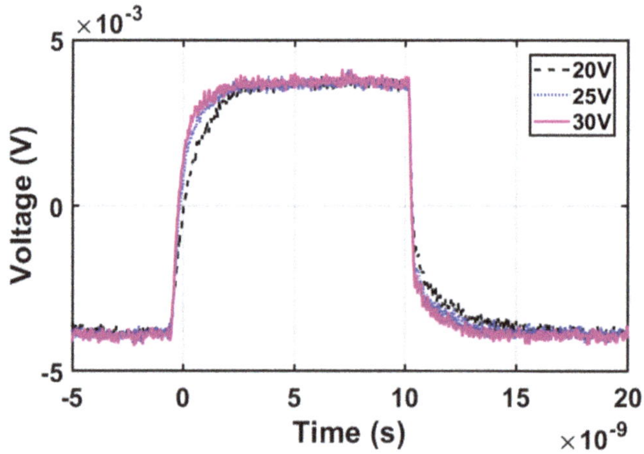

Figure 5.33. Measured step responses of PIN1. Reproduced from [22] CC BY 4.0.

[3] No influence of the opto window on the transient response and on the frequency response is expected.

Table 5.2. 10%–90% rise and fall times of the dot PIN photodiodes PIN1 and PIN2 in 0.18 μm HV CMOS for 675 nm. Reproduced from [22] CC BY 4.0.

Reverse voltage (V)	PIN1 Rise time (ns)	PIN1 Fall time (ns)	PIN2 Rise time (ns)	PIN2 Fall time (ns)
20	1.23	1.68	1.41	1.50
25	1.05	1.43	1.31	1.44
30	0.80	1.05	0.83	1.06

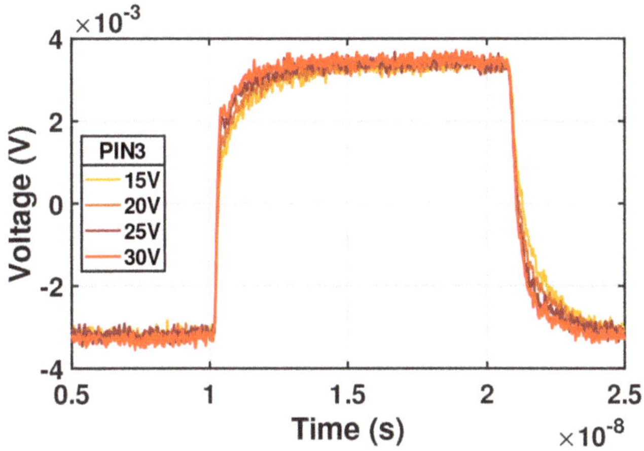

Figure 5.34. Measured step responses of PIN3. Reproduced from [24] CC BY 4.0.

The spot PIN-photodiodes PIN3 and PIN4 were also characterized concerning their transient response. The same equipment was used as for the dot PIN-photodiodes PIN1 and PIN2. Figure 5.34 presents the step responses of PIN3 from 15 V to 30 V reverse bias. At 15 V reverse bias, we can see a small contribution of carrier diffusion. This slow part of the carrier transport decreases with further widening of the depletion region for increasing reverse bias. From figure 5.35, we see that there is less contribution of carrier diffusion already at 15 V reverse voltage. Accordingly, figure 5.36 depicts a little advantage of PIN4 over PIN3 in the step response. For a more detailed comparison the extracted rise- and fall-time values are listed in table 5.3. These rise- and fall-time values were not corrected for the laser rise/fall times of 57 ps and 187 ps, i.e. they represent raw data. Table 5.3 shows that the rise and fall times of PIN3 fall below 1 ns at reverse biases larger than about 25 V. PIN4, however, distinguishes itself by rise and fall times of less than 1 ns in the whole voltage range from 15 V to 30 V.

Furthermore the dot PIN photodiodes PIN1 and PIN2 were characterized with respect to their frequency responses. A vector network analyzer Rhode&Schwarz ZNB8 was used to modulate the 675 nm laser diode. The devices were supplied with

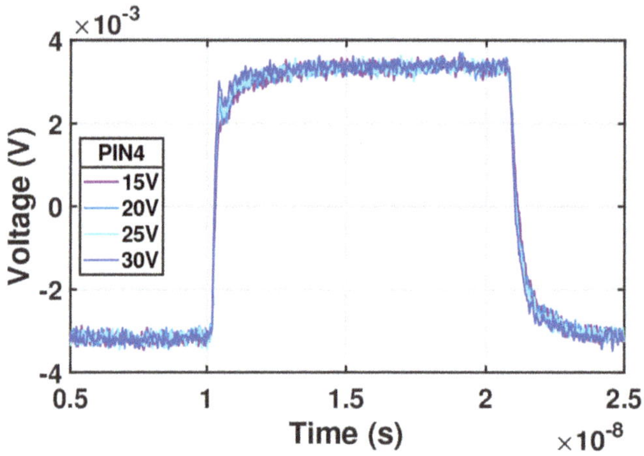

Figure 5.35. Measured step responses of PIN4. Reproduced from [24] CC BY 4.0.

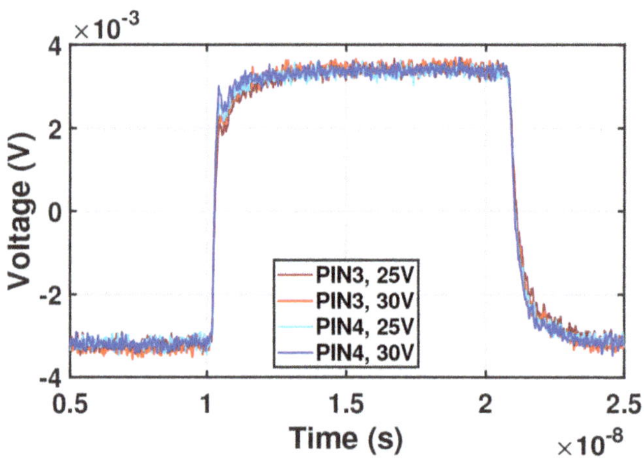

Figure 5.36. Comparison of measured step responses of PIN3 and PIN4. Reproduced from [24] CC BY 4.0.

the DC reverse voltage via a bias-tee. The measured frequency responses of PIN1 are presented in figure 5.37. The maximum −3 dB bandwidth is 310 MHz at 30 V reverse voltage. The measured bandwidth of PIN1 and PIN2 are listed in table 5.4. PIN1 has somewhat higher cut-off frequencies than PIN2 especially at 25 V and below.

The same equipment was used for measuring the frequency responses of the spot PIN-photodiodes PIN3 and PIN4. Figure 5.38 depicts the frequency responses of PIN3 from 5 V to 30 V reverse voltage. There is a pronounced contribution of slow carrier diffusion especially at 5 V reverse bias due to incomplete depletion. This slow carrier transport mechanism is somewhat less distinct in PIN4 (see figure 5.39). Figure 5.40 compares the frequency responses of PIN3 and PIN4 for the two highest reverse voltages. PIN4 outperforms PIN3 with respect to bandwidth. The extracted

Table 5.3. 10%–90% rise and fall times of the spot PIN photodiodes PIN3 and PIN4 in 0.18 μm HV CMOS for 675 nm. Reproduced from [24] CC BY 4.0.

Reverse voltage (V)	PIN3 Rise time (ns)	PIN3 Fall time (ns)	PIN4 Rise time (ns)	PIN4 Fall time (ns)
20	1.10	1.68	0.73	0.92
25	0.92	1.03	0.66	0.86
30	0.73	0.98	0.57	0.82

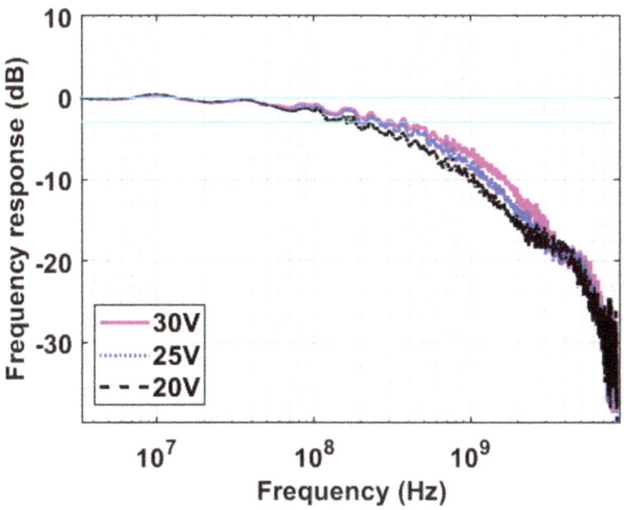

Figure 5.37. Measured frequency responses of PIN1. Reproduced from [22] CC BY 4.0.

Table 5.4. Measured −3 dB cut-off frequencies of PIN1, PIN2, PIN3, and PIN4 in 0.18 μm HV CMOS for 675 nm. Reproduced from [22, 24] CC BY 4.0.

Reverse voltage (V)	PIN1 (MHz)	PIN2 (MHz)	PIN3 (MHz)	PIN4 (MHz)
20	170	110	340	560
25	250	180	410	565
30	310	300	480	690

−3 dB cut-off frequencies of PIN3 and PIN4 are summarized and compared to the corresponding values of PIN1 and PIN2 in table 5.4. The bandwidth of PIN4 is already 520 MHz at 15 V reverse bias and is much larger (more than 150 MHz) than that of PIN3 in the voltage range 15 V to 30 V [24]. At 30 V, PIN4 achieves

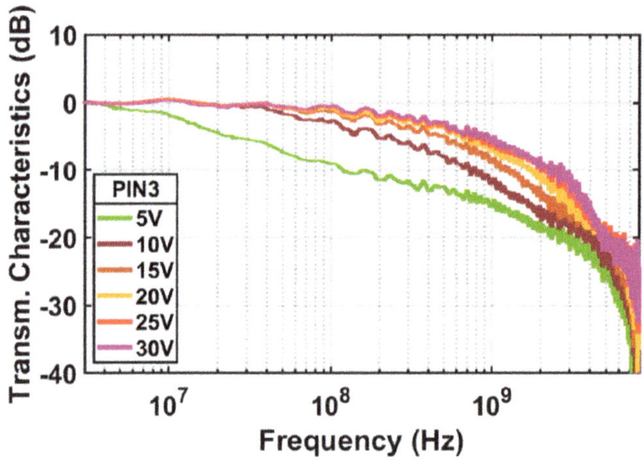

Figure 5.38. Measured frequency responses of PIN3. Reproduced from [24] CC BY 4.0.

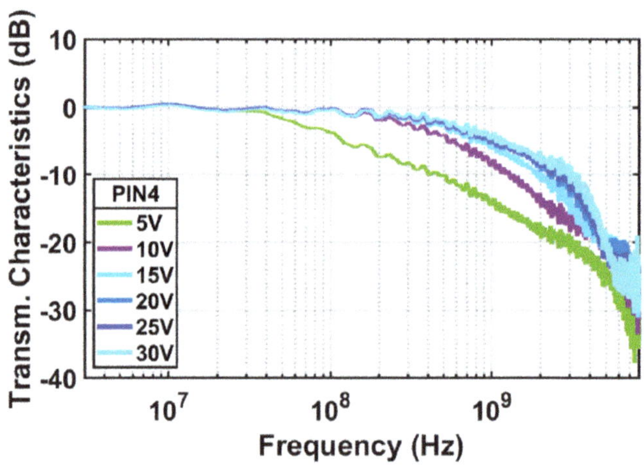

Figure 5.39. Measured frequency responses of PIN4. Reproduced from [24] CC BY 4.0.

690 MHz bandwidth, which should allow for data rates of up to 1 GHz. PIN3 and PIN4, both, are faster devices than PIN1 and PIN2 because of the larger cathode radius, which reduces the electric field line crowding at the P/N junction and leaves a higher electric field in the whole light-sensitive region (absorption region) of the spot PIN-photodiodes.

We can conclude that the dot PIN photodiodes PIN1 and PIN2 achieve capacitances being by factors of 8.2 and 9.2 smaller than that of a usual PIN photodiode with 10.5 aF μm^{-2} [9]. However, the price is a lower bandwidth of 300 MHz even at 30 V reverse bias compared to 625 MHz at 3 V reverse voltage [9]. The finger photodiodes in 0.6 μm BiCMOS [11] were characterized by a capacitance of 6.9 aF μm^{-2}. The capacitance of the field-crowding SPAD with a cathode radius of 0.6 μm [25] was 6.4 aF μm^{-2}. An usual PIN photodiode with an epi-layer thickness

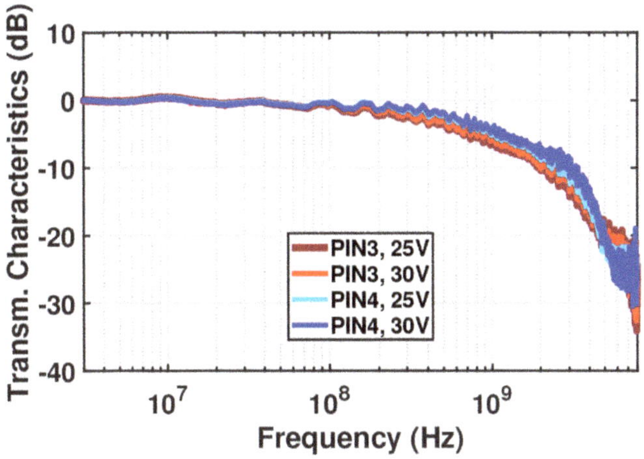

Figure 5.40. Comparison of measured frequency responses of PIN3 and PIN4. Reproduced from [24] CC BY 4.0.

of 40 μm [26] achieved a capacitance of 3 aF μm^{-2}, however the perimeter capacitance was neglected and such thick low-doped epi layers are difficult to grow.

Furthermore, the dot PIN photodiodes PIN1 and PIN2 join low capacitance with a large light-sensitive area and a bandwidth being sufficient for data rates up to a few hundred Mb s^{-1} at the price of a reverse bias up to 30 V. This voltage, however, is in the same range as the reverse voltage needed for SPADs. Unfortunately, the lowest capacitance and the highest bandwidth are not possible with the same device. PIN2 possesses the lowest capacitance but the smallest bandwidth.

The spot PIN-photodiodes PIN3 and PIN4 have still low capacitances. PIN4 achieves 2.08 aF μm^{-2} and PIN3 1.51 aF μm^{-2} at 24 V reverse voltage. Their advantage over PIN1 and PIN2 is their higher bandwidth at the same light-sensitive area. PIN4 with the largest capacitance possesses the highest bandwidth.

Concluding this section, we can say that the spot and dot PIN-photodiodes are very well appropriate for implementation in ultra-sensitive PIN and avalanche photodiode receiver (USPAR) integrated circuits. They are characterized by capacitances similar as in image sensors but at light-sensitive areas being up to three orders of magnitude larger. The bandwidth of the spot and dot PIN-photodiodes is also several orders of magnitude larger than the frame rate of image sensors.

5.3 Avalanche photodiodes

The PIN-photodiode CMOS technology was used to realize APDs with a thick absorption zone [27–29]. These APDs (see figure 5.41) actually are so-called reach-through APDs. The space-charge region pervades the P-well, where the multiplication zone is located below the N+ region, and fully depletes the low-doped epitaxial layer, which is the absorption zone, already (far) below the breakdown voltage. Due to the thick absorption zone this APD has a higher responsivity in the

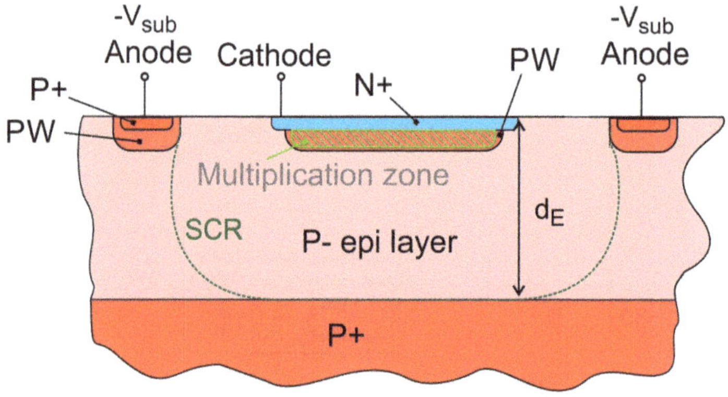

Figure 5.41. Cross-section of PIN-APD (PW: P-well).

red and near-infrared spectral range than thin P+/(deep) N-well APDs [30–35]. What also makes these thick APDs appropriate for ultra-sensitive optical receivers is their low capacitance, which is equal to that of the PIN photodiode shown above. It should be mentioned that the overlap of the N+ cathode over the P-well forms a so-called virtual guard ring to avoid premature edge breakdown, which would occur if the P-well was larger than the N+. With this virtual guard ring the breakdown voltage between N+ and P- epitaxial layer is (much) larger than the area breakdown voltage between N+ and P-well.

If the area of the P/N-junction (N+/P-well junction) of these APDs is reduced, the capacitance can be reduced. However, if the N+ radius and P-well radius become very small (keeping the overlap of N+ over the P-well constant as virtual guard ring), not enough multiplication zone will be left (see [36]). So the concept of the virtual guard ring cannot be used. The P-well has to surround the N+ dot, as shown in figure 5.42. If the N+ radius is small and the overlap of P-well over the N+ in lateral direction is about equal to its vertical overlap, the multiplication zone will be circular (actually like a half sphere). So, the edge breakdown happens everywhere at the N+/P-well junction. The breakdown voltage is therefore somewhat lower than that of the conventional device in figure 5.41.

Recently, another structure for a SPAD was suggested by IMEC [25]. Figure 5.43 shows the three-dimensional structure of this dot-SPAD fabricated in 0.13 μm CMOS. The n+ radius was 0.6 μm and the inner radius of the p+ anode region at the surface was 5 μm. Due to the small n+ radius crowding of the electric field lines at the n+p (i) junction occurs and the critical field strength for impact ionization is exceeded. In fact the authors claim a radius of 2 μm for the avalanche region. Furthermore, the n+ region was reported to be of half-spherical shape. In fact, the cross-section of the electric field shown in figure 5.43(b) reveals that the electric field distribution is not exactly spherical. At the silicon surface, which is actually the interface to an oxide layer, this is caused by the shallow n-type region. This low or medium doped (unfortunately not defined well in the paper) n-type region was implemented to reduce the dark count rate. Below the center of the device, the electric field is a little bit weaker, but it is large enough everywhere in the

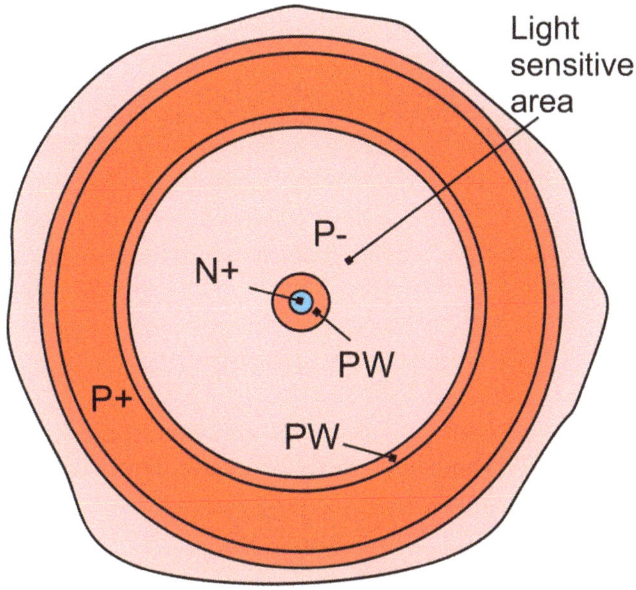

Figure 5.42. Top view of dot-PIN-APD (PW: P-well).

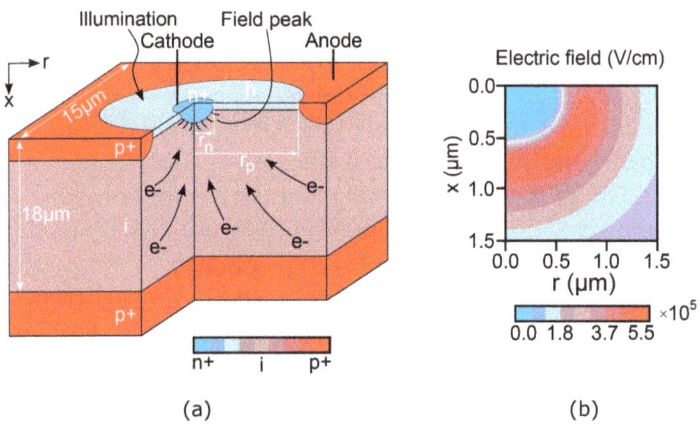

Figure 5.43. Structure of dot-SPAD (a) and simulated electric field at the n+ cathode (b) [25].

multiplication region to allow for multiplication of electrons drifting from all over the device. The epitaxial layer (i zone) had a thickness of 18 μm to enhance the PDP to 13% for 905 nm light in ranging applications. The median breakdown voltage of this dot-SPAD was 67.25 V at 25 °C. An n+n-well dot-SPAD had been suggested in [37], where a small dependence of its breakdown voltage on process nonuniformities of less than 0.6% over a wafer was reported.

A 'tip avalanche photodiode' (TAPD) was used as SPAD in a silicon photomultiplier (SiPM) [38] with the roots going back to [39], where a spherical P+/N APD in the Geiger mode was introduced. The devices in [38] had N+/P- junctions with n+ diameters of 0.6 μm, 0.8 μm and 1.0 μm. The P-doping concentration was 10^{14} cm^{-3}. The breakdown voltages were 43.4 V, 50.7 V and 53.9 V, respectively. It should be mentioned that the title of reference [38] is a bit misleading, since in terms of CMOS technologists, the TAPD is a planar device since the silicon surface is planar in this device. But this article is interesting since it describes the theory behind the field-crowding APDs and SPADs, i.e. the dependence of the electric field on the radius.

For sperical p/n-junctions the one-dimensional Poisson equation in spherical coordinates is [38, 40]:

$$\frac{1}{r^2}\frac{d}{dr}\left(r^2\frac{d\Psi(r)}{dr}\right) = -\frac{\rho r}{\varepsilon\varepsilon_0} \tag{5.3}$$

Assuming an abrupt N+/P- junction and complete ionization of donors and acceptors, the electric field can be obtained by integration of the Poisson equation in the N-type and P-type regions

$$E_{\mathrm{n}}(r) = \frac{qN_{\mathrm{D}}}{3\varepsilon\varepsilon_0}\frac{r^3 - r_{\mathrm{D}}^3}{r^2}$$

$$E_{\mathrm{p}}(r) = \frac{qN_{\mathrm{D}}}{3\varepsilon\varepsilon_0}\frac{r_j^3 - r_{\mathrm{D}}^3}{r^2} - qN_{\mathrm{A}}3\varepsilon\varepsilon_0\frac{r^3 - r_j^3}{r^2}, \tag{5.4}$$

where the boundary condition of a zero electric field outside the depletion region was used and where r_{D} and r_j are the radius of the undepleted n+ region and the radius of the N+/P junction, respectively. For our USPAR photodiodes, mainly the electric field in the p-type region is interesting and since N_{D} is 4 to 7 orders of magnitude (4 orders for the case of an N-well and 7 orders for the case of only N+ in the cathode) larger than N_{A}, the first term in the equation for $E_{\mathrm{p}}(r)$ dominates and the electric field decreases proportional to $1/r^2$ from the N+/P-junction.

According to our experience with APDs and SPAD [27–29, 41, 42], this N+/N-well dot structure also should work in the linear mode as APD. Figure 5.44 depicts the top view of the innovative dot-APD, which exploits the field crowding effect to form the multiplication zone. This structure lives from the small radii of N+ and N-well. But this makes the device's breakdown voltage less dependent on process variations and doping tolerances [37] than the device of figure 5.41. The shallow low- or medium-doped n-type region of [25] at the silicon surface in the intrinsic region is omitted in USPAR ICs.

The APD device was fabricated in the 0.18 μm HV CMOS technology using a PIN-photodiode wafer with a 24 μm thick low-doped epitaxial layer with a resistivity of 1000 Ωcm and is described in [43]. The radius of the N+ region is 0.37 μm and the N-well has a radius of 0.6 μm (see figure 5.44). The inner radius of the anode substrate contact ring (P+) is 19 μm. The P+ substrate acts as a second anode. Both anodes are at the same voltage. The device is covered by the standard isolation and passivation stack of the 0.18 μm technology.

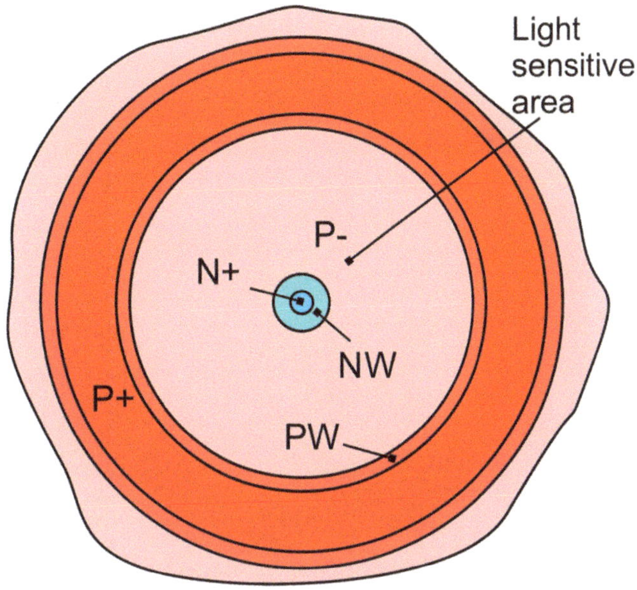

Figure 5.44. Top view of dot-PIN-N^+/N-well APD using the field-crowding effect (NW: N-well).

The electric field inside the device was calculated by TCAD simulations using ATLAS [15]. Figure 5.45 illustrates that the field distribution is quite spherical and the field vectors point nicely into radial direction. The maximum of the electric field is at the N-well/P-epi junction and decreases with about d^{-2} into the epitaxial layer. At the P+ substrate contact ring ($r = 19\ \mu$m), the electric field strength still has a value of about 3500 V cm^{-1} [43]. In the 'diagonal' direction (following the dashed line in figure 5.45(a)) the electric field strength is still about 3000 V cm^{-1} at $d = 25\ \mu$m. At a depth of 20 μm (at the transition from the P-epi layer to the P+ substrate) just vertically below the cathode, the electric field strength is still about 5000 V cm^{-1} [43]. The device, therefore, is well depleted at 69 V reverse bias and promises a high drift velocity, i.e. a high bandwidth. The whole low-doped epitaxial layer serves as a thick absorption zone and the photogenerated electron and hole pairs are separated by the electric field; the electrons drift with a high speed into the multiplication region. The width of the multiplication region, where the electric field exceeds 2×10^5 V cm^{-1}, is about 0.9 μm (see the inset in figure 5.45(b)). The device therefore promises a high responsivity (also for long wavelengths) in addition to the high bandwidth.

Measurements of responsivity and bandwidth were performed on a wafer prober whose thermochuck was regulated to 25 °C. The dark current and the photocurrents were determined with a Keithley B2987A electrometer. The light was coupled into the device with a 675 nm single-mode fiber whose core diameter was 5 μm and an optical attenuator was used to adjust the optical power P_{opt}, which was measured with a calibrated optical power meter [43].

Figure 5.46 presents the current–voltage characteristics in reverse direction for the dark case and for different incident optical powers. The reverse current is below

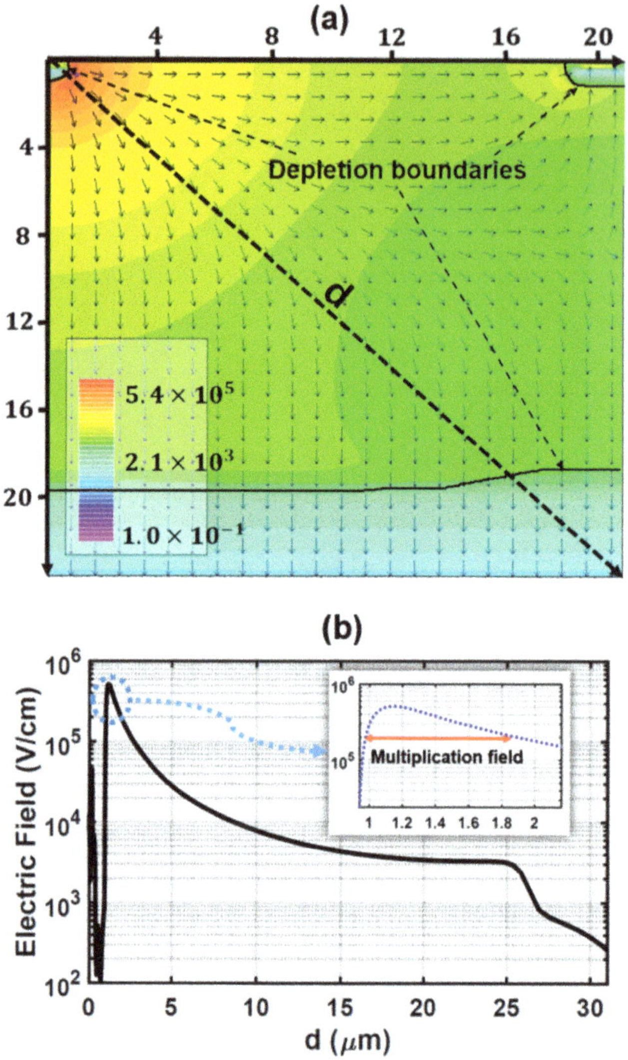

Figure 5.45. (a) Two-dimensional plot of the electric field of the dot-APD in 0.18 μm HV CMOS at a reverse bias of 69 V and (b) radial cross-section of the electric field along the dashed line in sub-figure (a). Reproduced from [43] CC BY 4.0.

1 pA up to more than 40 V reverse bias. From the dark-current curve we can read the breakdown voltage (at a dark current of 1 μA) of 70.7 V [43]. This voltage value is well within the isolation capability of the 0.18 μm HV CMOS process of 100 V.

The responsivity R was determined from the measured photocurrent characteristics by [43]

$$R(A/W) = \frac{I(V) - I_{\text{dark}}(V)}{P_{\text{opt}}}, \tag{5.5}$$

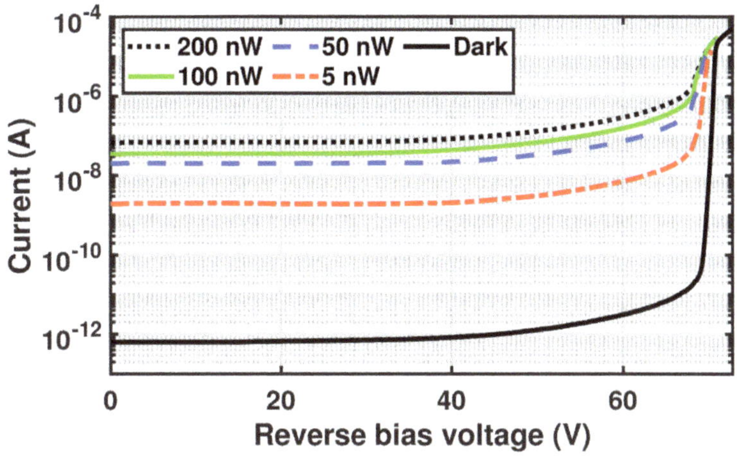

Figure 5.46. Reverse characteristics and photocurrents of field-crowding dot-APD in 0.18 μm HV CMOS at 675 nm. Reproduced from [43] CC BY 4.0.

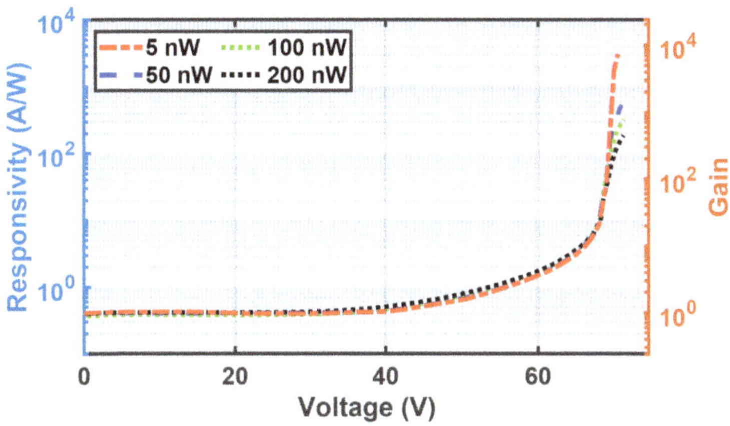

Figure 5.47. Measured responsivity (left, blue) and gain (right, red) of field-crowding dot-APD in 0.18 μm HV CMOS at 675 nm [43] CC BY 4.0.

where I represents currents. The gain was calculated with [43]

$$\text{Gain} = \frac{I(V) - I_{\text{dark}}(V)}{I(V_0) - I_{\text{dark}}(V_0)}, \tag{5.6}$$

where V_0 is a reference voltage for unity gain, i.e. where the device acts as a pure PIN photodiode without any gain.

The responsivity and gain in dependence on the reverse bias is plotted in figure 5.47 for 5 nW, 50 nW, 100 nW and 200 nW optical power. The unity-gain responsivity is 0.41 A W^{-1} for $V_0 = 1$ V. At 70.5 V, the maximum responsivity of 3050 A W^{-1} was obtained [43]. The responsivity and the gain are independent of the optical power up to a reverse voltage of 68.94 V, where the responsivity is below

30.6 A W^{-1} and the gain is lower than 76.5. Above this reverse voltage, the responsivity and gain for higher optical power are smaller than those for lower optical power. This is due to a saturation effect of the impact ionization process because of partial screening of the electric field by many charge carriers [44]. Table 5.5 lists the responsivity and gain values.

The frequency responses were measured with a Rhode&Schwarz ZNB8 vector network analyzer and using a high-voltage bias-tee having a frequency range from 5 MHz to 18 GHz for 675 nm at an optical power of 200 nW [43]. The results are plotted in figure 5.48. The maximum −3 dB-bandwidth of 1.6 GHz was obtained for 69 V reverse voltage, where the gain was 80 and the responsivity was 32 A W^{-1}. This bandwidth is 90% larger as that of an APD in 0.35 μm HV CMOS [45]. The reason for this is the large radius of about 20 μm of the half-sphere drift zone compared to a thickness of the drift zone of only approximately 8 μm in [45], which left a small contribution of slow carrier diffusion. The bandwidth at the reverse bias of 68 V, where the gain is 10 and the responsivity is 25 A W^{-1}, is almost the same as at 69 V. At 70 V reverse bias, where the gain is approximately 300 and the responsivity is 115 A W^{-1}, the bandwidth is 1.28 GHz, which is a bit smaller. This is because of the slightly larger avalanche build-up time needed for the higher gain. However, because of the high

Table 5.5. Responsivity and gain for the field-crowding dot-APD in 0.18 μm HV CMOS for 675 nm at 70.5 V reverse bias. Reproduced from [43] CC BY 4.0.

Quantity	5 nW	50 nW	100 nW	200 nW
R (A W^{-1})	3050	397	234	130
Gain	7470	980	610	325

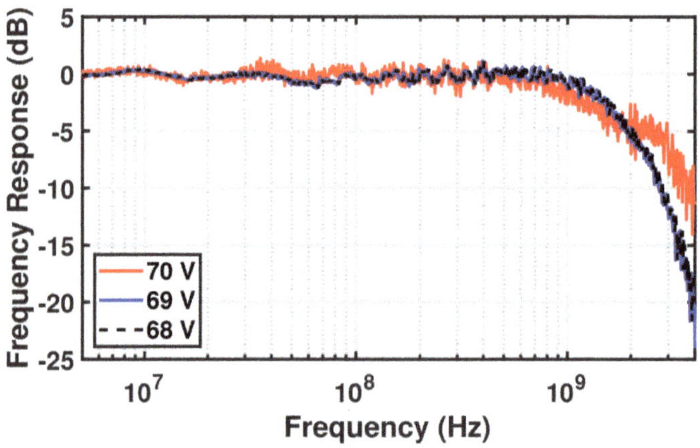

Figure 5.48. Normalized measured frequency responses of field-crowding dot-APD in 0.18 μm HV CMOS at 675 nm and 200 nW optical power. Reproduced from [43] CC BY 4.0.

optical power of 200 nW, the avalanche build-up time effect is not pronounced leading to this small drop of the bandwidth. The device possesses gain-bandwidth products of 128 GHz and 384 GHz for gain = 80 and bandwidth = 1. 6 GHz as well as gain = 300 and bandwidth = 1.28 GHz, respectively [43]. In fact, it is better and fairer to use the responsivity-bandwidth product to compare different APDs from literature, since the gain does not consider that in many devices a (considerable) part of the photogenerated carriers does not contribute to the photocurrent. The responsivity, however, is a measured quantity and allows a trustworthy and realistic comparison. The responsivity-bandwidth product of the PIN dot-APD for 675 nm and 200 nW is 51.2 A W^{-1} GHz at 69 V reverse voltage. At 70 V reverse voltage the responsivity-bandwidth product is 147.2 A W^{-1} GHz.

The field-crowding dot-APD can work with less optical power and at higher avalanche gain as well as higher bandwidth compared to other APDs [43]. Its breakdown voltage is comparable to the conventional reach-through APD in 0.35 μm HV CMOS [45] and to that of the field-crowding SPAD of 67.3 V [25]. Regarding the reponsivity, gain and bandwidth of the dot-APD in 0.18 μm HV CMOS, we can conclude that it should be well appropriate for USPAR ICs.

The electric field-line crowding (EFLC) APD was investigated in some more detail [46]. Figure 5.49(a) shows its three-dimensional structure as well as its photo taken after fabrication (b). The N-well radius r_w and the device radius r_a were varied.

The electric field inside the device was investigated with ATLAS TCAD simulations. Figure 5.50(a) depicts the electric field along the diagonal direction (see figure 5.45) at a reverse voltage of 69 V. For decreasing r_w the electric field strength increases in the multiplication zone as being visible especially in the inset of figure 5.50(a). The width of the multiplication region, where the electric field is stronger than 2×10^5 V cm^{-1}, is about 0.9 μm. At $d = 25$ μm, E is still about 3000 V cm^{-1}. However, it should be mentioned that a smaller N-well radius concentrates the electric field more to the center of the device and the electric field strength at a fixed d (e.g. at 15 μm) or at a fixed radius decreases. As we see from figure 5.50(b), the surface anode ring increases the electric field compared to a device without surface anode ring at least for $r > 10$ μm. Therefore, we have to conclude from these two properties that a large N-well radius r_w is needed, if we want a large device radius r_a and a high bandwidth of the EFLC-APD [46].

The frequency response and the bandwidth were also investigated with ATLAS TCAD simulations. Figure 5.51(a) compares the frequency responses for $r_a = 19$ μmin dependence on the N-well radius. The influence of the N-well radius on the bandwidth is quite small as we can conclude from the bandwidths 1.50, 1.54, and 1.59 GHz for $r_w = 0.6$, 0.7, and 0.8 μm, respectively. This is because the lateral width of the lateral drift region decreases only slightly with increasing N-well radius. The device radius r_a has a stronger impact on the bandwidth, as we can see in figure 5.51(b). For $r_w = 0.6$ μm, the bandwidths are 1.23, 1.35, and 1.50 GHz for $r_a = 19$, 15, and 12 μm, respectively. This dependence can be explained by the reduction of the lateral carrier drift distance for the decreasing devices radius r_a [46].

The N-well radius and the device radius also have an influence on the breakdown voltage. Figure 5.52(a) shows that the breakdown voltages (determined for a dark

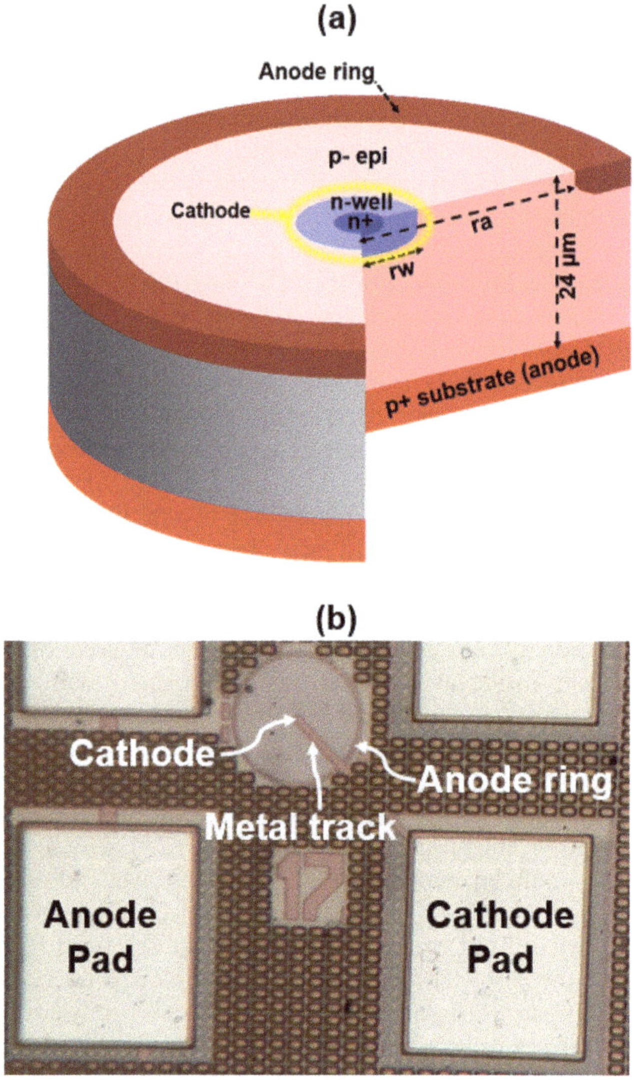

Figure 5.49. Structure of EFLC-APD (top) and microphotograph of fabricated device (bottom). Reproduced from [46] CC BY 4.0.

current of 1 μA) for $r_a = 19\,\mu$m are 70.7 V, ≈74 V, and ≈81 V for $r_w = 0.6$, 0.7, and 0.8 μm, respectively. For decreasing r_a, however, the breakdown voltages reduce from 70.7 V to 60.5 V for r_a decreasing from 19 μm to 12 μm [46]. This is because the space-charge region depletes a smaller volume and the field lines crowd stronger at the P/N-junction for decreasing device radius r_a [46].

Figure 5.53(a) shows that the responsivities for 850 nm and 675 nm are approximately equal. The responsivity at 850 nm for an avalanche gain $M = 1$ is 0.42 A W^{-1} at 1 V reverse bias, corresponding to a quantum efficiency of 61.4%. The maximum responsivity of 3050 A W^{-1} at 2 nW optical power and 70.5 V reverse

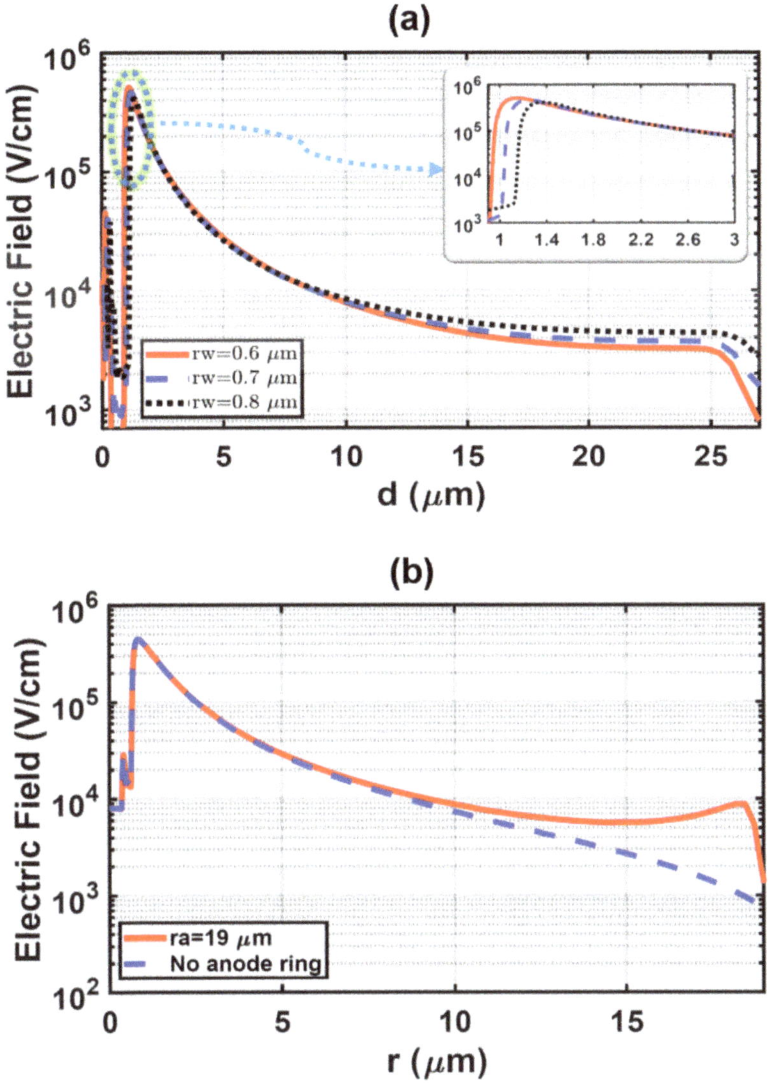

Figure 5.50. Diagonal cross-section of the electric field for three different N-well radii in the EFLC-APD with $r_a = 19\ \mu\text{m}$ (top) and radial cross-section of the electric field at the silicon surface of the EFLC-APD with $r_w = 0.6\ \mu\text{m}$ (bottom). All simulations were performed for an avalanche gain of 80, i.e. at reverse voltages of 69 V, 73.5 V and 80 V for $r_w = 0.6\ \mu\text{m}$, $r_w = 0.7\ \mu\text{m}$, and $r_w = 0.8\ \mu\text{m}$, respectively. Reproduced from [46] CC BY 4.0.

bias, corresponding to a gain of 7260, is obtained for the device with $r_w = 0.6\ \mu\text{m}$ and $r_a = 19\ \mu\text{m}$ [46]. The spectral response of this device is presented in figure 5.53(b), whereby the optical power was constant at 2 nW. The responsivity does not change much between 650 and 900 nm with a maximum for $M = 1$ at about 0.44 A W^{-1} at 770 nm. This is because of the thick absorption zone. Figure 5.53(c) depicts that for increasing optical power, the responsivity decreases, if the reverse voltage is made very large. At 70.5 V reverse voltage, the responsivities reduce to 196 A W^{-1} at 200 nW and

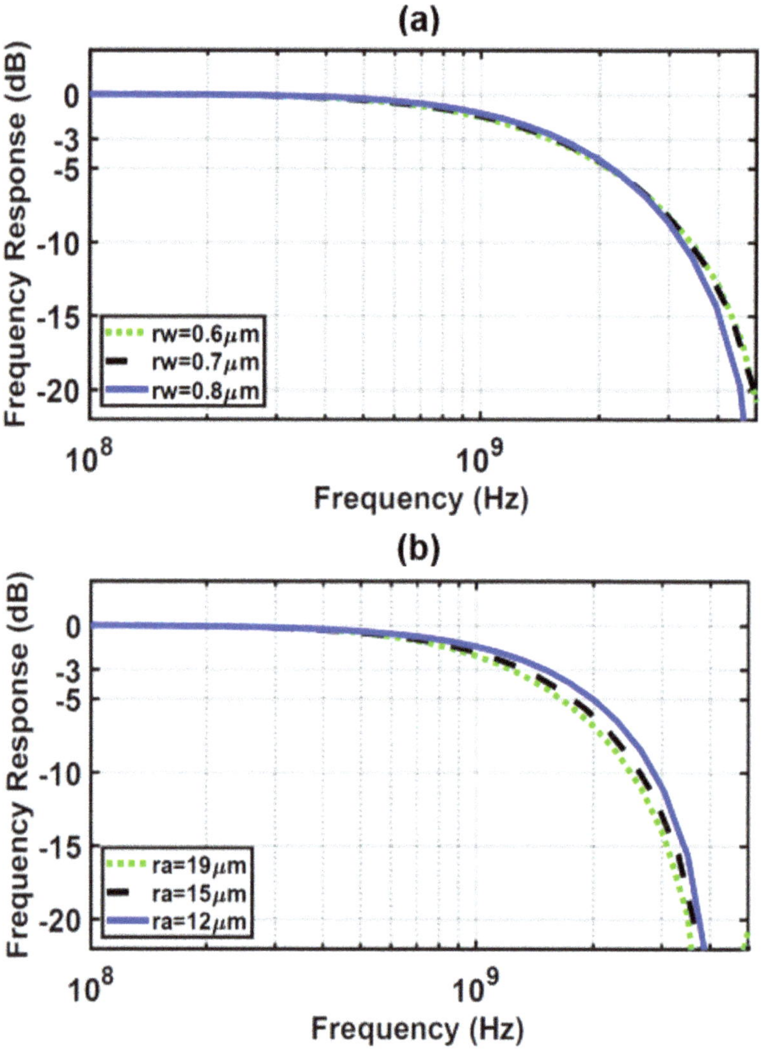

Figure 5.51. Simulated normalized frequency responses of EFLC-APD with $\lambda = 850$ nm for different N-well radii (a) and different device radii r_a (b). The reverse voltages were 69 V, 73.5 V and 80 V for $r_w = 0.6\,\mu$m, $r_w = 0.7\,\mu$m, and $r_w = 0.8\,\mu$m, respectively (a) and 60 V, 64 V and 69 V for $r_a = 12\,\mu$m, $r_a = 15\,\mu$m, and $r_a = 19\,\mu$m, respectively (b). Therefore, the avalanche gain was 80 in all six devices. Reproduced from [46] CC BY 4.0.

to 48 A W^{-1} at 1 μW optical power due to a saturation effect and partial screening of the electric field by the large amount of generated/amplified charges [46].

The measured frequency responses of the dot APD are compared in figure 5.54. Sub-figure (a) plots the curves for 675 nm and 850 nm for 200 nW optical power. The maximum bandwidth is 1.6 GHz at 69 V reverse bias for the device with $r_w = 0.6\,\mu$m and $r_a = 19\,\mu$m at a gain of 80 and $R = 32$ A W^{-1}. With 850 nm light and $M = 80$ ($R = 33$ A W^{-1}), the bandwidth is 1.2 GHz [46]. This is a remarkable

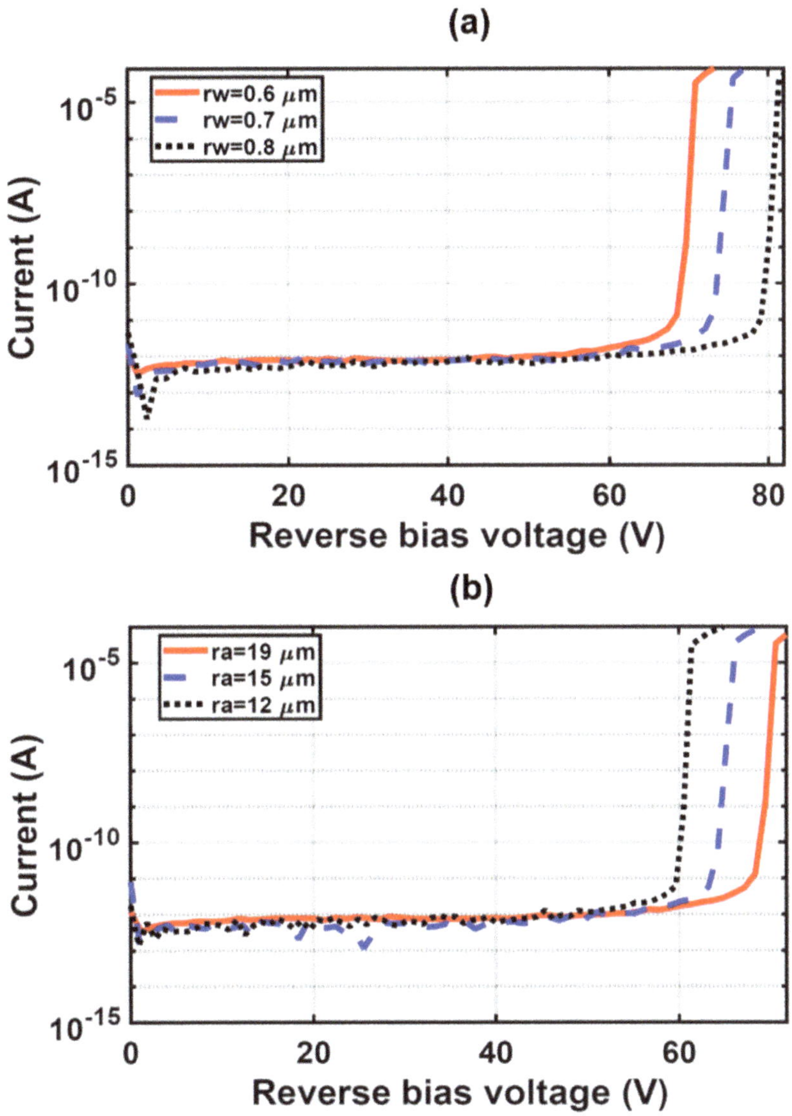

Figure 5.52. Measured dark reverse characteristics of EFLC-APD with different N-well radii with $r_a = 19\ \mu$m (a) and different device radii with $r_w = 0.6\mu$m (b). Reproduced from [46] CC BY 4.0.

bandwidth for the deep penetrating light with a wavelength of 850 nm, which is 350 MHz higher than the bandwidth reported in [45] for a planar APD in 0.35 μm high-voltage CMOS with a quite similar breakdown voltage.

If the device radius r_a is reduced from 19 μm to 12 μm, the bandwidth for 850 nm increases from 1.2 GHz to 1.5 GHz (figure 5.54(b)). For high avalanche gains, the avalanche build-up needs more time and the bandwidth reduces. This is demonstrated by figure 5.54(c) for the device with $r_w = 0.6\ \mu$m and $r_a = 12\ \mu$m. For the responsivity of 196 A W^{-1}, corresponding to an avalanche gain $M = 466$, the

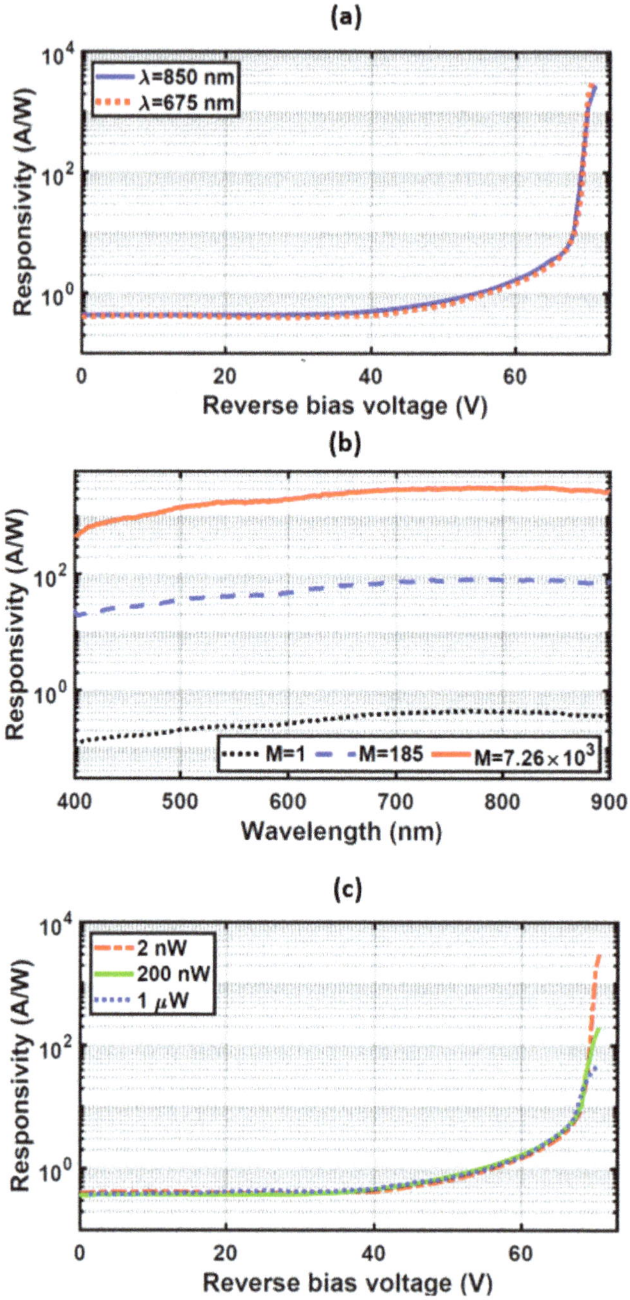

Figure 5.53. Measured responsivity of EFLC-APD for two wavelengths in dependence on reverse voltage (a), in dependence on the wavelengths (b), and in dependence on the reverse voltage for different optical power at 850 nm (c). Reproduced from [46] CC BY 4.0.

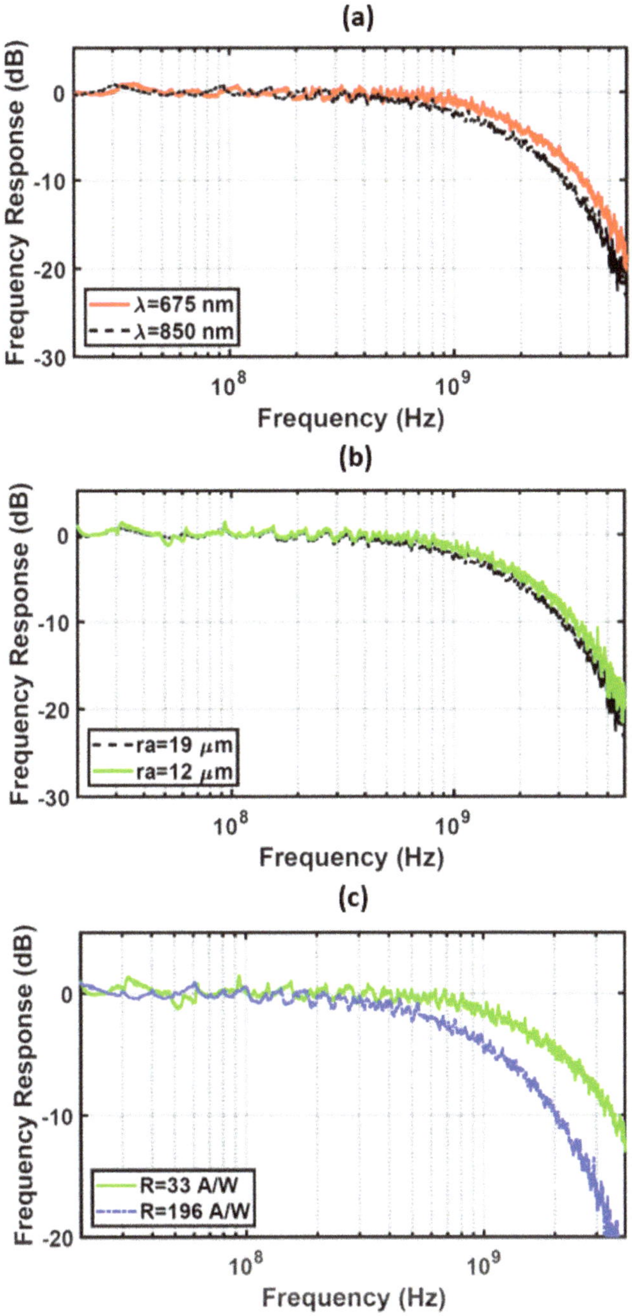

Figure 5.54. Normalized measured frequency responses of EFLC-APD with $r_w = 0.6\,\mu\text{m}$ and $r_a = 19\,\mu\text{m}$ for two wavelengths and an avalanche gain M of 80 (a), for two device radii r_a ($r_w = 0.6\,\mu\text{m}$, $\lambda = 850$ nm and $M = 80$) (b) and a device with $r_w = 0.6\,\mu\text{m}$, $r_a = 12\,\mu\text{m}$ and $\lambda = 850$ nm for two different responsivities (c). Reproduced from [46] CC BY 4.0.

bandwidth decreases to 760 MHz for an optical power of 200 nW. However, compared to the bandwidth of 1.5 GHz for the gain of 80, the bandwidth is only half pointing us to the conclusion that the avalanche build-up time effect is not strongly pronounced yet.

Figure 5.55 presents the dependence of bandwidth on the gain. The obtained bandwidths are 1.5 GHz, 1.06 GHz, and 0.76 GHz for M = 80, 300 and 466, respecticely. This figure compares the performance of the EFLC-APD to the gain-bandwidth limit of 1 THz for silicon [47]. For low gains the bandwidth is limited by the drift time in the thick absorption region. For high gains the avalanche build-up time dominates and the gain-bandwidth product is constant.

However, for many applications not just the responsivity (or gain) and bandwidth are important. The excess noise (see section 3.10) of APDs in the linear mode often is also an important specification. There are the McIntyre [48] and the dead-space [49–51] models for the excess noise factor F. However, these models were derived for uniform electric field and they cannot be applied to the EFLC-APD where the electric field is approximately hemispherical.

The excess noise factor of the EFLC-APD with r_w = 0.6 μm and r_a = 19 μm was determined experimentally. A special transimpedance amplifier with R_F = 28 kΩ and a bandwidth of 200 MHz was used [46]. Due to its limited dynamic range, many different optical power levels were necessary to cover a wide range of avalanche gains M. Figure 5.56 presents the obtained results. For M = 17.8, the excess noise factor F is 2, for M = 39 it is 4, and for M = 56.7 it is 6. Because the electric field is not uniform in the EFLC-APD, the ratio of the impact ionization coefficients cannot be determined using the McIntyre or dead-space models. The excess noise factor of the EFLC-APD, however, is about comparable to that of 6 for M = 50 and 6.5 for M = 60 of a planar APD in 0.35 μm HV CMOS [45]. A thick multiplication zone is known to be generally good for a low excess noise factor. Indeed, the width of the multiplication region of the EFLC-APD is about 0.9 μm compared to 0.73 μm (both for M = 60) of the planar 0.35 μm HV CMOS APD [46]. The field-line crowding

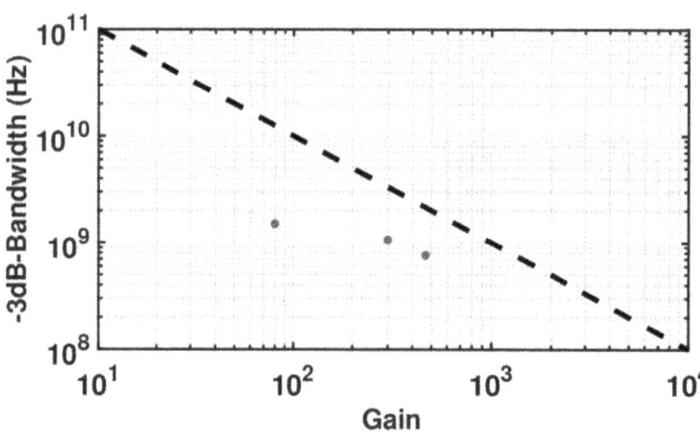

Figure 5.55. Bandwidth versus gain of EFLC-APD with r_w = 0.6 μm and r_a = 12 μm at 200 nW optical power. Reproduced from [46] CC BY 4.0. The dashed line illustrates the gain-bandwidth product of 1 THz [47].

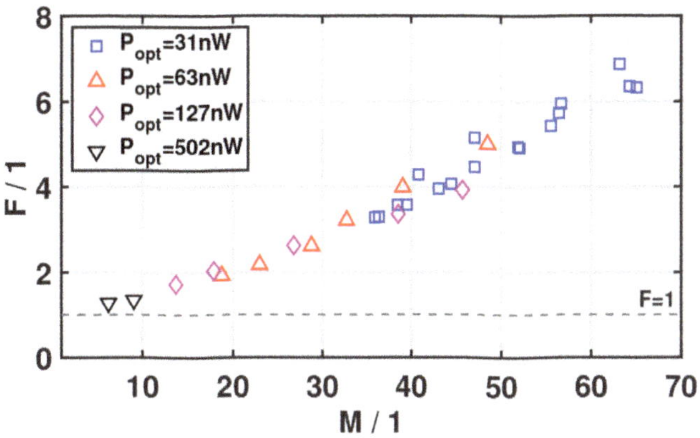

Figure 5.56. Measured excess noise factor of EFLC-APD at $\lambda = 642$ nm. Several optical power levels were necessary to cover the gain range due to a small dynamic range of the measurement system. Reproduced from [46] CC BY 4.0.

leading to crowding of multiplied carriers in the multiplication region, however, seems to reduce this advantage of the thicker multiplication region.

Concluding, the EFLC-APD is well appropriate for USPAR ICs. The excess noise of the device will make it even more interesting for application in ultra-sensitive receivers.

5.4 Multi-dot photodiodes for large photosensitive areas

In order to increase the light-sensitive area the inner radius of the P-well/P+ substrate contact would have to be increased. This, however, would reduce the electric field strength between cathode and P-substrate contact, reduce the carrier drift velocity and increase the carrier drift time. In other words the bandwidth of the photodiode would be reduced. The maximum inner radius of the P-well/P+ substrate contact ring has to be smaller than the lateral width of the space-charge region. Otherwise slow carrier diffusion will slow down the device considerably. Therefore, several dot-photodiodes have to be put together. With circular P-well/P+ substrate contacts, a loss of light sensitive areas would result between the P-well/P+ rings, even if these rings overlap. But with hexagons like in a honeycomb, no light sensitive area is lost.

Figure 5.57 shows the layout of a structure with seven hexagonal dot-PIN photodiodes. Each of the seven N+/N-well cathodes is surrounded by a hexagonal P-well/P+ substrate contact. The wide outer surrounding substrate contact reduces the series resistance to the P+ substrate through the low-doped epitaxial layer. The contacts and metal connections to the substrate contact are not shown. There are six metal lines necessary to connect all cathode dots with each other and a seventh metal line goes up to the input of the amplifier (not shown in figure 5.57), which is isolated by a deep N-well from the substrate. It should be mentioned that the slanting metal lines have to be constructed by many short horizontal, vertical and 45° lines, because other angles are not allowed in the layout according to the design rules. As a

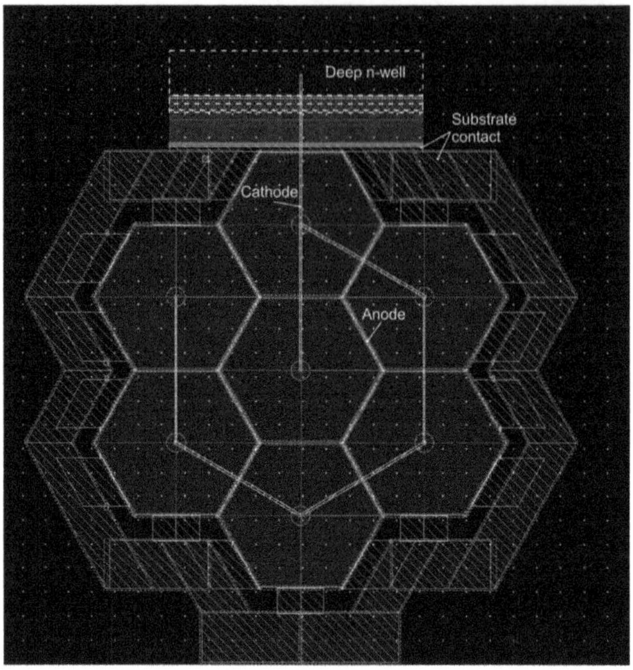

Figure 5.57. Layout of a 7-dot-PIN-photodiode with hexagonal substrate contact rings.

consequence these slanting metal lines cannot be realized in minimum width as the horizontal and vertical connections. The height of one honeycomb is 30 μm (inside the substrate contact) and the width at the widest location is 34 μm. This leads to a light-sensitive area of seven times 768 μm^2, i.e. to 5376 μm^2 in total. The capacitance of 13.14 fF for the shown metal lines in metal 3 (metal posts and P/N-junctions considered) was extracted from the layout. With top metal (4th metal layer in the 0.18 μm CMOS process used), the capacitance of 9.1 fF was obtained.

Another possibility is that several cathodes have to be arranged inside one P-well/P+ substrate contact ring as depicted in figure 5.58. For a hexagonal arrangement with the edge length a, the maximum drift distance along the surface is $\frac{\sqrt{3}}{2}a$. This distance can be used as width (radius) of the simulation region (in cylinder coordinates) in a conservative way to determine the maximum rise/fall times of the device-by-device simulations. A device simulation with such a simulation area determines a worst case rise/fall time. Since most of the light penetrates at other locations, where the drift distance to the cathode is smaller, the photodiode should show a faster response in practice.

The same argumentation is valid for the dot-APD. Therefore, several cathodes each with its own multiplication P-well have to be placed inside the P-well/P+ substrate contact, as depicted in figure 5.59, for increasing the light-sensitive area.

The same reasoning holds for the multi-dot-PIN-N-well APD using the field-crowding effect. The resulting structure for obtaining a larger light-sensitive area as that of the single-cathode device is illustrated in figure 5.60.

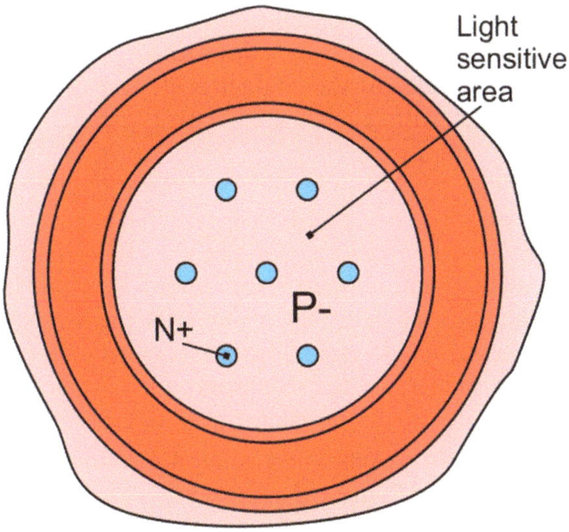

Figure 5.58. Top view of multi-dot-PIN-photodiode.

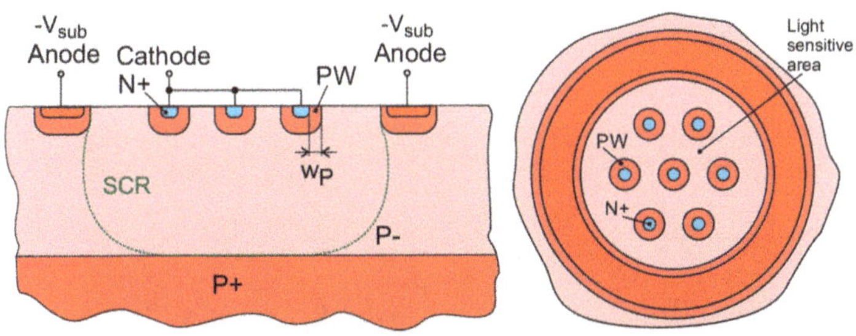

Figure 5.59. Cross-section of multi-dot-PIN-APD (left) and top view of multi-PIN-APD (right); (PW: P-well).

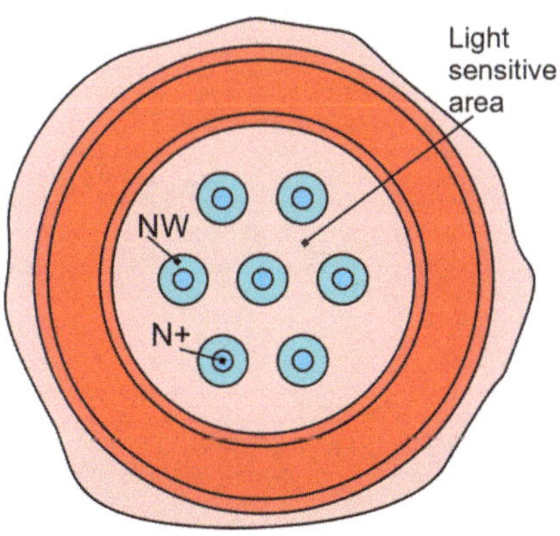

Figure 5.60. Top view of multi-dot-PIN-N-well APD using the field-crowding effect (NW: N-well).

When we observe the capacitance of multi-dot photodiodes, we can take the value of 0.9 fF mentioned above for each cathode dot with an N-well radius of 2 μm. Then we have to add the capacitance of the contact posts and of the metal lines, which connect the cathode dots with each other. These capacitances depend on the structure size of the technology used and on the distance of the cathode dots. For a 0.18 μm CMOS technology and metal 3, we can estimate about 1 fF for the connection between two dots. So, the devices with 7 cathode dots (considering 6 metal connections) possess a capacitance of about 13 fF, which is less than that of the PIN photodiode with 20 μm diameter (see figure 5.1). Their light sensitive area, however, is several times larger than that of this PIN photodiode.

Then, we investigated also multi-dot photodiodes with cathode dots at the corners of squares. In such an arrangement the maximum drift distance along the surface is $\sqrt{2}\frac{a}{2} = \frac{a}{\sqrt{2}}$. Figure 5.61 shows the layout of a 25-dot PIN photodiode in square arrangement. The circular parts of the substrate contact have to be approximated by short vertical, horizontal and 45° lines. The distance a of the cathode dots is 20 μm (from middle of cathode to middle of cathode). The distance between cathodes and substrate contact is $a/2$. The light-sensitive area therefore is $24a^2 + \pi(\frac{a}{2})^2$, which gives about 9914 μm^2. The minimum width of metal 3 and metal 4 lines in the 0.35 μm CMOS technology is 0.6 μm. The extraction of the metal capacitances determined values of 31.5 fF for the metal connections in metal 3 and 24 fF for the metal connections in metal 4. For a 15 μm pitch of the cathode dots and metal 4

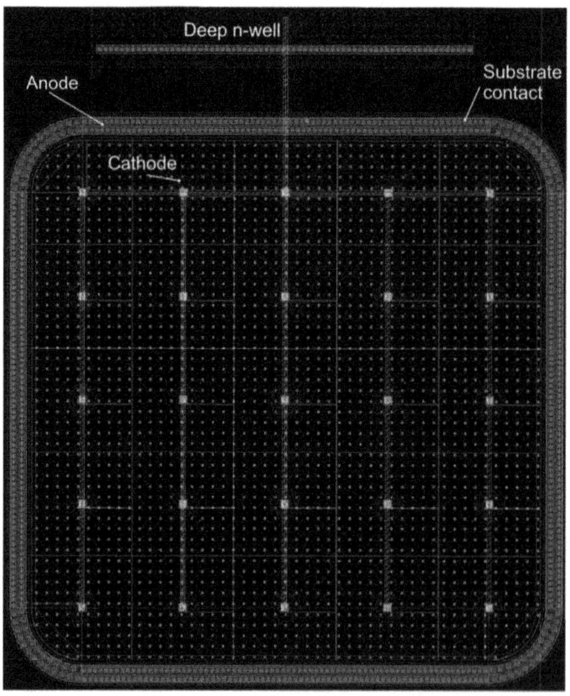

Figure 5.61. Layout of a 0.35 μm-PIN-photodiode-CMOS 25-dot-PIN-photodiode in square arrangement.

connections, their capacitance reduces to 17.5 fF (however, the light-sensitive area reduces also). The P/N-junction capacitances of the 25 cathode dots of about 0.9 fF each have to be added and the total capacitance of the 25-dot PIN photodiode is about 54 fF with metal 3 connections and 46.5 fF with metal 4 connections.

Figure 5.62 depicts the layout of a 4-dot-PIN photodiode in square arrangement. The circular parts of the substrate contact have to be approximated by short vertical, horizontal and 45° lines. The distance a of the cathode dots is 21 μm (from middle of cathode to middle of cathode). The distance between cathodes and substrate contact is $a/2$. The light-sensitive area therefore is $3a^2 + \pi(\frac{a}{2})^2$, which gives about 1669 μm^2. It should be mentioned that the total length of the metal connections could be reduced by about 20%, but then one horizontal connection would be needed inside the light sensitive area, which would require a smaller area of the optical window due to the design rule of minimum distance between metal 3 and optical window. Table 5.6 lists the capacitances and light-sensitive areas of PIN photodiodes with 4 to 36 cathode dots in square arrangement. It should be mentioned that using the top metal (4th metal layer in the 0.18 μm CMOS process used) instead of metal 3 for the metal connections of the dot cathodes would only slightly reduce the capacitance of the metal connections as the extraction from the corresponding layouts showed 20.6 fF compared to 21.8 fF. The reason for this is that the top metal in the 0.18 μm

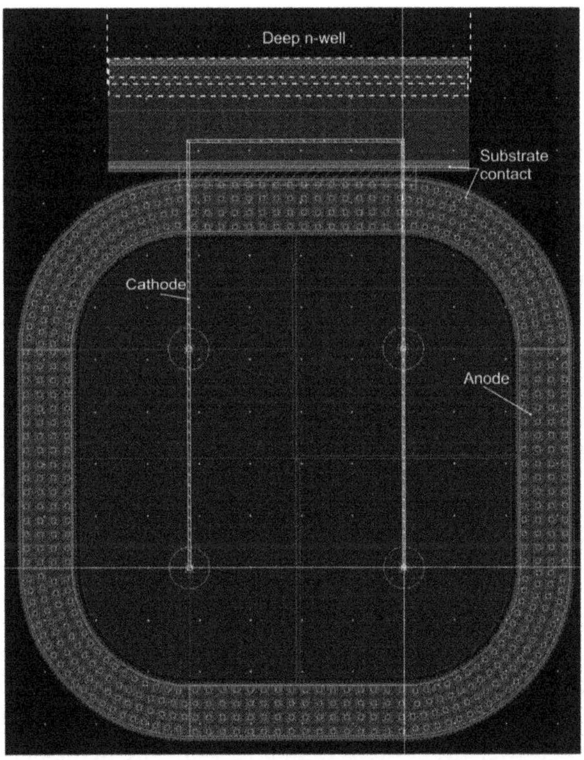

Figure 5.62. Layout of a 0.18 μm-PIN-photodiode-CMOS 4-dot-PIN-photodiode in square arrangement.

Table 5.6. Capacitance of multi-dot PIN photodiodes in square arrangement with a cathode pitch a of 21 μm. The values are for a 0.18 μm CMOS process and metal connections realized in metal 3.

Number of dots	Capacitance (fF)	Light-sensitive area (μm^2)
2×2	5.7	1 669
3×3	12.4	3 874
4×4	21.8	6 961
5×5	33.8	10 930
6×6	49.0	15 781

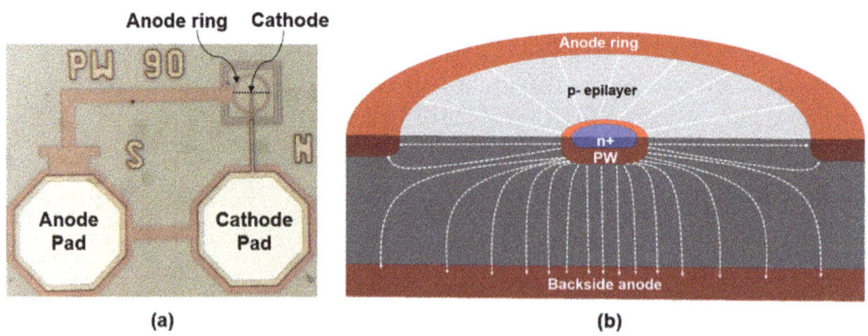

Figure 5.63. Chip microphotograph of a single-dot APD (a) and three-dimensional schematic drawing of single-dot APD (b) in 0.35 μm PIN-photodiode CMOS. Reproduced from [52] CC BY 4.0.

CMOS process used has a larger minimum width which leads to a certain increase of the capacitance again, although the oxide thickness between top metal and substrate is larger than that between metal 3 and substrate.

Planar N+/P-well APDs with large N+ cathodes need a virtual guard ring to avoid edge breakdown [36]. In USPAR, we developed the idea to exploit an APD based on only edge breakdown. To achieve this case, a small N+ cathode dot is necessary, which is surrounded by a P-well dot having in all (radial and vertical) directions the same thickness. This implies that the lateral width of the P-well has to be equal to its penetration depth. Of course, the N+ cathode needs a radius being equal to its depth. So, N+ and P-well need a hemisperical shape. Such a N+/P-well dot APD was realized in 0.35 μm CMOS technology [52]. Furthermore, it was extended to a multi-dot structure. But let us have a closer look first at the single-dot N+/P-well APD (see figure 5.63).

The N+ dot cathode in the center has a diameter of 1.7 μm and the diameter of the P-well, which forms the multiplication region, is 1.8 μm. The design rules of the process used did not allow a smaller diameter for a circular N+ cathode. These values are design values. We should, however, be aware that the P-well diffuses more strongly during the process than the N+ region and that the radial thickness of the P-well is (much) larger after the process than 0.05 μm. The hemispherical shape of the N+ region inside the P-well obviously will be somewhat violated, because the depth of the N+ region will not be as large as 0.85 μm (half of its diameter) in the 0.35 μm PIN-photodiode CMOS (XO035 of X-FAB Semiconductor Foundries) process used. But this seems not to have a large influence, as we will see below that this APD works quite well. In addition, simulations showed that for a P-well diameter of less than 0.9 μm, the electric field distribution becomes non uniform. The doping of the epitaxial layer is 2×10^{13} cm^{-3}. The thickness of this absorption layer is about 12 μm [52]. The surface P+/P-well anode ring has a diameter of about 22 μm. The P+ substrate forms a bottom anode in addition. The surface anode ring works in addition as a contact to the substrate.

The electric field inside this device was simulated with ATLAS [15]. Figure 5.64 presents the results at a reverse voltage of 24.5 V. The breakdown voltage is about 25.2 V [52]. According to the left part of subfigure (a), the light-sensitive area inside the surface anode ring is fully depleted and the electric-field distribution looks quite well hemispherically. The right part of this subfigure shows some deviations from the semispherical shape. At $d = 10$ μm (on the 'diagonal'), the electric field strengths is still 200 V cm^{-1} (see figure 5.64b). Subfigure (b) also depicts a maximum electric field of about 5×10^5 cm^{-3} at the N+/P-well junction. The width of the multiplication region at 2×10^5V cm^{-1}, where the impact ionization starts, is about 0.4 μm.

From the energy band diagram in figure 5.64(c) we see that the reverse voltage of 24.5 V the dot APD devides into about 7.5 V across the absorption region and 17 V across the multiplication region. The electrons and holes photogenerated in the absoption region drift into opposite direction. The electrons drift towards the

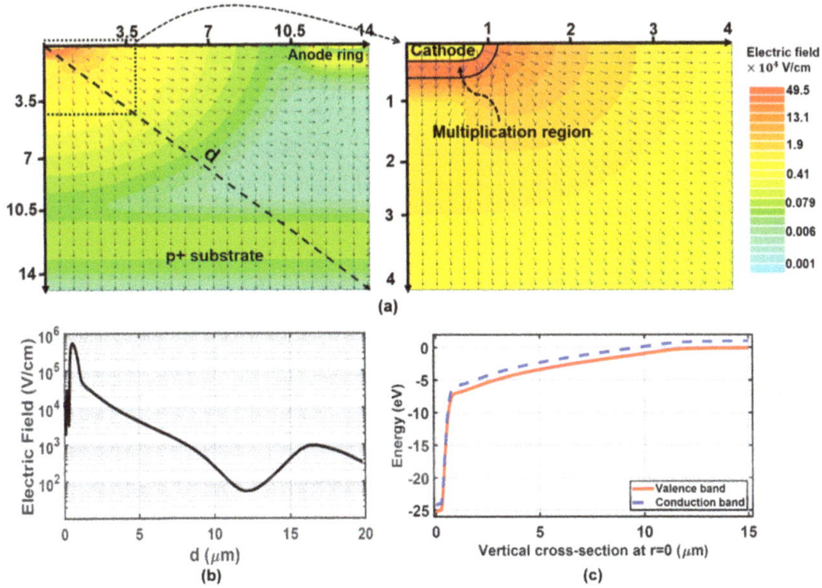

Figure 5.64. Simulated electric field in vertical-radial two-dimensional cross-section (a) of a single-dot APD in 0.35 μm PIN-photodiode CMOS at 24.5 V (left) with zoomed part (right). Electric field along the dashed line (d) in the top left subfigure (a). Energy band diagram in vertical direction at the center $r = 0$ μm (c). Reproduced from [52] CC BY 4.0.

multiplication region and have the full width of the multiplication region for impact ionization. The depletion into the depth of the absoption region is also quite good.

Due to the thick absorption zone and its full depletion, the responsivity of the single-dot APD is 0.27 A W^{-1} and its bandwidth amounts to 1.25 GHz [52]. The measured reverse characteristics in darkness are shown in figure 5.65a. The break-down voltage, where the reverse current equals 1 μA [53], is 25.2 V. As we know from dot PIN photodiodes that their capacitance is very low, we can expect a very small capacitance for the N+/P-well dot APD. The capacitance was simulated with ATLAS [15], because an accurate measurement is not possible even with a precision LCR meter. Figure 5.65(b) presents the simulated $C(V)$ curve of the single-dot N+/P-well APD. Its capacitance is 0.65 fF at the reverse voltage of 24.5 V. This value is much smaller than the capacitance of planar CMOS-integrated APDs [27, 45].

The performance of the single-dot N+/P-well APD in PIN-photodiode CMOS technology with respect to bandwidth and capacitance is quite impressive. However, its diameter is somewhat smaller than that of the single-dot PIN photodiodes

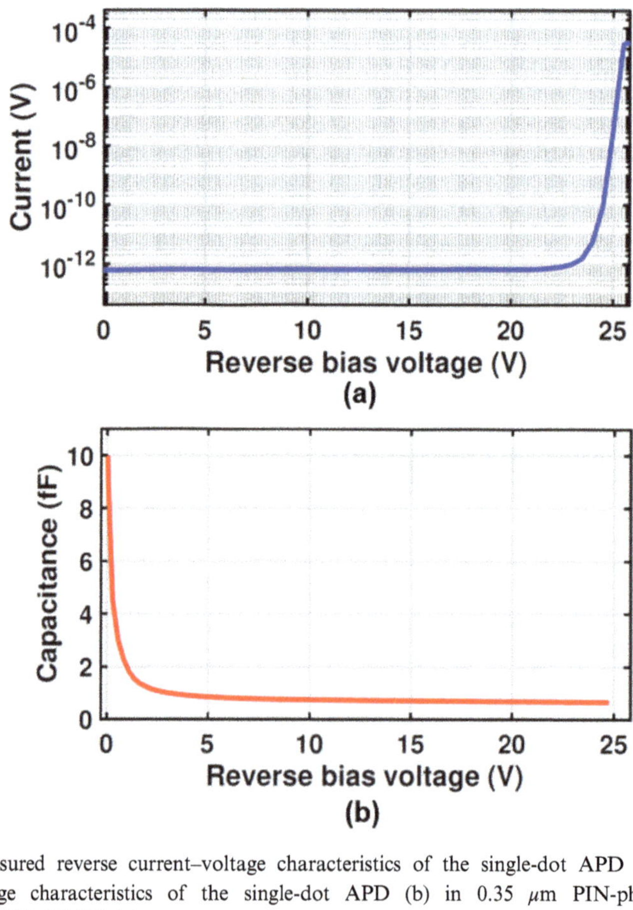

Figure 5.65. Measured reverse current–voltage characteristics of the single-dot APD (a) and simulated capacitance–voltage characteristics of the single-dot APD (b) in 0.35 μm PIN-photodiode CMOS. Reproduced from [52] CC BY 4.0.

described above. For many applications like e.g. plastic optical fiber receivers and optical wireless communication, a larger light-sensitive area is needed. But it is not possible to simply increase the radius of the surface ring anode, if we want to keep the high bandwidth, because the electric field strength decreases rapidly with larger radius. Then slow carrier diffusion of photogenerated carriers will come into play and the bandwidth will deteriorate considerably. The way to increased light-sensitive area at high bandwidth can only be via multi-dot structures. In the following, such a multi-dot N+/P-well APD will be introduced.

Figure 5.66 presents the layout, the microphotograph of the fabricated device and the schematic cross-section of the multi-dot N+/P-well APD. The N+/P-well dots are arranged in a square structure and the distance between two dots in horizontal and vertical direction is a (pitch = a in x- and y-direction in figure 5.66(a), left part). The metal lines (in the highest metal level 4) connecting the cathode dots have minimum width (0.6 μm in the 0.35 μm technology used) [52]. The multi-dot APD was covered by the complete isolation and passivation stack. The most important difference to the single-dot device shown in figure 5.63 is that there is no surface anode per N+/P-well dot. There is only one surface anode around the complete device. So, all field lines from the inner dots bend down to the bottom anode. Due to this difference, we cannot expect the same bandwidth as for the single-dot N+/P-well APD, when we assume the half pitch should correspond to the radius of the single-dot N+/P-well APD. Since a three-dimensional device simulation is still difficult and very CPU-time consuming, we have to approximate and estimate the behavior of the multi-dot APD. What we see from the cross-section in figure 5.66(b) is that a two-dimensional simulation with the simulation area marked with the yellow rectangle

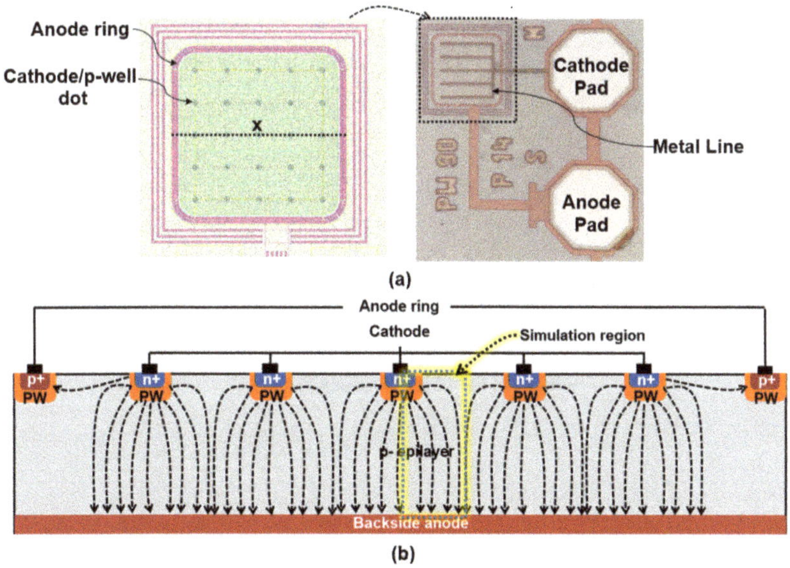

Figure 5.66. Layout (left) and microphotograph (right) of a multi-dot APD in 0.35 μm PIN-photodiode CMOS (a) and schematic cross-section (b) of the multi-dot APD (not to scale) along the dotted line x in the left part of subfigure (a). Reproduced from [52] CC BY 4.0.

due to symmetry reasons and the boundary conditions implemented in the device simulator will save a lot of computing time.

The longest drift path is present for carriers being photogenerated in the center of a square, as illustrated in figure 5.67. These carriers have to drift along the surface for a distance of the half diagonal of the square. From this center point we can expect the slowest response of the multi-dot APD. The maximum lateral drift distance is $\sqrt{2} \cdot a/2$. So, we can assume the yellow rectangle with a lateral extension of $\sqrt{2} \cdot a/2$, i.e. in diagonal direction instead in x-direction as drawn in figure 5.66 with a lateral extension of $a/2$.

Device simulations with ATLAS [15] were used to estimate the best array pitch a of the multi-dot APD. Using Cartesian coordinates with the two-dimensional simulation region represents an estimation because it implies a cathode finger structure, which will result in an overestimation of the bandwidth. However, not all carriers will be photogenerated in the center of a square. So, with a lateral width of $\sqrt{2} \cdot a/2$ for the simulation region we will underestimate the bandwidth of the multi-dot APD (considerably). Anyway, figure 5.68 collects the electric field distributions obtained for $a/2 = 13$, 10, 7, and 4 μm.

The electric field vectors behave as expected and shown in figure 5.66. The field lines extend along the silicon surface and bend down towards the P+ substrate. The strength of the electric field decreases for increasing d, if the half-pitch $a/2$ increases. Therefore, the bandwidth is expected to decrease for increasing half-pitch sizes. However, making the pitch size smaller, results in more dots needed to fill a certain

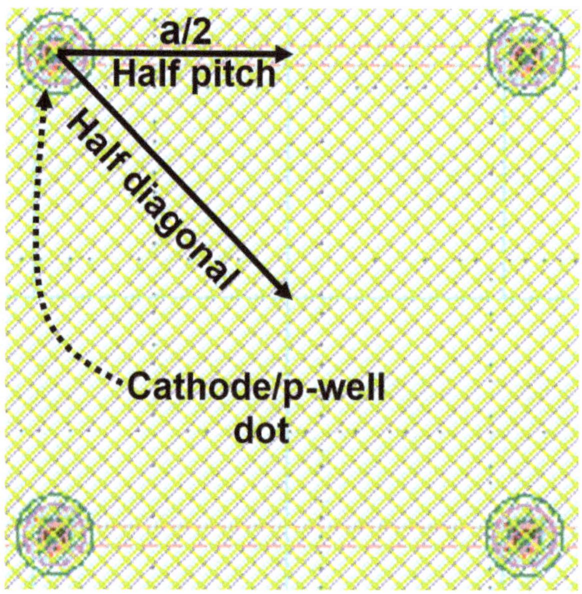

Figure 5.67. Zoomed top view of a multi-dot APD with four N+/P-well dots arranged in a square having the side length a. Reproduced from [52] CC BY 4.0.

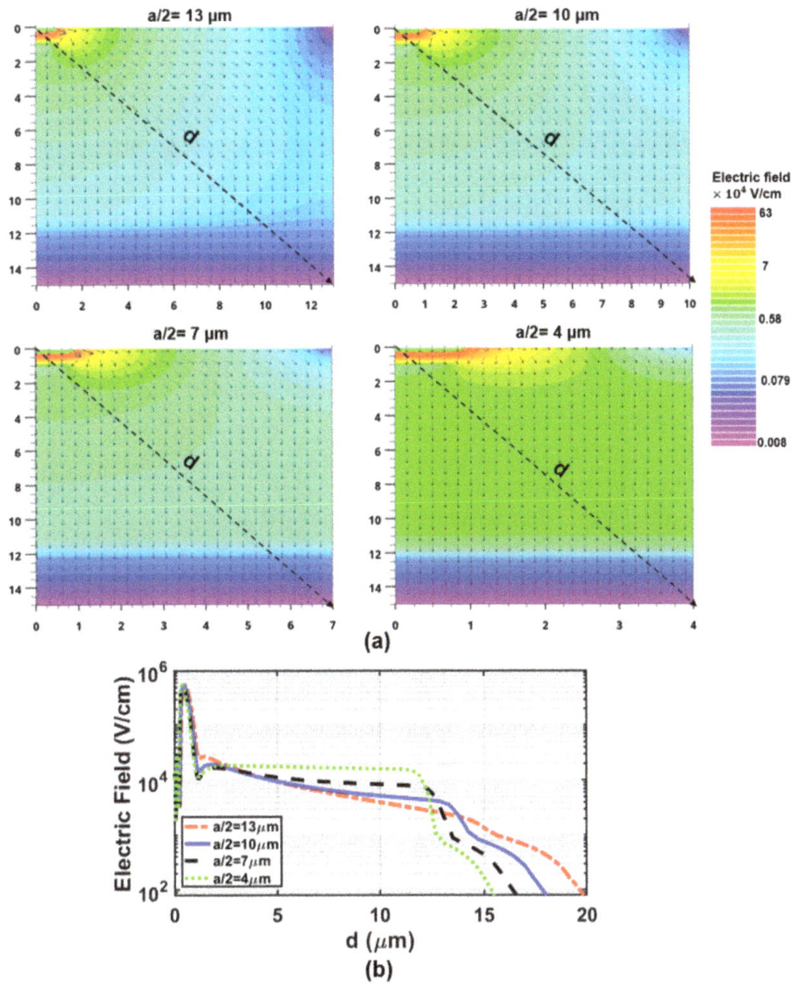

Figure 5.68. (a) Electric-field distributions of the multi-dot APD in the two-dimensional cross-section marked in figure 5.66(b) as 'simulation region' for the half-pitch sizes $a/2 = 13$ μm, 10 μm, 7 μm, and 4 μm at $M = 40$. (b) Electric field along the dashed line d in subfigure (a). Reproduced from [52] CC BY 4.0.

light-sensitive area and therefore in a larger capacitance as well as an increasing loss in detection efficiency due to reflection of incident light by more metal lines.

The frequency responses calculated with ATLAS for the four half-pitch sizes are presented in figure 5.69. As expected, the bandwidth increases for decreasing half-pitch sizes. The −3 dB-bandwidths are 1.25 GHz, 1.7 GHz, 1.85 GHz, and 2.0 GHz for $a/2 = 13$, 10, 7, and 4 μm, respectively. A reduction of $a/2$ from 13 to 10 μm increases the bandwidth from 1.25 GHz to 1.7 GHz, whereas a further reduction of $a/2$ results in less enhancement of the bandwidth. This fact can be explained with the long vertical drift distance from the about 12 μm thick absorption zone and the light penetration depth for $\lambda = 675$ nm. So, for further decreasing half-pitch size, the bandwidth will saturate at the bandwidth value of a conventional PIN photodiode

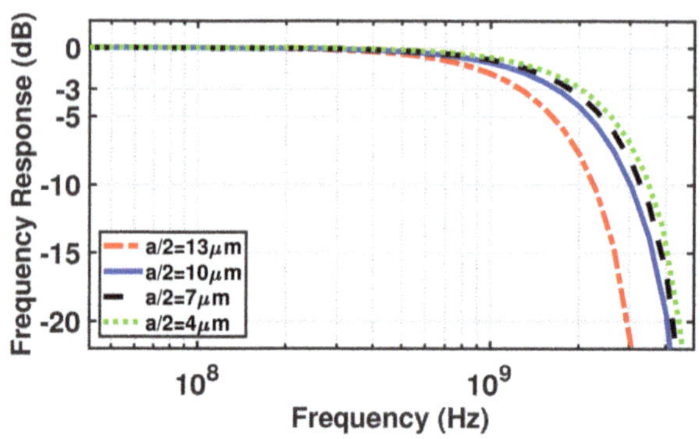

Figure 5.69. Simulated frequency responses for a wavelength of 675 nm and $M = 24$ for the half-pitch sizes of 13 μm, 10 μm, 7 μm, and 4 μm. Reproduced from [52] CC BY 4.0.

with a large cathode, respectively, at the bandwidth of a planar N+/P-well APD with a large cathode and P-well, due to vertical carrier drift time.

Having the results of these device simulations, now we can determine the best pitch size a. The longest lateral drift path is $\sqrt{2} \cdot \frac{a}{2} = \frac{a}{\sqrt{2}}$. Since $a/2 = 10\,\mu$m gave a good value for the bandwidth (be aware that this value has to be understood as the length of the half diagonal for the slowest response), $a = 14\,\mu$m (this results in a length of 10 μm of the half diagonal) should be a good choice for the pitch of the N+/P-well dots. According to the results of figure 5.69, the bandwidth for $a = 14\,\mu$m should be between 1.7 GHz and 1.85 GHz [52]. For the 5 × 5 multi-dot N+/P-well APD, we obtain then a light-sensitive area of almost $70 \times 70\,\mu\text{m}^2$ (exact value: $5a^2 - a^2(1 - \pi/4) = 4858\,\mu\text{m}^2$) with $a = 14\,\mu$m. It should be mentioned that the bandwidth and responsivity should not be affected if another number of N+/P-well dots is used as long as the pitch a is kept the same. The capacitance will scale with the number of dots. So, the multi-dot N+/P-well APD is easily scalable to larger (or smaller) light-sensitive areas, if observing an unchanged pitch.

There is another aspect of the different drift path lengths depending on the location where the photon is absorbed. Jitter may be the consequence of these different drift path lengths. The jitter is even more difficult to simulate precisely than simulating the transient response in three dimensions. Therefore, the jitter was estimated by simulations using the two-dimensional simulation area with the half-diagonal extension of 10 μm. In the first case, perpendicular light incidence was only present between $r = 0$ and $r = 0.5\,\mu$m. The transient response for this case is depicted in figure 5.70 as red curve and the blue dashed curve represents photogeneration only beween $r = 9.5\,\mu$m and $r = 10\,\mu$m. The rise time should correspond to the carrier drift time. In the first case the obtained drift time was 160 ps and in the second case it was 240 ps. The difference between both drift times can be taken as an approximation for the maximum possible jitter of the multi-dot N+/P-well APD with a pitch of 14 μm.

In the following, the measured results of the multi-dot N+/P-well APD are presented. The measurements were performed on a wafer prober whose thermo-chuck was controlled to 25°. The 675 nm light was fed from laser diode into the

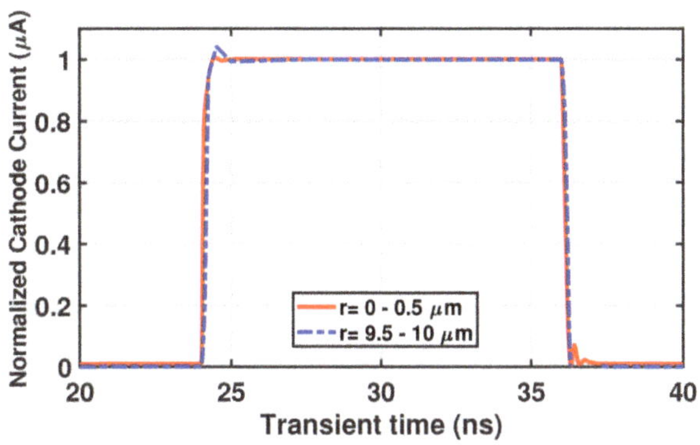

Figure 5.70. Simulated transient responses for a wavelength of 675 nm and $M = 24$ for the pitch of 14 μm, (half diagonal of 10 μm). Reproduced from [52] CC BY 4.0.

device via a 62.5 μm core-diameter multi-mode fiber. The dark reverse characteristics (see figure 5.71) were measured with a Keysight B2987 electrometer [52]. The breakdown voltage of the multi-dot N+/P-well APD is 24.5 V (at 1 μA reverse current, i.e. 0.7 V less than that of the single-dot N+/P-well APD described above). This seems reasonable, since the electric field has to bend down within 7 μm in the multi-dot APD instead of about 11 μm in the single-dot APD.

The capacitance of the multi-dot APD was measured with the precision LCR meter Agilent 4284A at 1 MHz and an AC amplitude of 100 mV. An open-structure was used to calibrate the measurement, i.e. to subtract the capacitance of the bondpad and of the measurement cables [52]. Figure 5.71(b) shows the measured $C(V)$curve. At 0 V, the capacitance is 220 fF which is 22 times larger than the capacitance of the single-dot APD (compare figure 5.65(b)) of about 10 fF. However, the capacitance of the multi-dot APD rapidly decreases to 27 fF for raising reverse voltage. 25 single-dots result in a capacitance of 25×0.65 fF = 16.25 fF leaving about 10.75 fF for the capacitance of the metal connections to all 25-dot cathodes. It should be mentioned that the low capacitance of the multi-dot APDs enables optical receivers and TIAs with a better sensitivity.

The measured responsivity and avalanche gain for 675 nm light are shown in figure 5.72(a). The low-field responsivity (i.e. for $M = 1$) of the multi-dot APD is 0.27 A W^{-1} at 1 V reverse bias. This corresponds to a quantum efficiency of about 50% [52]. Above about 20 V reverse bias the avalanche gain rises and at 24 V, the responsivity of 9.7 A W^{-1} corresponding to the avalanche gain $M = 36$ is achieved. For higher optical power, however, the gain drops. The spectral dependence of the responsivity is presented for $M = 1$ in figure 5.72(b). The optical power was controlled to 2 nW independent of the wavelength. For slightly different wavelengths from 675 nm, the responsivity could be better than 0.27 A W^{-1} due to the maxima of the transmission stemming from optical interference in the isolation and passivation stack. The absolute responsivity maximum of 0.37 A W^{-1} is present for 732 nm light.

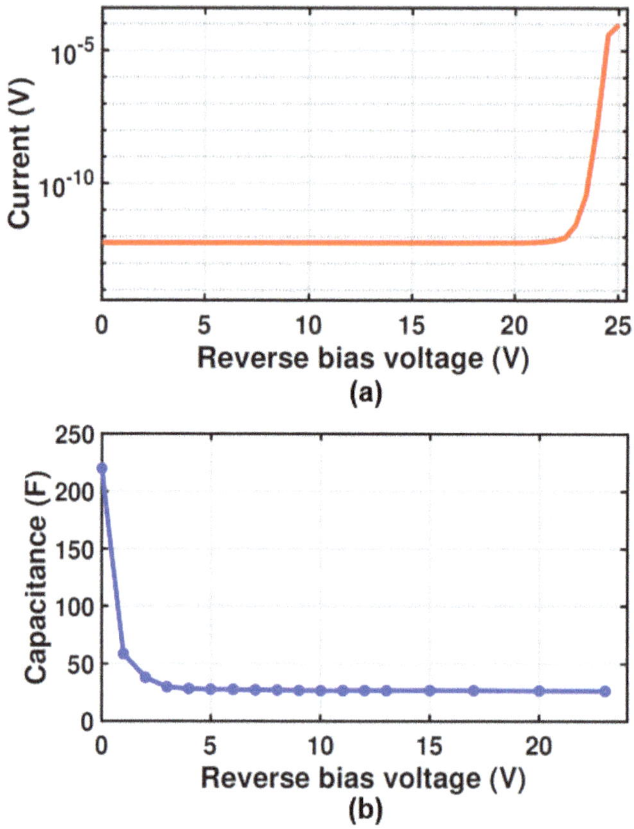

Figure 5.71. Measured dark reverse current–voltage characteristics (a) and measured capacitance (b) of 5 × 5 multi-dot APD. Reproduced from [52] CC BY 4.0.

The frequency response measured with a Rohde&Schwarz ZNB8 vector network analyzer is presented in figure 5.73. The reverse voltage of the multi-dot N+/P-well APD was 24 V, where the responsivity is 9.7 A W^{-1} and the gain is 36, and an optical power of 200 nW with 675 nm was used. The −3 dB bandwidth is 1.8 GHz. This measured value fits well into the range from 1.7 GHz to 1.85 GHz estimated above by device simulations.

Let us compare the properties of the multi-dot N+/P-well APD to the state of the art. Table 5.7 summarizes the most interesting APDs presented in literature. The thin APDs of [54] and [56] are characterized by high bandwidths, however, their responsivity is quite limited because of their small thickness. Compared to the planar APDs with N+/P-well areas of 5808 μm^2 [45] and 2827 μm^2 [29] with capacitances of 125 fF and 100 fF, respectively, the multi-dot N+/P-well APD with an area of 4858 μm^2 has only about a fifth of the capacitance. The current-assisted APD [55] has a smaller capacitance, however, its bandwidth is only 275 MHz. The EFLC-APD [43] with a light-sensitive area of 1134 μm^2 also has a lower capacitance, but its needs an

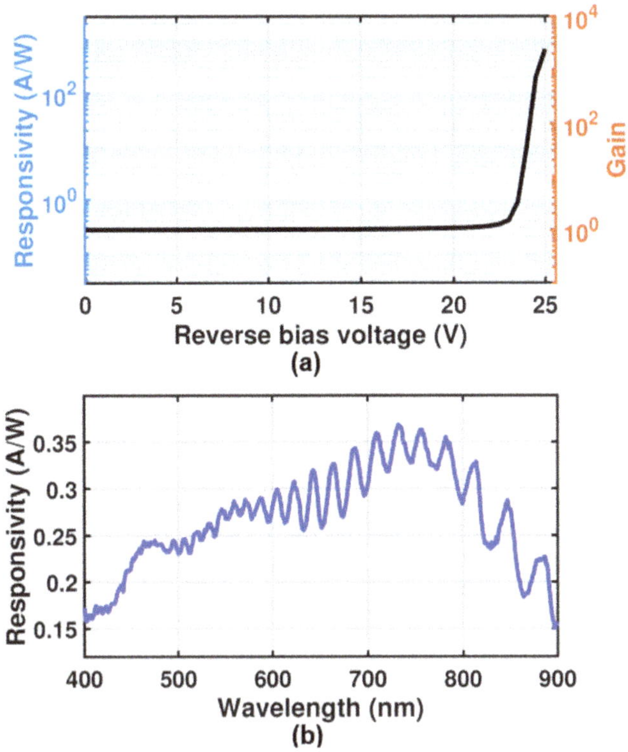

Figure 5.72. Measured responsivity (left) and gain (right) for 675 nm at 200 nW (a) and measured spectral responsivity for $M = 1$ and 2 nW optical power (b) of 5 × 5 multi-dot APD. Reproduced from [52] CC BY 4.0.

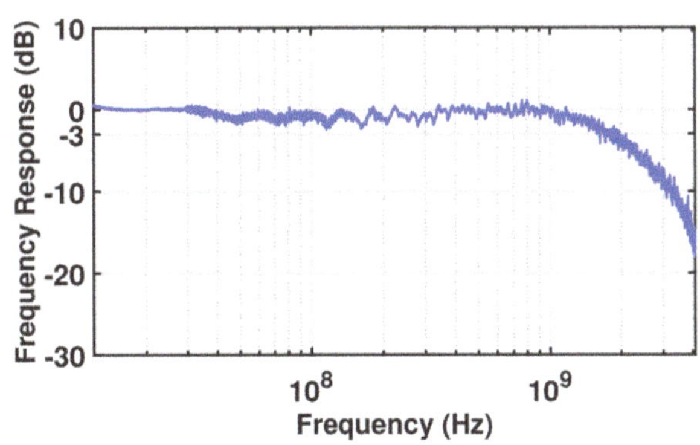

Figure 5.73. Normalized measured frequency response for 675 nm at 24 V and 200 nW of 5 × 5 multi-dot APD. Reproduced from [52] CC BY 4.0.

operating voltage of 69 V, whereas the multi-dot N+/P-well APD achieves a bandwidth of 1.8 GHz already at 24 V. The responsivity-bandwidth product of the EFLC-APD is almost three times higher than that of the multi-dot N+/P-well APD but at the expense of almost three times the operating voltage.

Table 5.7. Comparison of the performance of linear-mode APDs. Reproduced from [52]

Reference	[45]	[54]	[29]	[55]	[56]	[43]	[52]
Structure	n^+/pw RT*	p^+/nw APD	n^+/pw RT*	p^+/nw CA ⊕	Double pw/nw	n^+/nw EFLC #	n^+/pw MD &
Technology (nm)	350	250	350	350	45	180	350
Active area (μm^2)	5808	100	2827	1600	400	1134	4900
Capacitance (fF)	125	—	⩾100	1	—	⩽1	27
Operating voltage (V)	63	12.2	47	68	20.8	69	24
Optical power (μW)	0.5	1000	1	2	100	0.2	0.2
Wavelength (nm)	670	850	670	830	850	675	675
Gain	50	16.7	20	43.9	23	80	36
Responsivity (A/W)	20.5	0.2	7.4	13.17	0.56	32	9.7
Bandwidth (GHz)	0.85	5.6	2.3	0.275	8.4	1.6	1.8
R-BW product $\left(\frac{A}{W}\cdot GHz\right)$	17.4	1.1	17.02	3.6	4.7	51.2	17.46

* RT-APD: reach-through APD; ⊕ CA-APD: current-assisted APD; # EFLC-APD: electric field line crowding APD; & MD-APD: multi-dot APD.

Summarizing and concluding, multi-dot photodiodes were the next logical step for further reduction of capacitance after the finger photodiodes investigated in [11–14]. The multi-dot N+/P-well APD in 0.35 μm PIN-photodiode CMOS is easily scalable to larger or smaller light-sensitive areas. Its operating voltage is comparable to that of PIN-photodiode CMOS based APDs and SPADs (actually somewhat lower) and its bandwidth is high at a low capacitance. The multi-dot N+/P-well APD is well appropriate for ultra-sensitive optical receivers with larger light-sensitive area than that of single-dot photodiodes. The multi-dot APDs, however, are not only appropriate for USPAR ICs, they also allow the improvement of resistive-feedback TIAs with respect of their sensitivity without reduction of the light-sensitive area.

5.5 Isolation of circuits

Some of the photodiodes introduced above need very negative substrate potentials to deplete the thick absorption zone fully. For the dot-APDs exploiting the field crowding effect, the substrate can be at −60 V or even more. The breakdown voltage of the P-channel MOSFET inside the N-well towards substrate is much lower and the N-channel MOSFET is located inside a P-well, which is on substrate potential in a so-called twin-well process. Fortunately, there are triple-well CMOS processes, which offer a third well, the deep N-well. Both MOSFETs can be placed inside the deep N-well (see figure 5.74), which isolates the N-channel MOSFET from the substrate and increases the breakdown voltage of the P-channel MOSFETs. According to the standard design rules of the 0.35 μm PIN-photodiode CMOS

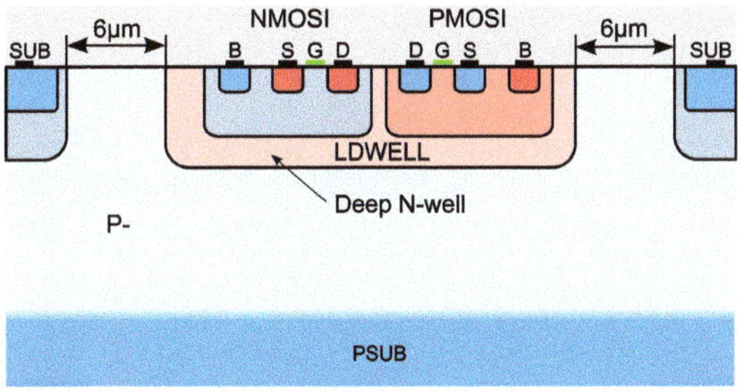

Figure 5.74. Cross-section of MOSFETs in 0.35 μm PIN-photodiode CMOS with increased distance to substrate contacts [57].

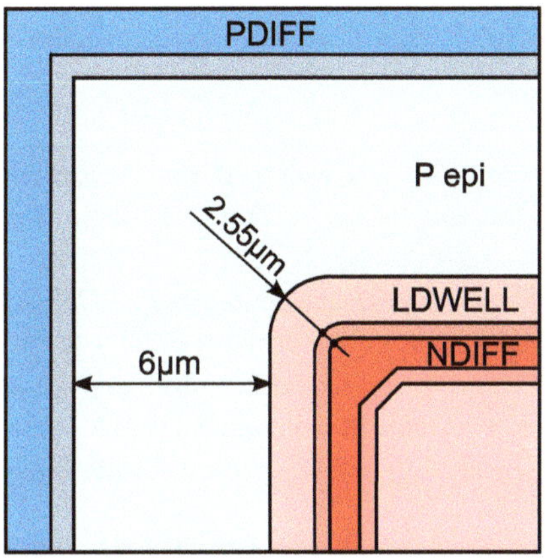

Figure 5.75. Modified corner layout of deep N-wells for better isolation of MOSFETs [57].

process, the breakdown voltage of this NMOSi transistor and of the PMOSi transistor is 40 V [41].

This breakdown voltage of the MOSFETs towards substrate can be increased by using a larger distance of 6 μm between deep N-well and the substrate contacts (see figure 5.74). In addition, the rectangular corner of the deep N-well was replaced by a circular shape as depicted in figure 5.75.

By these two isolation measures, the breakdown voltage of the NMOSi transistor towards substrate was increased to 130 V as the measured current to substrate shows (see figure 5.76). The breakdown voltage of the PMOSi transistor exploiting these layout modifications was 126 V.

These layout and design rule modifications can also be applied to USPAR receivers in 0.18 μm CMOS technology, which also uses low-doped epitaxial p-type wafers. In fact, the 0.18 μm CMOS process is then also a PIN-photodiode CMOS process. There is also a deep N-well for isolation of transistors. The breakdown voltages of the isolated MOSFETs towards substrate were not determined by measurements, but should also be much larger than 40 V.

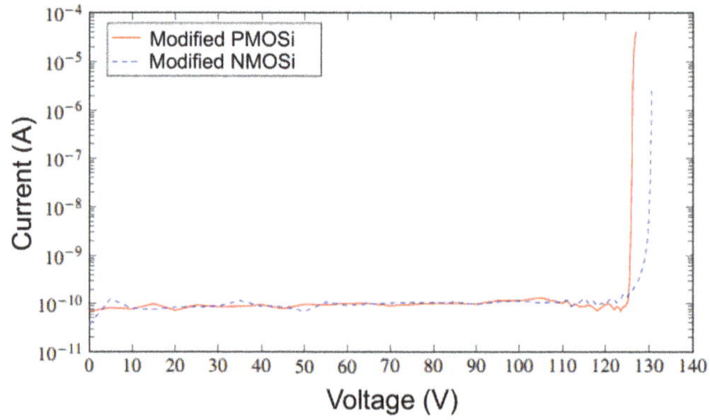

Figure 5.76. Breakdown characteristics towards substrate of MOSFETs with modified deep-N-well layout [57].

References

[1] Ebeling K J 1993 *Integrated Optoelectronics* (Berlin, Heidelberg: Springer)

[2] Säckinger E 2018 *Analysis and Design of Transimpedance Amplifiers for Optical Receivers* (Hoboken, NJ: Wiley)

[3] Zimmermann H 2018 *Silicon Optoelectronic Integrated Circuits* (Switzerland: Springer)

[4] Salvia J, Lajevardi P, Hekmat M and Murmann B 2009 A 56 MΩ CMOS TIA for MEMS applications *2009 IEEE Custom Integrated Circuits Conference* 199–202

[5] Ma J, Zhang D, Elgendy O A and Masoodian S 2021 A 0.19e- rms read noise 16.7 mpixel stacked quanta image sensor with 1.1 μm-pitch backside illuminated pixels *IEEE Electron Device Lett.* **42** 891–4

[6] Vlaskovic M, Zimmermann H, Meinhardt G and Kraft J 2020 Image sensor for spectral-domain optical coherence tomography on a chip *Electron. Lett.* **56** 1306–9

[7] Zimmermann H 2010 *Integrated Silicon Optoelectronics* (Heidelberg, Germany: Springer)

[8] Schneider K and Zimmermann H 2006 *Highly Sensitive Optical Receivers* (Berlin, Heidelberg, Germany: Springer)

[9] Förtsch M, Zimmermann H, Einbrodt W, Bach K and Pless H 2002 Integrated PIN photodiodes in high-performance BiCMOS technology *Digest. International Electron Devices Meeting* 801–4

[10] X-Fab Semiconductor Foundries 2019 Process & Device Specification XO035—0.35 μm Modular CMOS for Fast Optical Applications

[11] Gaberl W and Zimmermann H 2006 Low-capacitance integrated silicon finger photodetector *3rd IEEE Int. Conf. on Group IV Photonics* 122–4

[12] Ghazi A, Zimmermann H and Seegebrecht P 2002 CMOS photodiode with enhanced responsivity for the UV/blue spectral range *IEEE Trans. Electron Devices* **49** 1124–8

[13] Nemecek A, Oberhauser K, Zach G and Zimmermann H 2005 Fast and efficient integrated silicon PIN-finger photodetector from ultraviolet to up to near infrared *Proc. 2nd IEEE Conf. on Group IV Photonics* 108–10

[14] Zimmermann H, Marchlewski A, Gaberl W, Jonak-Auer I, Meinhardt G and Wachmann E 2009 Blue-enhanced PIN finger photodiodes in a 0.35 μm SiGe BiCMOS technology *IEEE Photon. Technol. Lett.* **21** 1656–8

[15] Silvaco 2023 Atlas user's manual, available: https://www.silvaco.com accessed: 23 March 2023

[16] Zimmermann H, Müller B, Hammer A, Herzog K and Seegebrecht P 2002 Large-area lateral p-i-n photodiode on soi *IEEE Trans. Electron Devices* **49** 334–6

[17] Avset B, Ellison J, Evensen L, Hall G, Hansen T-E, Roe S and Wheadon R 1989 A silicon drift photodiode *IEEE Trans. Nucl. Sci.* **36** 295–9

[18] Fiorini C, Gola A, Zanchi M, Longoni A, Lechner P, Soltau H and Struder L 2006 Gamma-ray spectroscopy with labr$_3$:ce scintillator readout by a silicon drift detector *IEEE Trans. Nucl. Sci.* **53** 2392–7

[19] Bertuccio G *et al* 2016 X-ray silicon drift detector-CMOS front-end system with high energy resolution at room temperature *IEEE Trans. Nucl. Sci.* **63** 400–6

[20] Mahmoudi H, Poushi S S K, Steindl B, Hofbauer M and Zimmermann H 2021 Optical and electrical characterization and modeling of photon detection probability in CMOS single-photon avalanche diodes *IEEE Sens. J.* **21** 7572–80

[21] Poushi S S K, Mahmoudi H, Steindl B, Hofbauer M and Zimmermann H 2020 Comprehensive modeling of photon detection probability in CMOS-based single-photon avalanche diodes *2020 IEEE SENSORS* 1–4

[22] Goll B, Schneider-Hornstein K and Zimmermann H 2023 Dot PIN photodiodes with a capacitance down to 1.14 aF μm^{-2} *IEEE Photon. Technol. Lett.* **35** 301–4

[23] https://en.wikipedia.org/wiki/Capacitance#Capacitance_of_simple_systems, 'Capacitance,' accessed 20 Nov. 2022

[24] Goll B, Schneider-Hornstein K and Zimmermann H 2023 Ultra-low capacitance spot PIN photodiodes *IEEE Photonics J.* **15** 1–6

[25] van Sieleghem E, Süss A, Boulenc P, Lee J, Karve G, de Munck K, Cavaco C and van Hoof C 2021 Near-infrared enhanced silicon single-photon avalanche diode with a spherically uniform electric field peak *IEEE Electron Device Lett.* **42** 879–82

[26] Sakic A *et al* 2010 Versatile silicon photodiode detector technology for scanning electron microscopy with high-efficiency sub-5 keV electron detection *2010 International Electron Devices Meeting* 31.4.1–4

[27] Steindl B, Gaberl W, Enne R, Schidl S, Schneider-Hornstein K and Zimmermann H 2014 Linear mode avalanche photodiode with 1 GHz bandwidth fabricated in 0.35 μm CMOS *IEEE Photon. Technol. Lett.* **26** 1511–4

[28] Gaberl W, Schneider-Hornstein K, Enne R, Steindl B and Zimmermann H 2014 Avalanche photodiode with high responsivity in 0.35 μm cmos *SPIE Opt. Eng.* **53** 043105-1–4

[29] Steindl B, Jukic T and Zimmermann H 2017 Optimized silicon CMOS reach-through avalanche photodiode with 2.3 GHz bandwidth *SPIE Optical Engineering* **56** 110501–110501-3

[30] Rochas A *et al* 2002 Low-noise silicon avalanche photodiodes fabricated in conventional CMOS technologies *IEEE Trans. Electron Devices* **49** 387–94

[31] Chou F-P, Wang C-W, Li Z-Y, Hsieh Y-C and Hsin Y-M 2013 Effect of deep n-well bias in an 850 nm si photodiode fabricated using the cmos process *IEEE Photon. Technol. Lett.* **25** 659–62

[32] Youn J-S, Lee M-J, Park K-Y and Choi W-Y 2012 10 Gb s^{-1} 850 nm CMOS OEIC receiver with a silicon avalanche photodetector *IEEE J. Quantum Electronics* **48** 229–36

[33] Iijama K *et al* 2010 Hole-injection type and electron-injection type silicon avalanche photodiodes fabricated by standard 0.18 μm CMOS process *IEEE Photon. Technol. Lett.* **22** 932–4

[34] Atef M, Polzer A and Zimmermann H 2013 Avalanche double photodiode in 40 nm standard CMOS technology *IEEE J. Quantum Electronics* **49** 350–6

[35] Shimotori T, Maekita K, Maruyama T and Iiyama K 2012 Characterization of APDs fabricated by 0.18 μm CMOS process in blue wavelength region *Proc. Opto-Electronics and Communications Conf. (OECC)* 509–10

[36] Poushi S S K, Mahmoudi H, Hofbauer M, Steindl B, Schneider-Hornstein K and Zimmermann H 2021 Experimental and simulation study of fill-factor enhancement using a virtual guard ring in n+p-well CMOS single-photon avalanche diodes *Opt. Eng.* **60** 67105–11

[37] van Sieleghem E, Süss A, Boulenc P, Rosmeulen M and van Hoof C 2019 A NIR-enhanced silicon BSI SPAD with low sensitivity to process fulctuations and 15 μm pitch *Proc. Single Photon Workshop (Milan)* 149

[38] Engelmann E, Schmailzl W, Iskra P, Wiest F, Popova E and Vinogradov S 2021 Tip avalanche photodiode–a new generation silicon photomultiplier based on non-planar technology *IEEE Sens. J.* **21** 6024–34

[39] Sigmund H 1968 Photoelectric properties of spherical avalanche diodes in silicon *Infrared Phys.* **8** 259–64

[40] Baliga B J and Ghandhi S K 1976 Analytical solutions for the breakdown voltage of abrupt cylindrical and spherical junctions *Solid-State Electron.* **19** 739–44

[41] Zimmermann H, Steindl B, Hofbauer M and Enne R 2017 Integrated fiber optical receiver reducing the gap to the quantum limit *Sci. Rep.* **7** 2652

[42] Steindl B, Hofbauer M, Schneider-Hornstein K, Brandl P and Zimmermann H 2018 Single-photon avalanche photodiode based fiber optical receiver up to 200 Mb s^{-1} *J. Sel. Top. Quantum Electron.* **24** 3801308

[43] Kohneh Poushi S S, Goll B, Schneider-Hornstein K, Hofbauer M and Zimmermann H 2022 CMOS integrated 32 A W^{-1} and 1.6 GHz avalanche photodiode based on electric field-line crowding *IEEE Photonics Technol. Lett.* **34** 945–8

[44] Karar A *et al* 1995 Investigation of avalanche photodiodes for EM calorimeter at LHC *SCAN-9510272, Technical report* pp 932–4

[45] Steindl B, Enne R, Schidl S and Zimmermann H 2014 Linear mode APD with high responsivity integrated in high-voltage CMOS *IEEE Electron Device Lett.* **35** 897–9

[46] Kohneh Poushi S S, Gasser C, Goll B, Hofbauer M, Schneider-Hornstein K and Zimmermann H 2023 A near-infrared enhanced field-line crowding based CMOS-integrated avalanche photodiode *IEEE Photonics J.* **15** 1–9

[47] Decoster D and Harari J 2013 *Optoelectronic Sensors* (New York: Wiley)

[48] McIntyre R J 1966 Multiplication noise in uniform avalanche diodes *IEEE Trans. Electron Devices* **13** 164–8

[49] Hayat M, Sargeant W and Saleh B 1992 Effect of dead space on gain and noise in si and gaas avalanche photodiodes *IEEE J. Quantum Electron.* **28** 1360–5

[50] Hayat M, Saleh B and Teich M 1992 Effect of dead space on gain and noise of double-carrier-multiplication avalanche photodiodes *IEEE Trans. Electron Devices* **39** 546–52

[51] Hossain M M, Ray S, Cheong J S, Qiao L, Baharuddin A N A P, Hella M M, David J P R and Hayat M M 2017 Low-noise speed-optimized large area CMOS avalanche photodetector for visible light communication *J. Light. Technol.* **35** 2315–24

[52] Kohneh Poushi S S, Goll B, Schneider-Hornstein K, Hofbauer M and Zimmermann H 2023 Area and bandwidth enhancement of an n+/p-well dot avalanche photodiode in 0.35 μm CMOS technology *Sensors* **23** 3403

[53] Koyama A *et al* 2015 Development of pixelated linear avalanche integration detector using silicon on insulator technology arXiv:150706721 https://arxiv.org/abs/1507.06721

[54] Lee M-J, Lee J-M, Rücker H and Choi W-Y 2015 Bandwidth improvement of CMOS-APD with carrier-acceleration technique *IEEE Photon. Technol. Lett.* **27** 1387–90

[55] Jegannathan G, Ingelberts H, Boulanger S and Kuijk M 2019 Current assisted avalanche photo diodes (CAAPDs) with separate absorption and multiplication region in conventional CMOS *Appl. Phys. Lett.* **115** 132101

[56] Zhi W, Quan Q, Yu P and Jiang Y 2020 A 45 nm CMOS avalanche photodiode with 8.4 GHz bandwidth *Micromachines* **11** 65

[57] Jukic T 2017 Empfänger mit integrierter Lawinenfotodiode *PhD Thesis* TU Wien

Chapter 6

Ultra-sensitive PIN photodiode receivers

In this chapter ultra-sensitive optical receivers with low-capacitance photodiodes based on PIN-photodiode CMOS technology are investigated. These receivers employ integrators exploiting CMOS inverters and common-source amplifiers. Both integrator types are described and compared. The importance of the integration efficiency will be elucidated. In addition, reset-less integrating receivers without and with punch-through device (PTD) are introduced. Analytical expressions for gain, noise and sensitivity are derived.

6.1 Inverter integrator in 0.35 μm CMOS

The group of the author investigated inverter-based TIAs in optical receivers [1–4] and used integrating-type correlators with inverters as amplifiers for 3D time-of-flight sensors at about 10 MHz, because inverters have a high gain and cause a low input noise current [5, 6]. Due to the good experience with inverters, we employ them also the in ultra-sensitive PIN and avalanche photodiode receiver (USPAR). The circuit of the USPAR employing an inverter as integrator is shown in figure 6.1.

In the following, the equations describing the gain and noise are put together. The voltage gain of the inverter A_{inv} is

$$A_{\text{inv}} = \frac{g_{m,\text{n}} + g_{m,\text{p}}}{g_{\text{ds,n}} + g_{\text{ds,p}}}, \tag{6.1}$$

where $g_{m,\text{n}}$ and $g_{m,\text{p}}$ are the transconcuctances and $g_{\text{ds,n}}$ and $g_{\text{ds,p}}$ are the output conductances of the NMOS and PMOS transistor, respectively.

The noise can be calculated starting from the drain current noise of NMOS and PMOS transistor (see equation (3.8)). Here we neglect the $1/f$ noise. The squared drain current noises of NMOS and PMOS transistor have to be added to obtain at the inverter output

doi:10.1088/978-0-7503-5437-0ch6

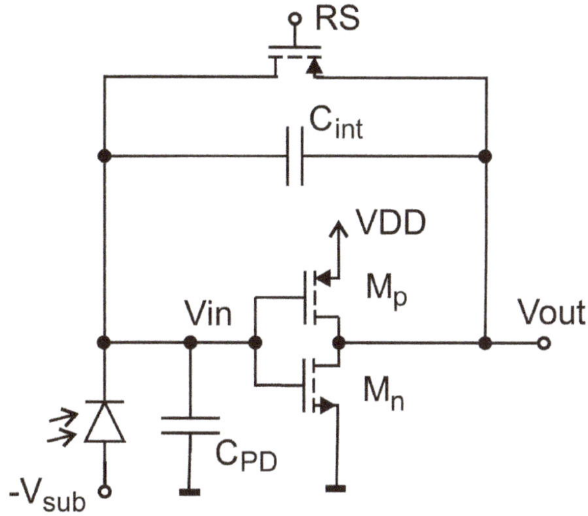

Figure 6.1. Integrating optical receiver exploiting an inverter as amplifier.

$$\overline{i^2_{\mathrm{n,out}}} = 4k_{\mathrm{B}}T\left(\frac{2}{3}g_{m,\mathrm{n}}\right)\Delta f + 4k_{\mathrm{B}}T\left(\frac{2}{3}g_{m,\mathrm{p}}\right)\Delta f = \frac{8}{3}k_{\mathrm{B}}T(g_{m,\mathrm{n}} + g_{m,\mathrm{p}})\Delta f. \tag{6.2}$$

Via the total transconductance of the inverter $g_m = g_{m,\mathrm{n}} + g_{m,\mathrm{p}}$, the output-node noise can be transformed to an input voltage noise of the inverter (note the definition of the transconductance $g_m = \frac{\Delta i_{\mathrm{D}}}{\Delta v_{\mathrm{GS}}}$):

$$\overline{v^2_{\mathrm{n,\,in}}} = \frac{\overline{i^2_{\mathrm{n,out}}}}{g_m^2} \tag{6.3}$$

and the squared input noise voltage is

$$\overline{v^2_{\mathrm{n,\,in}}} = \frac{8}{3}\frac{k_{\mathrm{B}}T}{(g_{m,\mathrm{n}} + g_{m,\mathrm{p}})}\Delta f. \tag{6.4}$$

The noise at the output is then obtained by applying the inverter gain A_{inv}:

$$\overline{v^2_{\mathrm{n,out}}} = A^2_{\mathrm{inv}}\,\overline{v^2_{\mathrm{n,\,in}}} = \frac{8k_{\mathrm{B}}T}{3}\frac{g_{m,\mathrm{n}} + g_{m,\mathrm{p}}}{(g_{\mathrm{DS,n}} + g_{\mathrm{DS,p}})^2}\Delta f \tag{6.5}$$

Another possibility is to multiply the output noise current (equation (6.2)) with the total output resistance $r_{\mathrm{inv,out}}$ of the inverter

$$\frac{1}{r_{\mathrm{inv,out}}} = \frac{1}{r_{\mathrm{DS,n}}} + \frac{1}{\mathrm{r}_{DS,p}} \tag{6.6}$$

Instead of the inverter's output resistance we can use the inverter's output conductance

$$g_{\mathrm{inv,out}} = g_{\mathrm{DS,n}} + g_{\mathrm{DS,p}}, \tag{6.7}$$

where $g_{\mathrm{DS,n}} = r_{\mathrm{DS,n}}^{-1}$ and $g_{\mathrm{DS,p}} = r_{\mathrm{DS,p}}^{-1}$ with the output resistances $r_{\mathrm{DS,n}}$ and $r_{\mathrm{DS,p}}$ of the NMOS and PMOS transistors. Now, we obtain the output noise voltage:

$$\overline{v_{\mathrm{n,out}}^2} = \frac{\overline{i_{\mathrm{n,out}}^2}}{g_{\mathrm{inv,out}}^2} = \frac{8}{3}\frac{k_{\mathrm{B}}T\,(g_{m,\mathrm{n}} + g_{m,\mathrm{p}})}{(g_{\mathrm{DS,n}} + g_{\mathrm{DS,p}})^2}\Delta f, \tag{6.8}$$

which is the same result as in equation (6.5).

The transconductances and the output conductances of the transistors can be determined with the circuit simulator (e.g. with SPECTRE in the CADENCE design framework [7]). Of course, the noise can be directly calculated in circuit simulation, but here we aim towards an analytical model for the sensitivity, which can be achieved in USPAR. For this analytical model we need the signal voltage at the output of the integrator, which is the output of the inverter. Here, we have to be aware that the simple calculation with $U = Q/C_{\mathrm{int}} = n_{\mathrm{e}}q/C_{\mathrm{int}}$ (n_{e} is the number of electrons generated in the photodiode by the signal light) is only valid for an amplifier gain of infinity. For a limited gain A_{inv}, the integration capacitance C_{int} has to be charged to $(A_{\mathrm{inv}} + 1) \times U_{\mathrm{in}}$ to fulfil the gain condition $U_{\mathrm{out}} = -A_{\mathrm{inv}}U_{\mathrm{in}}$. Therefore, we obtain the integration efficiency

$$\eta_{\mathrm{int}} = \frac{A_{\mathrm{inv}}C_{\mathrm{int}}}{(A_{\mathrm{inv}} + 1)C_{\mathrm{int}} + C_{\mathrm{PD}}}. \tag{6.9}$$

The output signal voltage is then $U_{\mathrm{out}} = \eta_{\mathrm{int}}n_{\mathrm{e}}q/C_{\mathrm{int}}$. Note that the inverter gain A_{inv} has to be inserted into equation (6.9) as a positive number.

For a classical (digital) inverter, $W_{\mathrm{n}} = 1.0\ \mu$m and $W_{\mathrm{p}} = 2.5\ \mu$m was chosen, since the mobility of holes is about 2.5 times smaller than that of electrons and about the same transconductance should be present for both transistors in the inverter. Circuit simulations showed a gain $A_{\mathrm{inv}} = 16.6$ for the 0.35 μm PIN-photodiode CMOS technology. The bandwidth was 288 MHz with a common-source amplifier having a width of 5 μm as a load. The DC operating point of the inverter was at 1.58 V (VDD = 3.3 V). The transconductances were $g_{m,\mathrm{n}} = 210\ \mu$S and $g_{m,\mathrm{p}} = 189\ \mu$S. The output conductances were $g_{\mathrm{DS,n}} = 7.4\ \mu$S and $g_{m,\mathrm{p}} = 16.6\ \mu$S. The root-mean square (rms) output noise voltage calculated with the equations above is $v_{\mathrm{n,out}} = 1.45$ mV. According to the process and device specifications manual [8], the integration capacitance of this inverter is 0.7 fF.

With a photodiode capacitance (more precisely: input node capacitance) of 20 fF, the resulting sensitivity (considering reset noise, see section 3.12) was −48.8 dBm at a data rate of 250 Mb s^{-1} for BER $= 2 \times 10^{-3}$ ($Q = 3$, see section 3.6). The number of photogenerated electrons in a '1'-bit being necessary for this sensitivity was 290. The reduction of the input node capacitance to 10 fF diminished the number of required electrons to 200 corresponding to a sensitivity of −50.45 dBm. Assuming perfect double sampling for elimination of the reset noise in addition reduced the number of necessary electrons to 125 and the sensitivity was improved to −52.5 dBm.

There is, however, the possibility of further reduction of the amount of necessary electrons, if we depart from classical, digital inverter design. The p-channel MOSFET with $W_p = 2.5 \times W_n$ contributes to $C_{int} = 3.5 \times C_{GD,n}$, but it only doubles the effective transconductance. Therefore, an improvement of the sensitivity follows, if $W_p < 2.5 \times W_n$. The DC operating point then goes down from VDD/2 closer towards GND, but this does not matter since the output voltage is positive (the anode of the PIN photodiode is at a negative bias and the cathode is connected to the input of the integrator).

The realization of the integration capacitor for USPAR would somehow be prohibited, if a polysilicon–polysilicon capacitor, a metal–metal (MIM) capacitor or another predefined capacitor described in the process-device-specification manuals of ASIC foundries were to be used. Such capacitors usually start from 10 fF upwards, which does not enable ultra-sensitive optical receivers. The reason for this lower limit of the capacitance values are their parasitic capacitances towards substrate and the parasitic capacitances of the metal lines connecting them to other circuit components, which change the effective capacitance. To keep the error due to these parasitics low, the lower limit is set to 10 fF.

But fortunately we do not need an explicit separate integration capacitor, because the MOSFETs possess gate–drain (and gate–source) overlap capacitances (see figure 3.22), which are included in the transistor models of the design kits. These overlap capacitances constitute parasitics for most CMOS circuits, but the gate–drain overlap capacitances of the inverters's NMOS and PMOS transistors can be most advantageously exploited as integration capacitor. Figure 6.2 depicts the

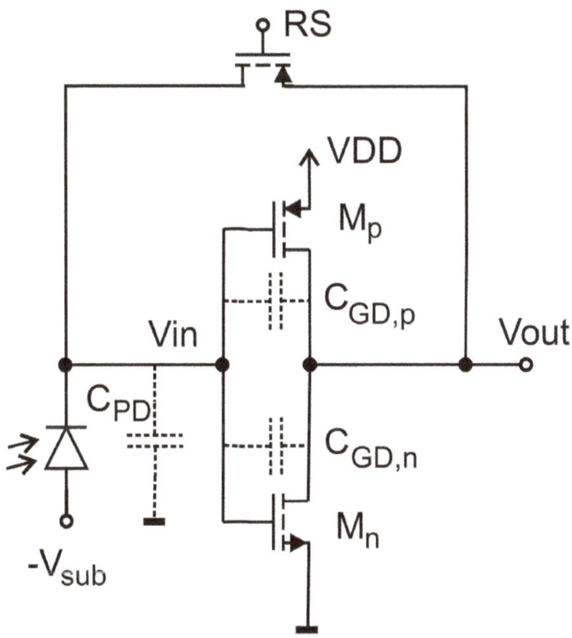

Figure 6.2. Inverter integrator with gate–drain overlap capacitances as integration capacitance.

resulting integrator circuit. In such a way, the parasitics of a separate integration capacitor are avoided and another realization with equal or fewer parasitics does not exist.

For the USPAR integrator, the gate–drain overlap capacitances are a huge piece of luck, as we saw just above. But we still have to solve another issue, which is related with the reset transistor. The reset transistor also possesses gate–drain and gate–source overlap capacitances, as depicted in figure 6.3.

The clock feed-through in the reset transistor (see figure 6.3) injects a huge amount of charges into the input node and into the output node of the integrating amplifier. For a width of the reset transistor of 1.4 μm the gate–drain overlap capacitance $C_{OV} = C_{GS} = C_{GD}$ (see figure 3.22) is 0.28 fF for the 0.35 μm PIN-photodiode CMOS process with a specific gate–drain overlap capacitance of 0.2 fF μm. According to

$$Q_c = C \cdot U, \tag{6.10}$$

the injected charge Q_c is 4.62×10^{-16} As for a voltage step at the gate of VDD/2 = 1.65 V. After dividing this charge by the electron charge 1.6×10^{-19} As, we obtain 2887 injected electrons. This is a dramatically large amount of injected charges if we are conscious of the number of photo-electrons to be detected in USPAR. The number of electrons to be detected in USPAR is on the order of 100

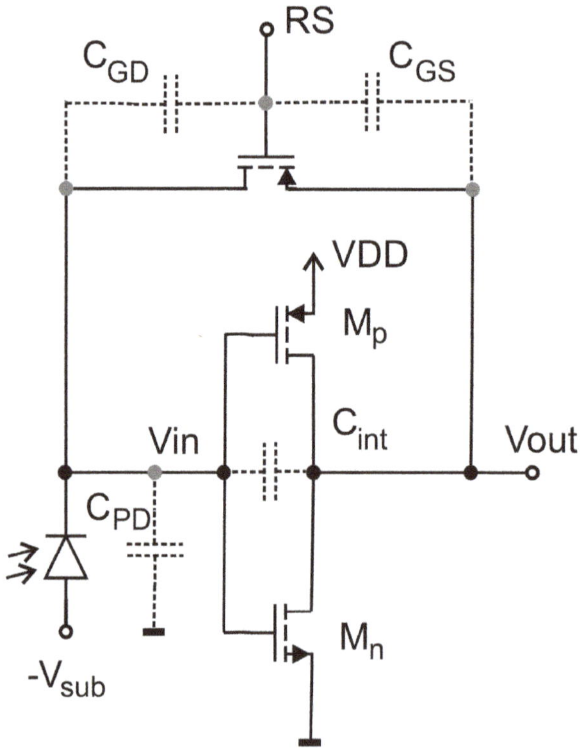

Figure 6.3. Inverter integrator with clock feed-through of reset transistor.

and possibly below. The consequence of this comparison is that the signal voltage cannot be recovered without additional measures, because the 'error' voltage is about 30 times larger than the signal voltage. Even, if a minimum-size transistor with a gate width of 0.7 μm is taken, the 'error' voltage is still 15 times larger than the signal voltage.

The idea of the dummy switch (see figure 3.24) can be applied to compensate this charge injected by the reset transistor. The compensating transistor needs half the width of the reset transistor and we need the compensating transistor twice, for the input node and for the output node of the integrating amplifier (see figure 6.4). In this circuit, the compensation transistors M_{comp} are connected as MOS capacitors. The minimum width of 0.7 μm in the 0.35 μm CMOS technology is used for a compensating transistor (as a consequence, the width of the reset transistor was chosen as 1.4 μm or two minimum size transistors connected in parallel can be used for better matching). Ideally, the injected charges would be canceled exactly, if the voltage step at the gates of the compensating transistors has the same height (1.65 V) and if it is of opposite direction compared to the voltage step at the gate of the reset transistor. But in reality, there is mismatch between the transistors, especially because they are of relatively small size. According to this mismatch, the 'error' voltage can still be of the same magnitude as the signal voltage. In addition, it is necessary that the voltage steps are applied simultaneously to the gates of the reset and compensation transistors. Otherwise, large positive and negative voltage spikes may be visible at the output of the integrator.

The inverter, of course, possesses an input capacitance $C_{\mathrm{in}} = C_{\mathrm{GS,n}} + C_{\mathrm{GS,p}}$, the metal line to the photodiode adds another (parasitic) capacitance C_{L} and the reset

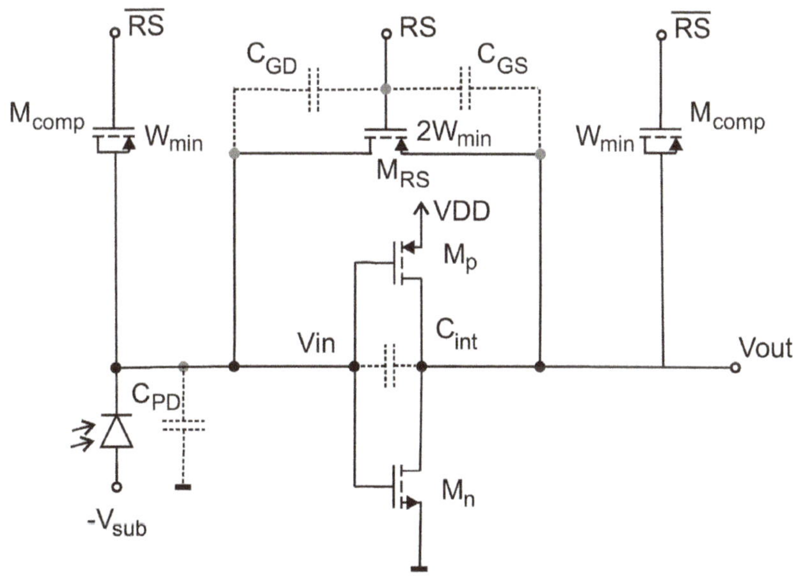

Figure 6.4. Inverter integrator with compensation transistors for clock feed-through of reset transistor.

transistor as well as the compensation transistor at the input append also capacitances towards VDD, i.e. towards small-signal ground:

$$C_T = C_{PD} + C_L + C_{GS,n} + C_{GS,p} + C_{DB,RS} + C_{GD,RS} + C_{GS,comp} + C_{GD,comp} + C_{DB,comp} + C_{SB,comp}, \tag{6.11}$$

where $C_{GS,n}$ and $C_{GS,p}$ are the gate–source overlap capacitances of the NMOS and PMOS transistors in the inverter. These overlap capacitances are equal to their gate–drain overlap capacitances. A part of the complete gate capacitance also adds in the models used by CADENCE. The reset transistor contributes its drain-bulk capacitance $C_{DB,RS}$ and its gate–drain overlap capacitance $C_{GD,RS}$. The compensation transistor located at the input contributes its gate–source and gate–drain overlap capacitances $C_{GS,comp}$ and $C_{GD,comp}$ as well as its drain-bulk and source-bulk capacitances $C_{DB,comp}$ and $C_{SB,comp}$.

These capacitances and C_{PD} lead to a total input-node capacitance C_T, which has to be used in the equation for the integration efficiency (6.9) instead of C_{PD}:

$$\eta_{int} = \frac{A_{inv} C_{int}}{(A_{inv} + 1)C_{int} + C_T}. \tag{6.12}$$

In addition to be able to determine the necessary optical input power, we have to consider the quantum efficiency of the photodiode ($\eta_{PIN} = \eta_d$) to calculate the signal voltage at the output of the integrator because the number of photogenerated electrons n_e is equal to $\eta_{PIN} \times n_{PH}$. Be aware that the signal output voltage can be obtained exploiting $U_{out} = Q_c/C_{int}$, where the charge Q is the number of electrons times the electron charge ($n_e \cdot q = n_{PH} \cdot \eta_{PIN} \cdot q$). Therefore, the signal output voltage is

$$U_{out} = \eta_{PIN}\eta_{int}n_{PH}q/C_{int}. \tag{6.13}$$

For the determination of the BER we need the total output noise of the integrator. But we have to consider also the kT/C-noise (reset noise) and the photon noise. The photon (shot) noise is determined by the square root of the number of photons.

The BER depends on the Q-factor or noise distance Q (see section 3.6). It is also possible to obtain Q from [9, 10]

$$Q = \frac{\langle i_1 \rangle - \langle i_0 \rangle}{\sigma_{i,1} + \sigma_{i,0}}, \tag{6.14}$$

where σ_1 and σ_0 represent the rms noise current of logical 1 and 0, respectively. For a very large extinction ratio of the light source and in darkness, we can neclect $\langle i_0 \rangle$ (there is no signal for the logical zero). Here, we compare the output signal voltage and the rms noise voltage σ_1 at the output of the inverter (note that we must not neglect σ_0) and use

$$Q = \frac{\langle u_1 \rangle}{\sigma_1 + \sigma_0}. \tag{6.15}$$

We furthermore have to consider the inverter noise $v_{n,out}$ and the reset noise $v_{n,RS}$ both in σ_1 and σ_0, but the photon noise $v_{n,PH}$ contributes only to σ_1 and not to σ_0 (extinction ratio of light source very large or infinity).

$$Q = \frac{U_{out}}{\sqrt{v_{n,out}^2 + v_{n,\,RS}^2 + v_{n,\,PH}^2} + \sqrt{v_{n,out}^2 + v_{n,\,RS}^2}} \tag{6.16}$$

We can take $v_{n,\,RS}^2$ from equation (3.69). The voltage noise at the integrator output from photon noise is

$$v_{n,\,PH}^2 = n_{PH}\eta_{PIN}^2\eta_{int}^2 q^2/C_{int}^2. \tag{6.17}$$

$$Q = \frac{\eta_{PIN}\eta_{int}n_{PH}q/C_{int}}{\sqrt{v_{n,out}^2 + k_BT/C_{int} + n_{PH}\eta_{PIN}^2\eta_{int}^2 q^2/C_{int}^2} + \sqrt{v_{n,out}^2 + k_BT/C_{int}}} \tag{6.18}$$

Now, we can plot the noise distance Q over the number of incident photons. The width of the NMOS transistor was varied from 1.0 μm to 0.7 μm, which is the minimum allowed gate width in the 0.35 μm CMOS technology. The width of the PMOS transistors was changed proportionally. The inverter integrator's bandwidth was 223 MHz (for $W_n = 0.7$ μm)[1] and the quantum efficiency of the PIN photodiode was 0.9. The inverter gain was the same for all four variations and the transconductances were scaled as the widths. The capacitance of the photodiode was 3.5 fF inclusive of the capacitance of the metal line to the integrator input. The input capacitance of the inverter was 4 fF μm$^{-1} \times W_n$ and the contribution of the reset transistor and of the left compensation transistor to the input-node capacitance was 3 fF (the widths of reset transistor and compensation transistors were constant). Figure 6.5 shows the noise distance Q for inverter noise, reset noise and photon noise.

For $W_n = 1.0$ μm, 168 photons are necessary for $Q = 3$. 166 photons are sufficient with $W_n = 0.9$ μm, 162 photons with $W_n = 0.8$ μm and 158 photons with $W_n = 0.7$ μm.

For the number of incident photons we can calculate the optical input power in a logical 1.

$$P_{opt,1} = n_{PH}h\nu B, \tag{6.19}$$

where $h\nu$ is the photon energy (h is Planck's constant) and B is the data rate ($B = 1/T$, T = duration of a bit). The photon frequency is $\nu = \frac{c}{\lambda}$ with the light velocity c and the wavelength λ. Therefore, the average optical power (assuming equally frequent zeros and ones) is

$$\langle P_{opt}\rangle = \frac{n_{PH}hcB}{2\lambda}. \tag{6.20}$$

[1] This was the minimum bandwidth for W_n from 0.7 μm to 1.0 μm. This bandwidth was taken for all four gate width cases, since the bandwidth could be adjusted by a separate low pass.

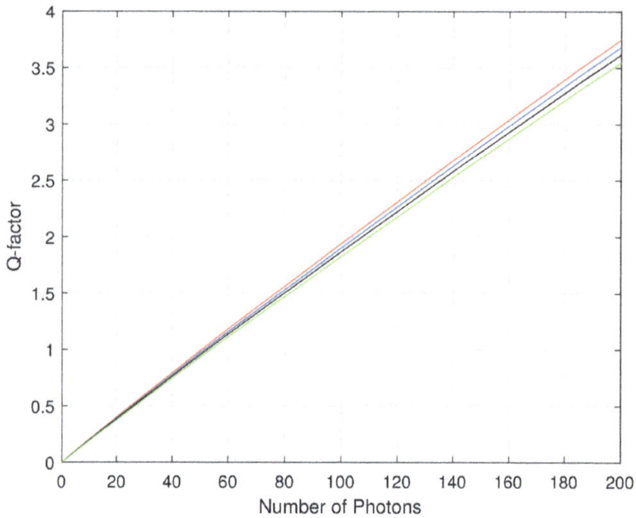

Figure 6.5. Q-factor of inverter integrator in 0.35 μm PIN-photodiode CMOS ($W_p = 2.5 \times W_n$, green: $W_n = 1.0$ μm, black: $W_n = 0.9$ μm, blue: $W_n = 0.8$ μm, red: $W_n = 0.7$ μm) with reset noise ($T = 300$ K).

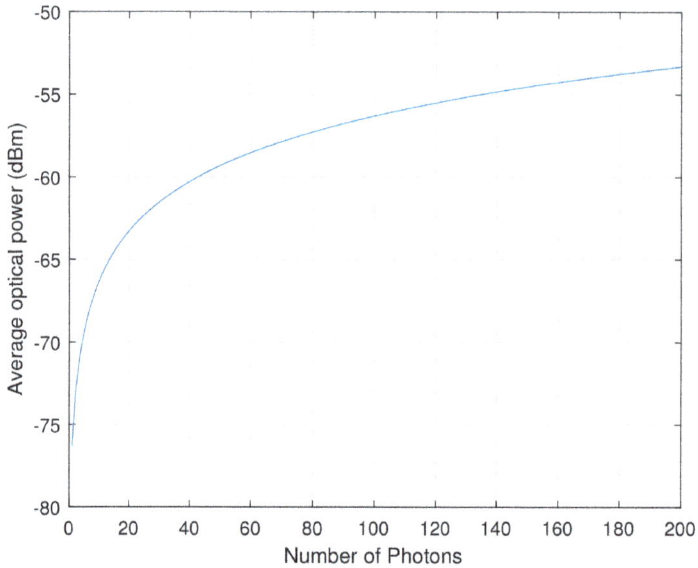

Figure 6.6. Sensitivity for a data rate of 150 Mb s^{-1} and a wavelength of 635 nm.

The sensitivity ($\langle P_{\text{opt}} \rangle$ for $Q = 3$) can be read from figure 6.6 and ranges from –54.0 dBm to –54.3 dBm for W_n from 1.0 μm to 0.7 μm (168 photons to 158 photons).

If double sampling eliminates the reset noise completely, we can set the temperature to zero ($T = 0$) in equation (6.18). The results are plotted in figure 6.7. We can read from these curves that 85 photons are necessary for $Q = 3$, if $W_n = 1.0$ μm. The number of photons reduces to 83, to 81 and 80 for $W_n = 0.9\ \mu$m, $W_n = 0.8\ \mu$m and $W_n = 0.7\ \mu$m, respectively[2]. We also see that this classical inverter is quite at its limit, because the number of photons necessary for $Q = 3$ does not change much for a reduction of the gate widths by 30%.

Since the necessary number of photons for $Q = 3$ is much smaller when the reset noise is completely omitted (compare figures 6.5 and 6.7), we should compare the reset noise to the noise of the inverter. Figure 6.8 presents the rms reset noise voltage ($\sqrt{k_B T / C_{int}}$) for the integration capacitances of the inverters with four different gate widths. When we compare these rms reset noise voltages to the rms inverter output noise voltages shown in figure 6.9, we notice that the rms reset noise voltages are considerably larger (almost two times). To be exact, it should be mentioned that the rms reset noise voltage of the integration capacitance will be allocated with an amount of $A/(A + 1)$ to the output of the inverter and to an amount of $1/(A + 1)$ to the input of the inverter, because the gain behavior of the inverter must be fulfilled. Since $A/(A + 1)$ is almost 1, we keep equation (6.18) as it is with respect to the $k_B T / C_{int}$-noise. Concluding, it is absolutely necessary to use double sampling for enhancing the sensitivity of the USPAR integrated circuits.

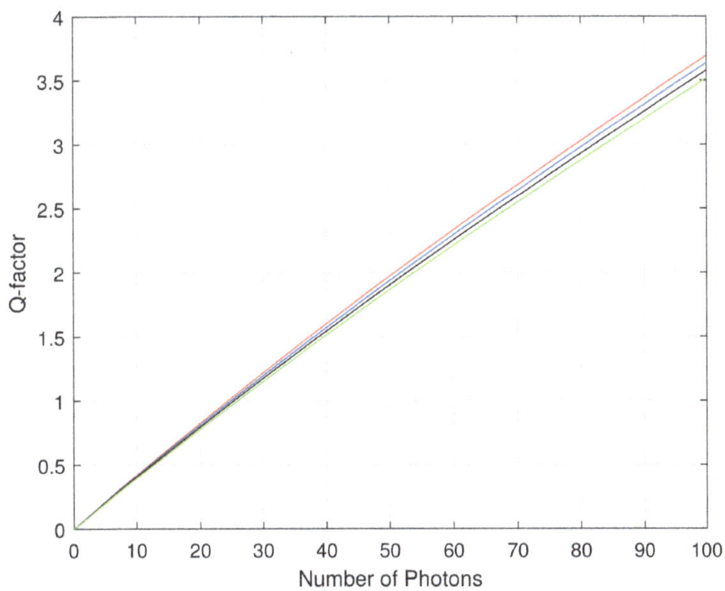

Figure 6.7. Q-factor of inverter integrator in 0.35 μm PIN-photodiode CMOS ($W_p = 2.5 \times W_n$, green: $W_n = 1.0\ \mu$m, black: $W_n = 0.9\ \mu$m, blue: $W_n = 0.8\ \mu$m, red: $W_n = 0.7\ \mu$m) assuming perfect double sampling, i.e. without reset noise.

[2] Below we will see that real double sampling will raise the thermal noise voltage by a factor of $\sqrt{2}$, see figure 6.19 and equations (6.22), (6.23)

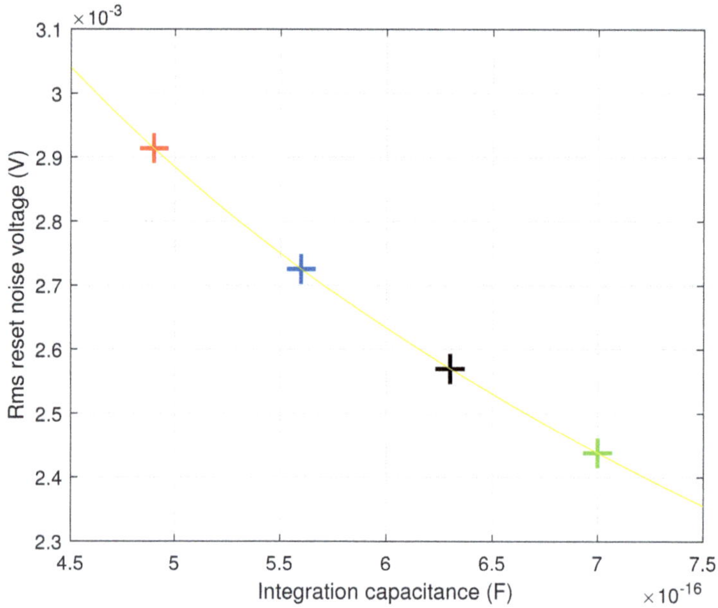

Figure 6.8. Reset noise ($T = 300$ K) for the integration capacitances of the inverter integrators in 0.35 μm PIN-photodiode CMOS ($W_p = 2.5 \times W_n$, green: $W_n = 1.0$ μm, black: $W_n = 0.9\mu$m, blue: $W_n = 0.8$ μm, red: $W_n = 0.7$ μm).

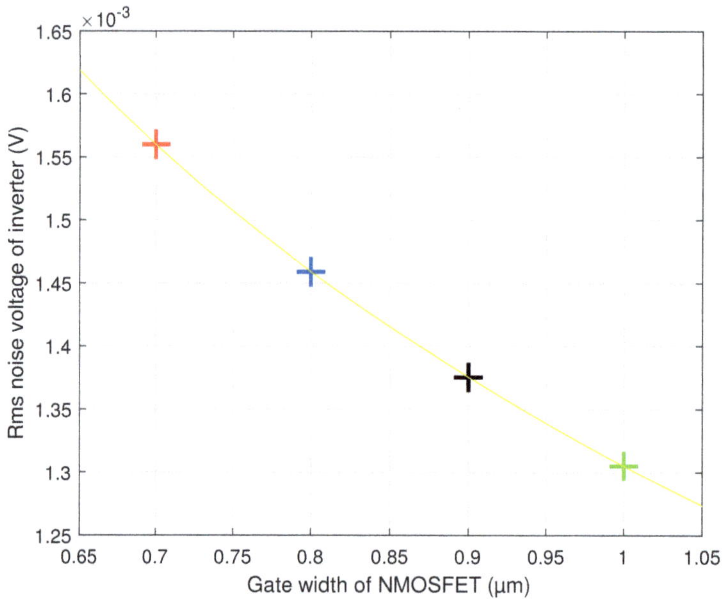

Figure 6.9. Rms output noise voltage of inverter integrators in 0.35 μm PIN-photodiode CMOS ($T = 300$ K, $W_p = 2.5 \times W_n$, green: $W_n = 1.0$ μm, black: $W_n = 0.9$ μm, blue: $W_n = 0.8$ μm, red: $W_n = 0.7$ μm).

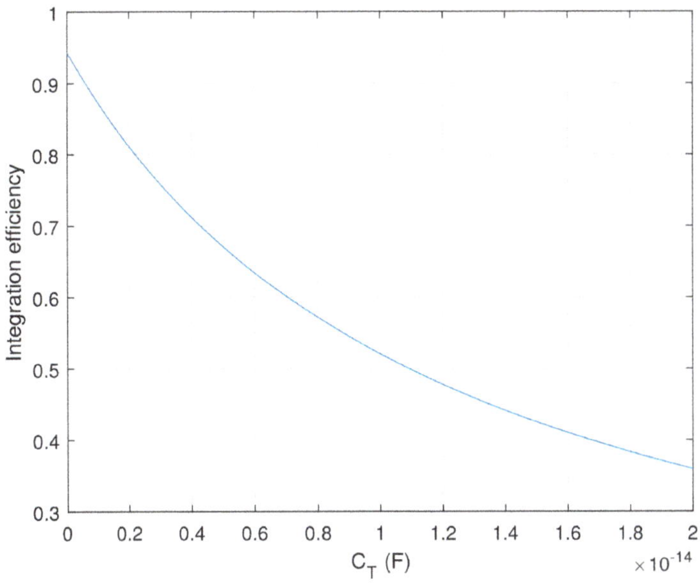

Figure 6.10. Integration efficiency for $A = 16.6$ and $C_{\text{int}} = 0.7$ fF.

The importance of the integration efficiency shall be illustrated with figure 6.10. The larger the input-node capacitance inclusive photodiode is, the smaller the integration efficiency. For a large amplifier gain A, the influence of the input-node capacitance decreases and for $C_{\text{T}} \to 0$ the integration efficiency approaches $A/(A + 1)$. This value is 16.6/17.6 for above example and can be read from the curve in figure 6.10 at $C_{\text{T}} = 0$. For the example of the inverter above, C_{T} is around 10 fF, for which the integration efficiency is near 0.5.

A higher integration efficiency can be achieved for a larger gain of the integrating amplifier. Then the influence of the input-node capacitance is less pronounced. Figure 6.11 plots the integration efficiency for $C_{\text{int}} = 0.7$ fF and $C_{\text{T}} = 10$ fF over the gain A.

Actually the gain A depends on the frequency making the integration efficiency of equation (6.12) also frequency dependent:

$$\eta_{\text{int}}(\omega) = \frac{A_{\text{inv}}(\omega)C_{\text{int}}}{(A_{\text{inv}}(\omega) + 1)C_{\text{int}} + C_T}. \tag{6.21}$$

Therefore, we should consider the frequency response of the inverter integrator. The frequency response of the integrator was simulated with SPECTRE inside the CADENCE integrated circuit design framework [7] using the design kit of the 0.35 μm PIN photodiode CMOS technology. A common-source amplifier stage consisting of an NMOS transistor with a gate width of 5 μm and of a resistor of 1 kΩ at its drain was connected to the output of the inverter ($W_{\text{n}} = 1.0$ μm, $W_{\text{p}} = 2.5$ μm). A small-signal analysis in the frequency range from 10 Hz to 10 GHz with an AC photocurrent as input source was performed for this purpose and the output voltage

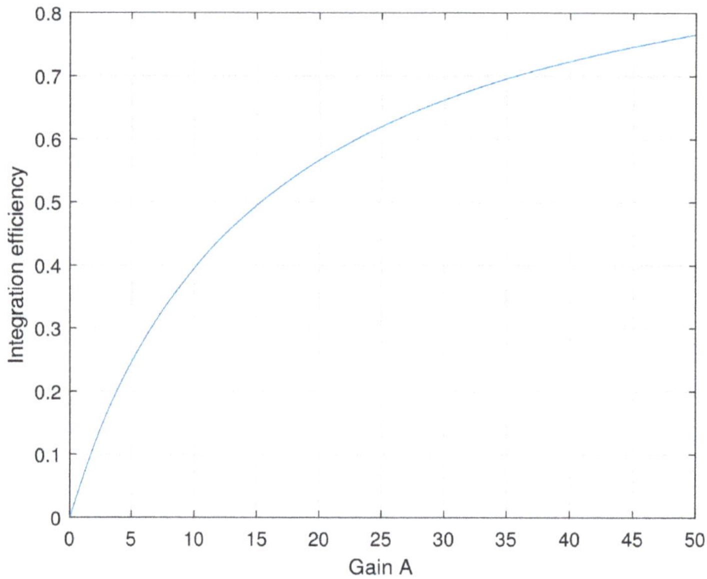

Figure 6.11. Integration efficiency for $C_{\mathrm{int}} = 0.7$ fF and $C_{\mathrm{T}} = 10$ fF.

of the inverter was devided by the input voltage of the inverter to obtain the gain. Figure 6.12 shows the obtained frequency response. It reveals an one-pole frequency roll-off (at 400 MHz the gain is 10 and it drops to 1 at about 4 GHz). The −3 dB bandwidth is 288 MHz The low-frequency gain is 16.6. The frequency response of an inverter with $W_{\mathrm{n}} = 0.7$ μm and $W_{\mathrm{p}} = 1.75$ μm showed a −3 dB bandwidth of 223 MHz.

Figure 6.13 shows the frequency dependence of the gain and of the integration efficiency for a one-pole amplifier. For a drop of the gain A of 30% from 0 Hz to the −3 dB bandwidth $f_{\mathrm{g}} = 223$ MHz the integration efficiency decreases by about 16%. The dependence of the integration efficiency on the frequency is therefore not strong, but a certain error will occur, if we use the DC integration efficiency (calculated with the DC gain A_0).

Therefore, the transient response of the integrator was investigated for a few data rates to get a better understanding of the possible error when using the DC gain of the inverter for the calculation of the integration efficiency. Circuit simulations were performed again with SPECTRE and the design kit of the 0.35 μm CMOS technology. The amount of photogenerated charges was held constant for the different data rates. This means that the photocurrent was varied accordingly for the different data rates, i.e. different integration periods. For 50 Mb s^{-1}, one bit lasts 20 ns, the integration time is 6.67 ns, the reset time and the times for sampling are both also 6.67 ns. Figure 6.14 depicts the simulated transient response to light pulses with a duration of 6.67 ns. The photocurrent was switched between zero and 120 nA.

The transients at the output of the inverter integrator (violet) and at the output of the second, common-source (CS) amplifier stage (blue) are shown in this figure. The second amplifier stage just served as a load to the inverter and as a kind of buffer for

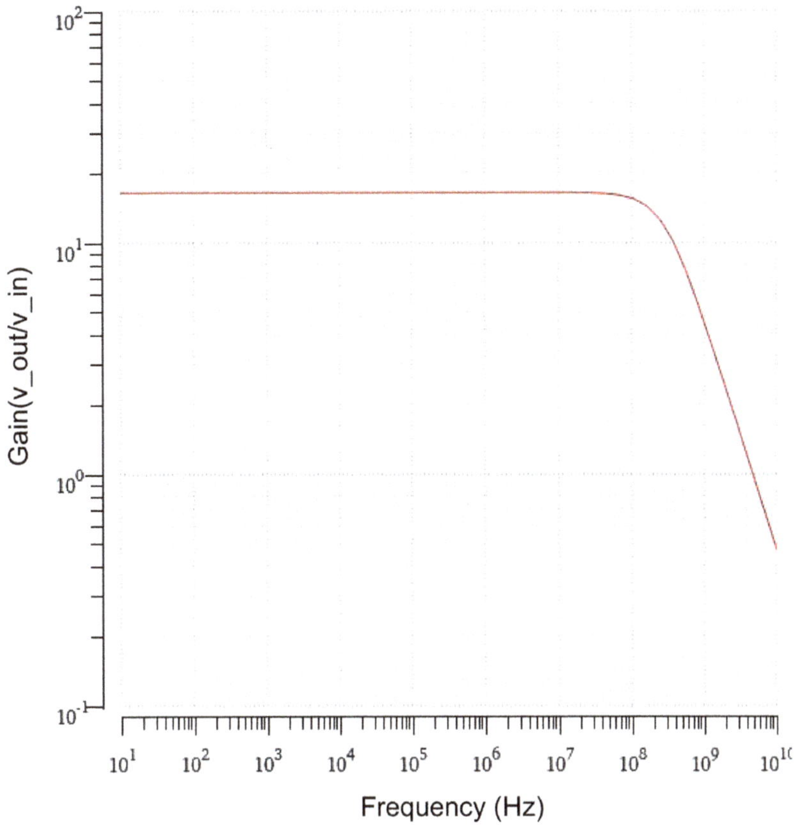

Figure 6.12. Frequency response of the inverter in 0.35 μm CMOS ($W_n = 1.0$ μm, $W_p = 2.5$ μm).

the sensitive integrator output. This second amplifier stage is certainly not optimized. From the two curves of the output voltages of the inverter integrator and of the second CS amplifier stage we see that the operating point is influenced by charge injection during switching the PMOS reset transistor RS off (during the positive edge of the brown gate signal of the reset (RS) transistor). At zero time, there was no positive edge of the gate signal of the RS transistor. We see that the voltage step due to injection charge is about 220 mV at the inverter output, which is quite large. In order to obtain well visible signal voltages by charge integration in presence of charge injection, it was necessary to apply the high photocurrent of 120 nA, which corresponds to a photo-charge of 8×10^{-16} As together with the 6.67 ns duration. This photo-charge corresponds to 5000 electrons in one bit, which is much larger than the target value of USPAR. The output voltage of the integrator was determined from the violet curve just before the start of the green light pulse and before the start of the brown reset pulse by subtraction (be aware that the reset starts at the negative edge of the brown reset signal). The result of the integration at 50 Mb s^{-1} is 0.3704 V.

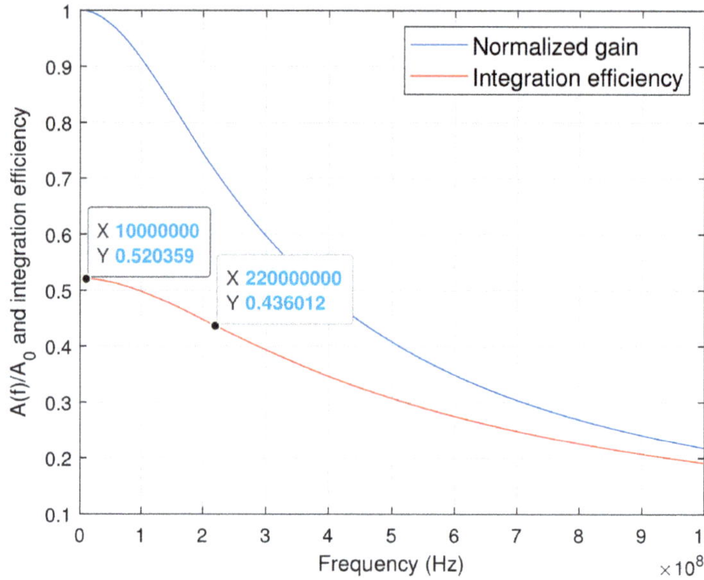

Figure 6.13. Frequency dependence of integration efficiency for $C_{\text{int}} = 0.7$ fF and $C_{\text{T}} = 10$ fF for a bandwidth $f_{\text{g}} = 223$ MHz.

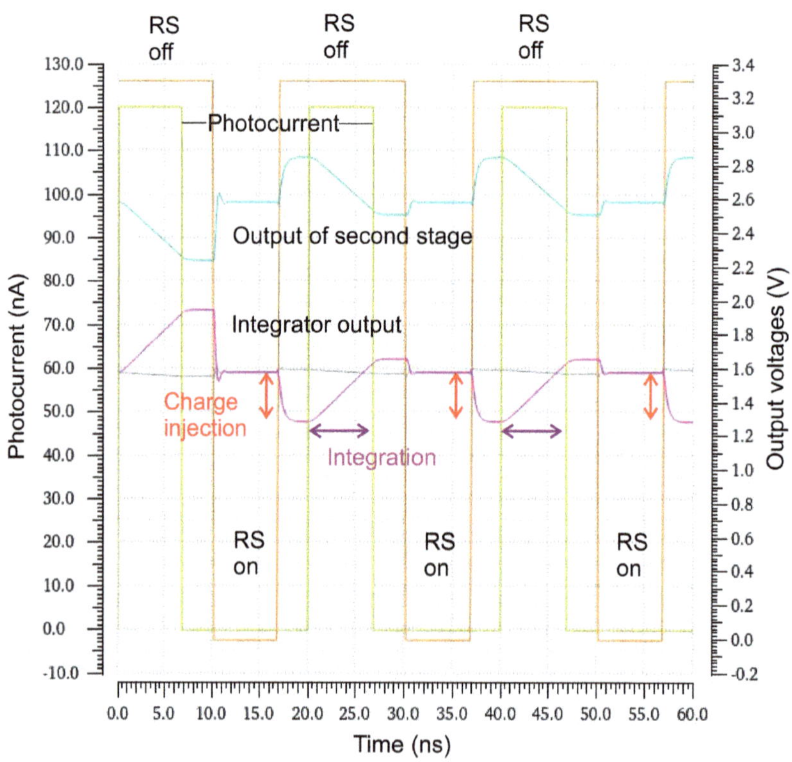

Figure 6.14. Transient response of the inverter integrator in 0.35 μm CMOS ($W_{\text{n}} = 1.0$ μm, $W_{\text{p}} = 2.5$ μm) for a data rate of 50 Mb s^{-1}.

For 100 Mb s^{-1}, the duration of a bit is 10 ns and the integration time and the reset time is 3.33 ns (see figure 6.15). The photocurrent, therefore, was switched between zero and 240 nA. The obtained result of the integration is 0.3667 V. Finally, for 150 Mb s^{-1}, one bit lasts 6.67 ns and the integration and reset take 2.22 ns each (see figure 6.16). For the integration result, we obtain 0.3423 V. This value is 7.6% smaller than the value for 50 Mb s^{-1}. To enable a better comparison, the integration results are plotted in figure 6.17 for the three data rates. The integration results are almost equal for 50 Mb s^{-1} and 100 Mb s^{-1}. But at 150 Mb s^{-1}, the integration result is already noticeably smaller, although the −3 dB bandwidth of the inverter is 288 MHz. In addition, we see from figure 6.16 that the data rate should not be increased above 150 Mb s^{-1} because the reset will then not be completed anymore.

The average optical input power is depicted over the number of photons for a data rate of 150 Mb s^{-1} and a wavelength of 635 nm in figure 6.18. For 80 photons (i.e. for the minimum size NMOS transistor in the inverter from above) we can read the sensitivity −57.2 dBm.

Let us compare this value of −57.2 dBm to the sensitivity of −54.3 dBm, which was obtained above for inverter noise, reset noise and photon noise. If double sampling can eliminate the reset noise completely, double sampling can improve the

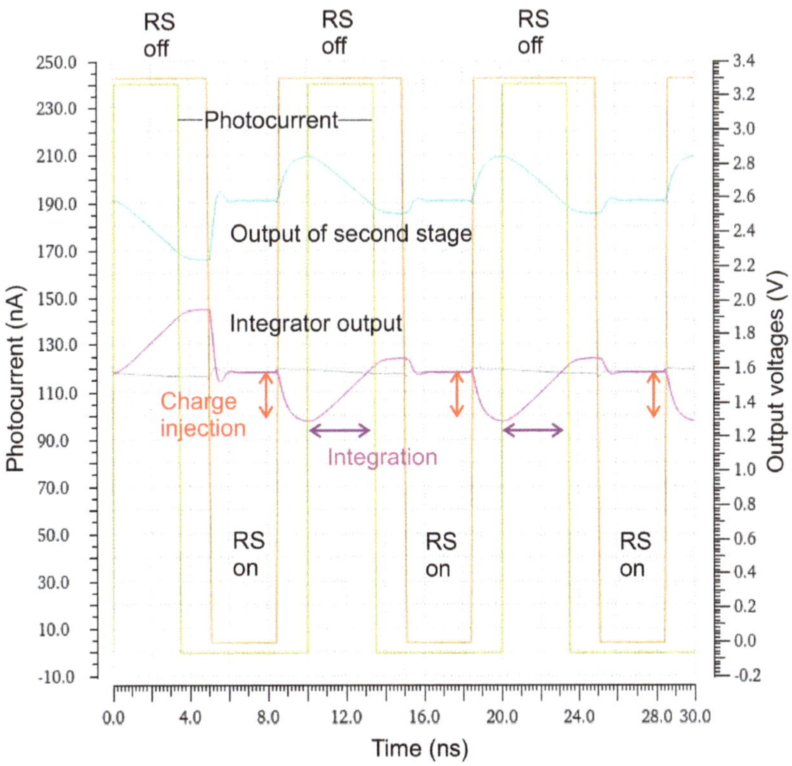

Figure 6.15. Transient response of the inverter integrator in 0.35 μm CMOS ($W_n = 1.0\ \mu$m, $W_p = 2.5\ \mu$m) for a data rate of 100 Mb s^{-1}.

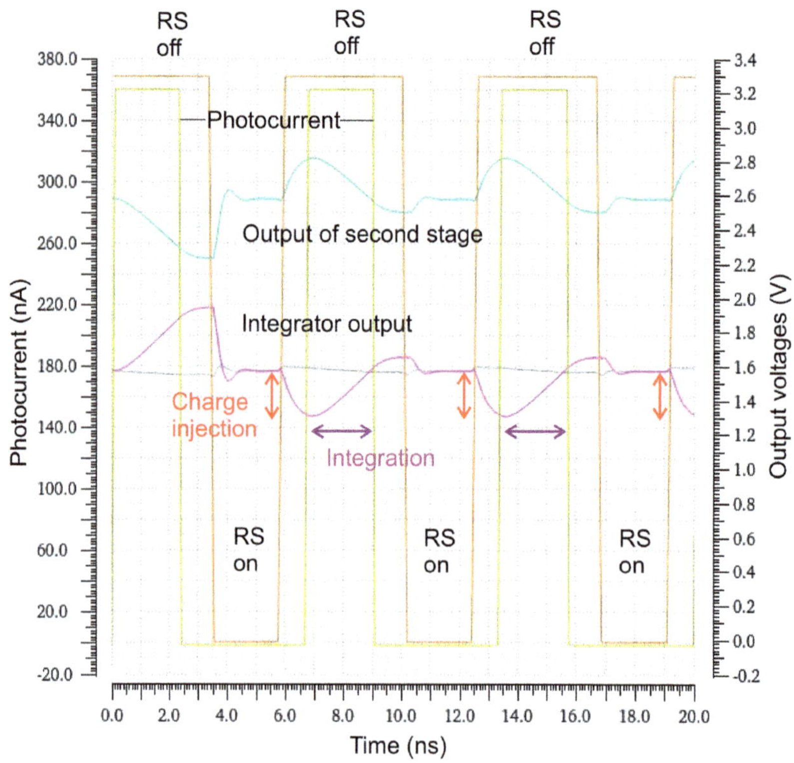

Figure 6.16. Transient response of the inverter integrator in 0.35 μm CMOS (W_n = 1.0 μm, W_p = 2.5 μm) for a data rate of 150 Mb s^{-1}.

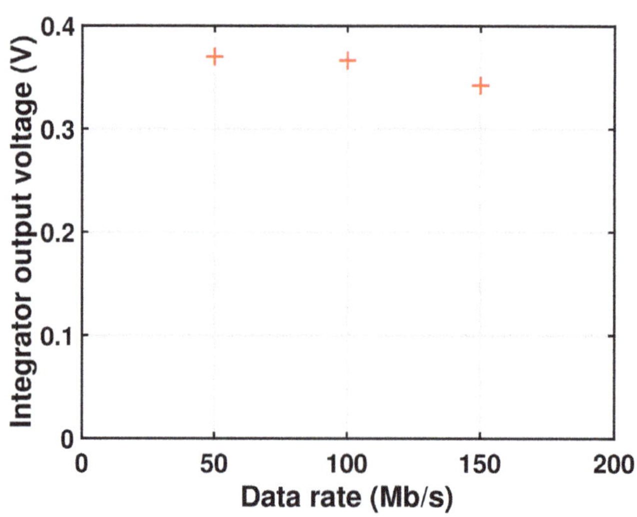

Figure 6.17. Output voltage of the inverter integrator in 0.35 μm CMOS (W_n = 1.0 μm, W_p = 2.5 μm) in dependence on data rate.

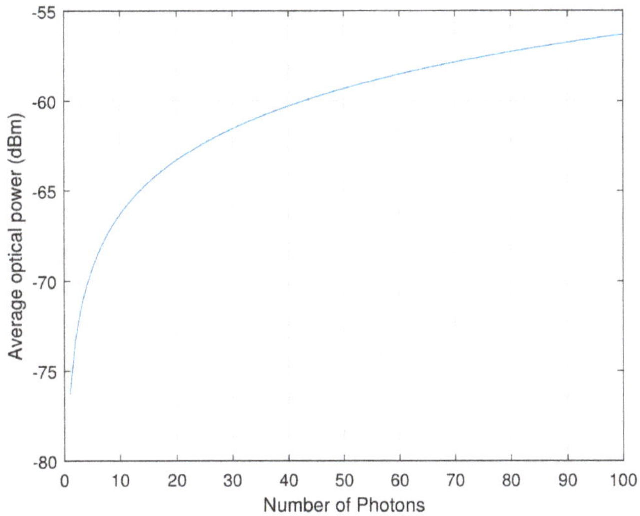

Figure 6.18. Sensitivity for a data rate of 150 Mb s^{-1} and a wavelength of 635 nm.

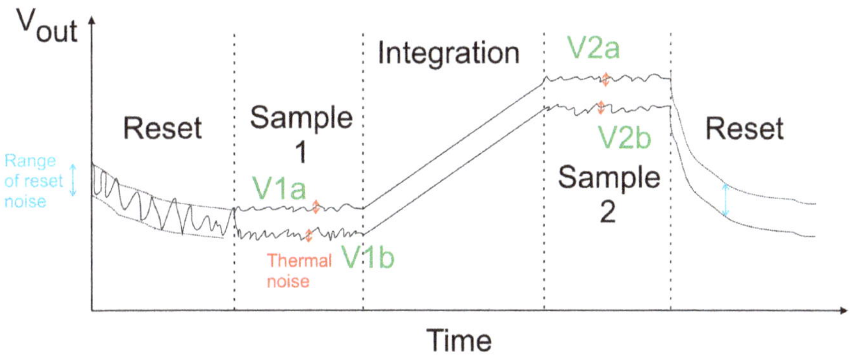

Figure 6.19. Illustration for double sampling.

sensitivity by almost 3 dB. However, if double sampling is used the difference of two signal voltages has to be composed to obtain the signal voltage. Figure 6.19 explains the context.

During reset, the RS transistor (see figure 6.1) is conducting and its channel resistance causes thermal noise. This (momentary) noise voltage is not known at the moment when the RS transistor is switched off at the end of the reset phase. According to section 3.12, the voltage, which is superimposed to the DC operating point of the inverter, is between 0 and $\sqrt{(k_B T/C_{int})}$. This range defines the uncertainty range of the voltage $V1$, which can lie between $V1a$ and $V1b$ ($V1a$–$V1b = \sqrt{(k_B T/C_{int})}$). During charge integration of the signal (for a transferred '1' bit), this uncertainty range is kept and the signal voltage after the integration phase has the same uncertainty range $V2a$–$V2b = \sqrt{(k_B T/C_{int})}$, if we neglect the thermal noise of the inverter which is superimposed in addition. If an eye diagram is constructed, the eye closes to $V2b$–$V1a$. This eye opening is smaller than $V2a$–$V1a$ and $V2b$–$V1b$, which correspond to the actual signals. Double sampling can be used to come close to these actual signals, if the reset noise is (much) larger than the thermal noise. With the first sampling, the reset noise is measured and the result of the second sampling gives the signal plus the same reset noise. Therefore, the difference of the two sampled voltages does not contain the reset noise anymore. The reason, why the actual signal cannot be reconstructed exactly, is the thermal noise of the inverter (in our case). In fact, because the signal is the difference $V2a$–$V1a$ or $V2b$–$V1b$, we have to add two squared noise voltages:

$$\overline{v_n^2} = \overline{v_{n2a}^2} + \overline{v_{n1a}^2}. \tag{6.22}$$

Since for the target sensitivities of USPAR the signals are in the millivolt range, the current in the inverter does almost not change and the thermal noise voltages are equal ($v_{n2a} = v_{n1a} = v_{n2b} = v_{n1b}$) and we obtain $\overline{v_{\mathrm{n,out}}^2} = 2\,\overline{v_{n2a}^2}$. Therefore, we have to implement a factor of 2 for the noise of the '1' and noise of the '0' in equation (6.18). Of course, we can omit the reset noise in this equation, if double sampling is used, and we obtain:

$$Q = \frac{\eta_{\mathrm{PIN}}\eta_{\mathrm{int}}n_{\mathrm{PH}}q/C_{\mathrm{int}}}{\sqrt{2v_{\mathrm{n,out}}^2 + n_{\mathrm{PH}}\eta_{\mathrm{PIN}}^2\eta_{\mathrm{int}}^2 q^2/C_{\mathrm{int}}^2} + \sqrt{2v_{\mathrm{n,out}}^2}} \tag{6.23}$$

Now, we can expect that more photons are necessary for a Q-factor of 3, as displayed in figure 6.7, but according to the comparison of reset noise and thermal noise voltages above, still fewer photons, as indicated by figure 6.5, should be sufficient for $Q = 3$. Indeed, figure 6.20 confirms this expectation. For $Q = 3$, 109,

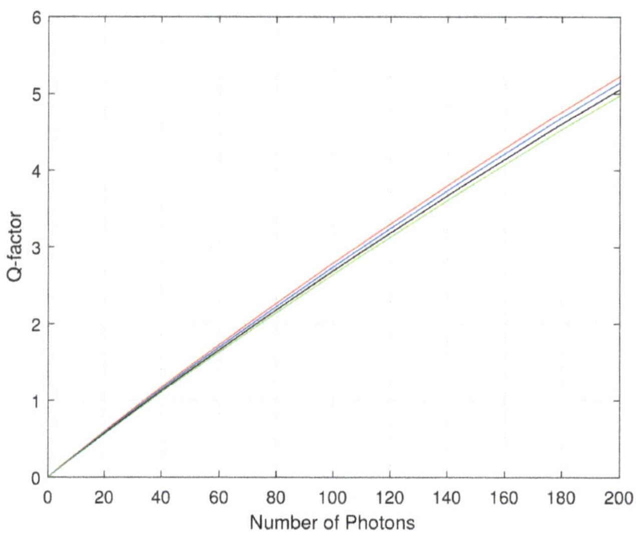

Figure 6.20. Q-factor of inverter integrator in 0.35 μm PIN-photodiode CMOS ($W_\mathrm{p} = 2.5 \times W_\mathrm{n}$, green: $W_\mathrm{n} = 1.0$ μm, black: $W_\mathrm{n} = 0.9$ μm, blue: $W_\mathrm{n} = 0.8$ μm, red: $W_\mathrm{n} = 0.7$ μm) without reset noise supposing real double sampling.

111, 113 and 115 photons are necessary for W_n equal to 0.7 μm, 0.8 μm, 0,9 μm and 1.0 μm, respectively. These photon numbers compare to 80, 01, 83 and 85 (see figure 6.7), respectively, for unrealistic, perfect double sampling and to 158, 162, 166 and 168 (see figure 6.5) for taking account reset noise completely. The photon numbers from 109 to 115 correspond to sensitivities from −55.9 dBm to −55.7 dBm for a data rate of 150 Mb s^{-1} and a wavelength of 635 nm. The improvement over the case without double sampling is still about 1.5 dB.[3]

6.2 Integrator with common-source amplifier in 0.35 μm CMOS

When we play the game of reducing the width of the PMOS transistor in the inverter to the very end to reduce the value of the integration capacitor, we can split the connection of the gates of the NMOS and PMOS transistors. Figure 6.21 shows the resulting circuit. The bias voltage V_b should be set to the same DC input voltage as the inverter had (1.58 V for W_n = 1 μm and W_p = 2.5 μm). The NMOS transistor works as common-source amplifier with the PMOS transistor as constant current source. Only the gate–drain overlap capacitance of the NMOS transistor acts as integration capacitor now. On the other side, only the transconductance of the NMOS transistor contributes to the gain and to the bandwidth. The gain of this common-source integrator in 0.35 μm PIN-photodiode CMOS is only 8.75 due to

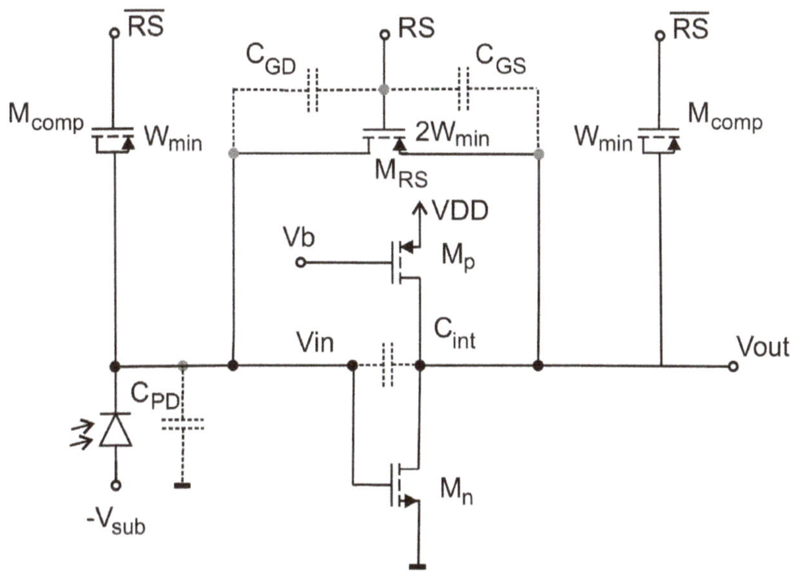

Figure 6.21. Integrator using a common-source stage with compensation transistors M_{COMP} for clock feed-through of reset transistor M_{RS}.

[3] For precision modelling, it may be necessary to consider the dependence of the noise voltage at the output of the integrator on the photodiode and input-node capacitance (see section 3.9).

circuit simulations. Due to this lower gain, the simulated bandwidth was about the same as for the inverter with the gain of 16.6.

The drain current noise of NMOS and PMOS transistor in the common-source amplifier is the same as for the inverter. The output resistance is also the same and therefore the output noise voltage is the same for inverter and common-source amplifier. The integration capacitance with the gate–drain overlap capacitance of the NMOS transistor with a width of 1 μm is 0.2 fF. Nevertheless, the integration efficiency is lower than that of the inverter integrator although the input-node capacitance C_{T} is also lower.

$$\begin{aligned} C_{\mathrm{T}} = {} & C_{\mathrm{PD}} + C_{\mathrm{L}} + C_{\mathrm{GS,n}} + C_{\mathrm{DB,RS}} \\ & + C_{\mathrm{GD,RS}} + C_{\mathrm{GS,comp}} + C_{\mathrm{GD,comp}} + C_{\mathrm{DB,comp}} + C_{\mathrm{SB,comp}} \end{aligned} \tag{6.24}$$

is somewhat smaller compared to that of the inverter integrator ($C_{\mathrm{GS,p}}$ is not present here compared to equation (6.11)). Figure 6.22 shows the integration efficiency for the gain of the common-source amplifier $A = 8.75$ in 0.35 μm CMOS. For $W_{\mathrm{n}} = 1$ μm and the photodiode capacitance of 3.5 fF inclusive of the line capacitance to the integrator input, C_{T} is about 7.5 fF, for which we can read an integration efficiency of somewhat below 0.2 from figure 6.22.

Although the integration efficiency is about 2.5 times lower than for the inverter integrator, the integration capacitance is 3.5 times lower, which should lead to a sensitivity advantage. Figure 6.23 shows, how the integration efficiency depends on the gain A for $C_{\mathrm{int}} = 0.2$ fF and $C_T = 7.5$ fF. According to this figure, a gain of 40 would raise the integration efficiency to about 0.5.

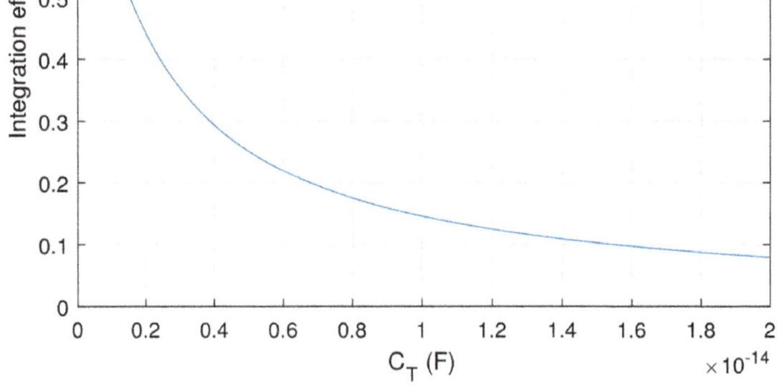

Figure 6.22. Integration efficiency for $A = 8.75$ and $C_{\mathrm{int}} = 0.2$ fF.

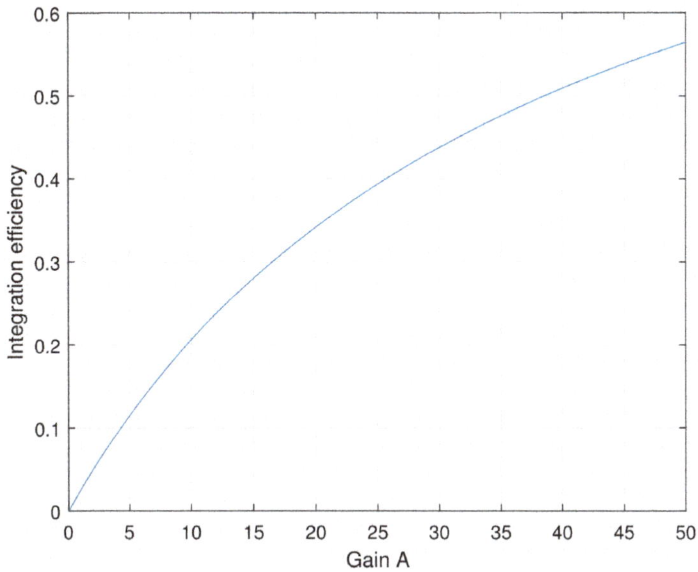

Figure 6.23. Integration efficiency for $C_{\text{int}} = 0.2$ fF and $C_T = 7.5$ fF in dependence on the gain.

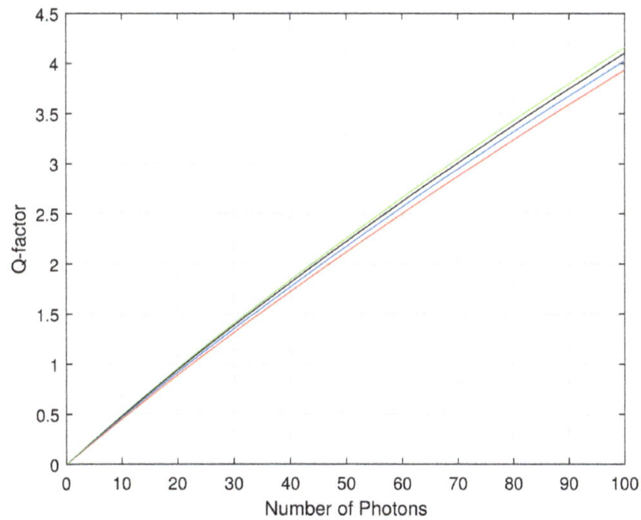

Figure 6.24. Q-factor of common-source integrator in 0.35 μm PIN-photodiode CMOS ($W_\text{p} = 2.5 \times W_\text{n}$, green: $W_\text{n} = 1.0$ μm, black: $W_\text{n} = 0.9$ μm, blue: $W_\text{n} = 0.8$ μm, red: $W_\text{n} = 0.7$ μm) without reset noise assuming perfect double sampling.

For the common-source integrator 69 to 74 photons are sufficient for a Q-factor of 3 (see figure 6.24). It may look a bit strange since for $W_\text{n} = 1$ μm the number of necessary photons is 69, for $W_\text{n} = 0.9$ μm 70 photons, for $W_\text{n} = 0.8$ μm 72 photons and for $W_\text{n} = 0.7$ μm 74 photons are needed for $Q = 3$. The reason for this behavior is the low gain and the low integration efficiency due to the rather large input-node capacitance.

Let us investigate this undesired outcome. To be able to understand what is going on, we compare the electronic noise voltage at the amplifier output, the noise voltage at the amplifier output caused by photon (shot) noise and the signal output voltage. Figure 6.25 shows that the electronic output noise voltage increases with decreasing gate width of the NMOS transistor. (This finding is the same, if the dependence of the output noise voltage on the photodiode capacitance (see section 3.9) is considered.) This increase is almost 20% from $W_n = 1\ \mu m$ to $W_n = 0.7\ \mu m$.

The noise at the amplifier output due to photon noise (see figure 6.26) increases only slightly because of the rather low integration efficiency. This increase is about 10% from $W_n = 1\ \mu m$ to $W_n = 0.7\ \mu m$ (around 70 photons).

The exiting question is how the signal output voltage changes with the gate width. This dependence is depicted in figure 6.27. The signal output voltage increases by only approximately 11% from $W_n = 1\ \mu m$ to $W_n = 0.7\ \mu m$ although the integration capacitor decreases by 30%. For a perfect integration efficiency, we therefore would expect an increase of the signal output voltage by about 40% (the output voltage is reciprocal to C_{int}, i.e. $1/0.7 = 1.4$). Since the increase in electronic noise is 20%, the increase of the signal output voltage of only 11% leads to a decrease in the Q-factor and more photons are necessary to achieve $Q = 3$, if W_n reduces from 1 μm to 0.7 μm. The conclusion is that the integration efficiency is too low. The reason for this is the dominance of the input node capacitance and the low gain A of the CS amplifier.

To determine the sensitivity of the common-source integrator we can look into figure 6.18. For $W_n = 1\ \mu m$, i.e. 69 photons (for $Q = 3$) in a logical 1, the sensitivity is −57.9 dBm. For 74 photons ($W_n = 0.7\ \mu m$) the sensitivity is −57.6 dBm.

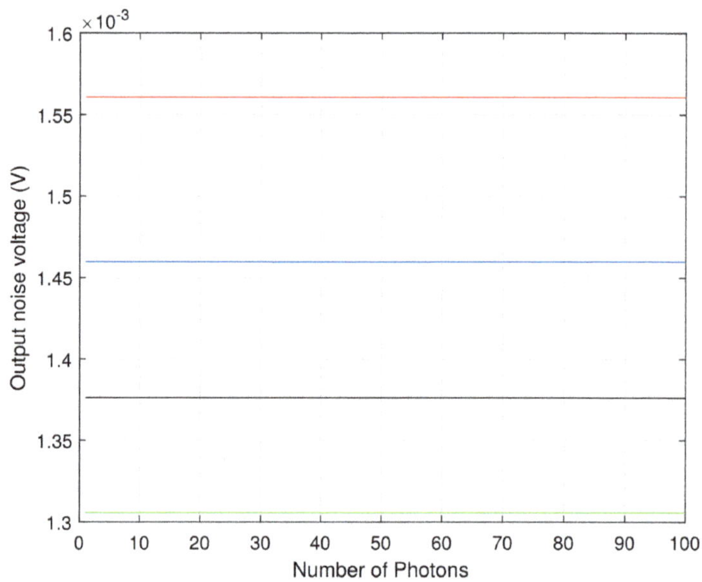

Figure 6.25. Rms electronic output noise of the common-source integrator in 0.35 μm PIN-photodiode CMOS ($W_p = 2.5 \times W_n$, green: $W_n = 1.0\ \mu m$, black: $W_n = 0.9\ \mu m$, blue: $W_n = 0.8\ \mu m$, red: $W_n = 0.7\ \mu m$).

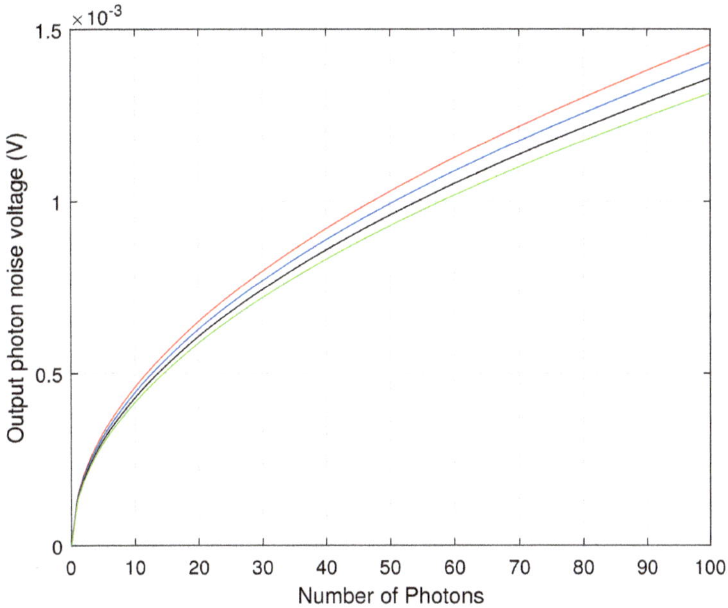

Figure 6.26. Rms output noise voltage of common-source integrator in 0.35 μm PIN-photodiode CMOS due to photon noise ($W_p = 2.5 \times W_n$, green: $W_n = 1.0$ μm, black: $W_n = 0.9$ μm, blue: $W_n = 0.8$ μm, red: $W_n = 0.7$ μm).

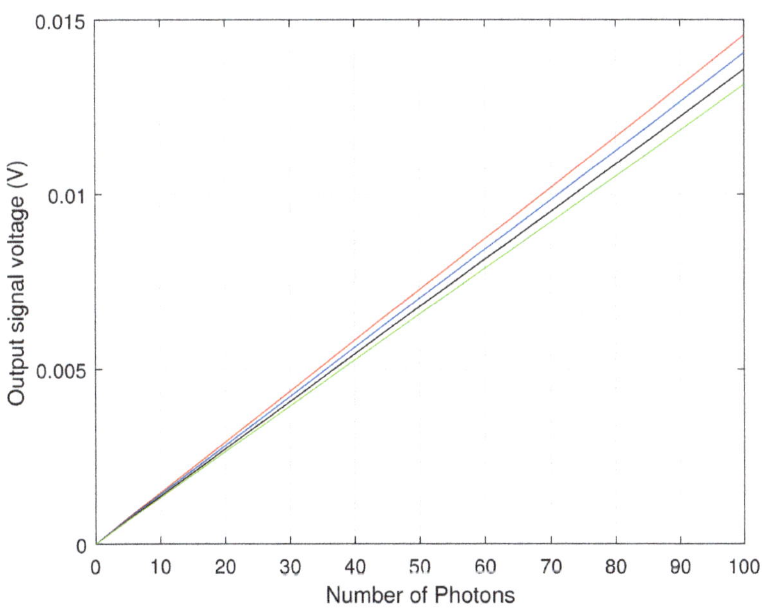

Figure 6.27. Signal output voltage of the common-source integrator in 0.35 μm PIN-photodiode CMOS ($W_p = 2.5 \times W_n$, green: $W_n = 1.0$ μm, black: $W_n = 0.9$ μm, blue: $W_n = 0.8$ μm, red: $W_n = 0.7$ μm).

We should also have a look onto the Q-factor for real double sampling calculated with equation (6.23). Figure 6.28 presents the Q-factor for real double sampling. For $Q = 3$ and $W_n = 0.7$ μm, 101 photons are necessary, for $W_n = 0.8$ μm 98 photons, for $W_n = 0.9$ μm 96 photons and for $W_n = 1.0$ μm 94 photons are required.

Since the sensitivity improves for increasing W_n from 0.7 μm to 1.0 μm, we can try to find the optimum W_n. Figure 6.29 contains the Q-factor curves for gate widths of the NMOS transistor from 1.0 μm to 3.4 μm. The lowest number of photons for $Q = 3$ is 65 and is achieved for $W_n = 1.8$ μm and $W_n = 2.6$ μm. For $W_n = 3.4$ μm, the necessary number of photons is 66. The optimum range for the gate width of the NMOS transistor, therefore, is from 1.8 μm to 2.6 μm. It should be noted that the bandwidth of the common-source amplifier increases with W_n and to keep the noise bandwidth constant, a low-pass filter has to be applied in the post amplifier.

For real double sampling, the Q-factor is presented in figure 6.30. The necessary photon numbers for $Q = 3$ are 94, 89, 89 and 91 for gate widths of the NMOS transistor W_n of 1.0 μm, 1.8 μm, 2.6 μm and 3.4 μm, respectively. The lowest photon number applies to $W_n = 1.8$ μm and $W_n = 2.6$ μm. The optimum range for the gate width of the NMOS transistor is also from 1.8 μm to 2.6 μm, if real double sampling is considered.

The best achievable sensitivity of the common-source integrator in 0.35 μm PIN-photodiode CMOS seems to be—58.16 dBm (for the minimum photon number of 65) for 150 Mb s^{-1} and 635 nm light, if perfect double sampling and perfect compensation of charge injection can be accomplished. For real double sampling,

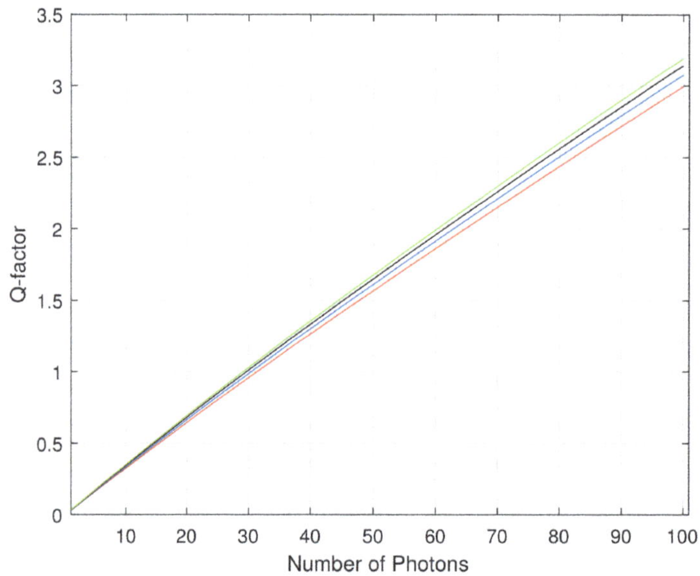

Figure 6.28. Q-factor of common-source integrator in 0.35 μm PIN-photodiode CMOS ($W_p = 2.5 \times W_n$, green: $W_n = 1.0$ μm, black: $W_n = 0.9$ μm, blue: $W_n = 0.8$ μm, red: $W_n = 0.7$ μm) without reset noise and assuming real double sampling.

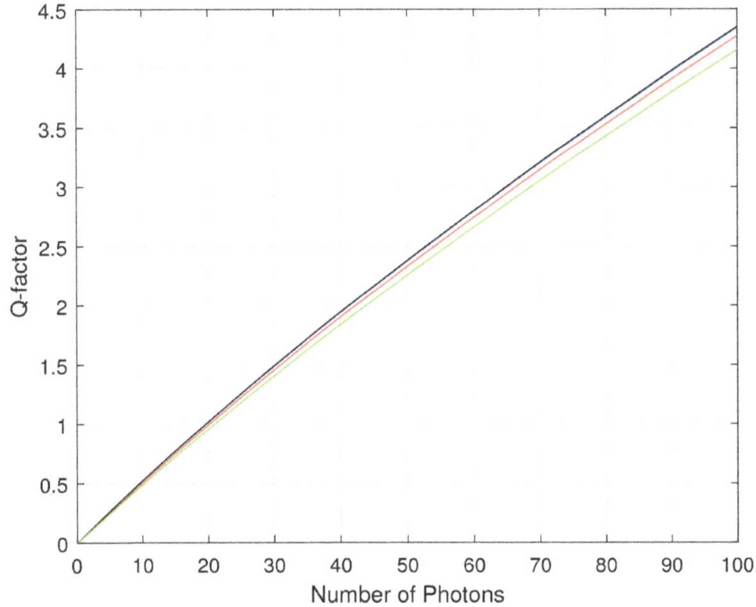

Figure 6.29. *Q*-factor of common-source integrator in 0.35 μm PIN-photodiode CMOS ($W_p = 2.5 \times W_n$, green: $W_n = 1.0$ μm, black: $W_n = 1.8$ μm, blue: $W_n = 2.6$ μm, red: $W_n = 3.4$ μm) without reset noise and assuming perfect double sampling.

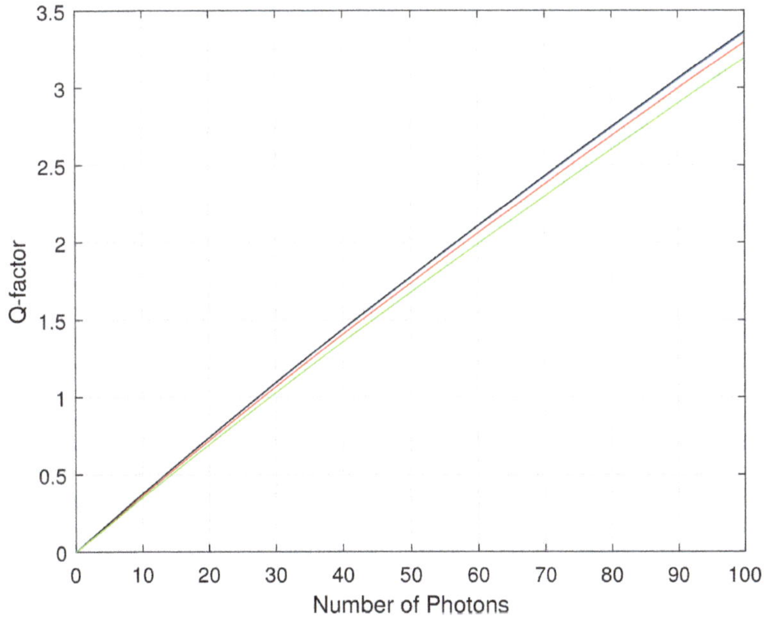

Figure 6.30. *Q*-factor of common-source integrator in 0.35 μm PIN-photodiode CMOS ($W_p = 2.5 \times W_n$, green: $W_n = 1.0$ μm, black: $W_n = 1.8$ μm, blue: $W_n = 2.6$ μm, red: $W_n = 3.4$ μm) without reset noise and assuming real double sampling.

the sensitivity reduces to −56.8 dBm. Let us conclude that the common-source integrator is somewhat better than the inverter integrator. However, due to the low gain and low integration efficiency, we should look for another technology, if we want to improve the sensitivity further.

6.3 Inverter integrator in 0.18 μm CMOS

When we look for another technology, we are limited to CMOS processes offering integrated PIN photodiodes. There is a 0.18 μm high-voltage (HV) CMOS technology, where just recently a PIN photodiode (see chapter 5) was integrated. Actually, this 0.18 μm HV CMOS technology offers a so-called low-power (LP) process module. This means that the threshold voltage of the LP MOSFETs is larger and the currents are lower than those of the standard MOSFETs with the standard threshold voltage. The photodiode capacitance was also 0.9 fF. But the narrower metal lines lead to a capacitance of 2.2 fF for photodiode and metal connection (with a minimum width metal-4 line having a length of 35 μm). The gain of an inverter is about 40 and the gate–drain overlap capacitance is 0.33 fF μm^{-1} for the NMOS transistor and 0.32 fF μm^{-1} for the PMOS transistor. These large values per μm seem to be a bad fact, but the minimum transistor width is 0.22 μm (compared to 0.7 μm in 0.35 μm CMOS) and a smaller integration capacitance is possible than in 0.35 μm CMOS.

Let us start with an inverter with $W_n = 0.44$ μm and $W_p = 2 \times W_n$, i.e. $W_p = 0.88$ μm, which has a simulated gain of 38 and a bandwidth of 160 MHz. The integration capacitance is about 0.43 fF. For the total input node capacitance $C_T = 20$ fF and assuming perfect double sampling (i.e. no reset noise), 150 photons are necessary for $Q = 3$. The corresponding sensitivity for 635 nm and BER = 2×10^{-3} is —55.6 dBm.

Figure 6.31 shows the dependence of the Q-factor of this inverter with $W_p = 2 \times W_n$ for W_n in the range 0.88 μm to 0.22 μm. The capacitance of the photodiode inclusive metal line to the amplifier input is 2.2 fF. The inverter's input capacitance (scales with W_n) and the capacitances of the reset and compensation transistor (independent of W_n) at the input are 3.5 fF (for $W_n = 0.44$ μm). The quantum efficiency of the PIN photodiode is assumed as 0.9. For $W_n = 0.44$ μm, 95 photons are necessary for $Q = 3$ leading to a sensitivity of −57.6 dBm (see figure 6.32). 80 photons are sufficient for $W_n = 0.22$ μm corresponding to a sensitivity of −58.3 dBm. Comparing these results to the sensitivity and Q-factor of the inverter in 0.35 μm CMOS technology (see figure 6.7) we are astonished, because the inverter in 0.18 μm technology does not need fewer photons for $Q = 3$.

The change of the dependence of the Q-factor on the number of photons for implementing real double sampling can be seen in figure 6.33 in comparison to figure 6.31. The minimum necessary photon number for $W_n = 0.22$ μm is then 109 instead of 80. The necessary photon numbers for $Q = 3$ and the corresponding sensitivities are listed in table 6.1. The best sensitivity for real double sampling is −57 dBm instead of −58.3 dBm for ideal double sampling. For the inverter integrator in 0.35 μm CMOS using real double sampling, the minimum necessary number of

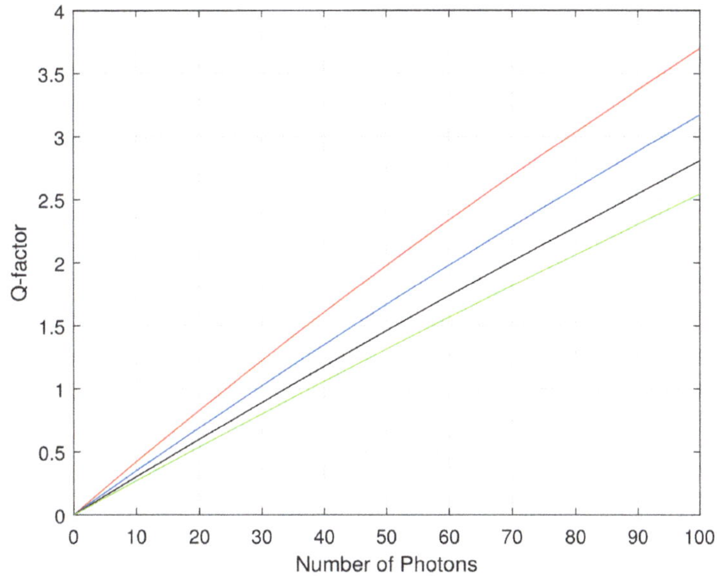

Figure 6.31. Q-factor of inverter integrator in 0.18 μm PIN-photodiode CMOS ($W_p = 2.0 \times W_n$, green: $W_n = 0.88$ μm, black: $W_n = 0.66$ μm, blue: $W_n = 0.44$ μm, red: $W_n = 0.22$ μm) assuming perfect double sampling, i.e. without reset noise.

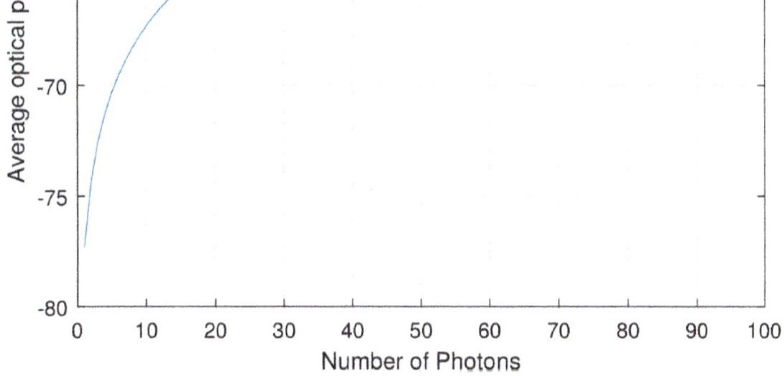

Figure 6.32. Sensitivity for a data rate of 118 Mb s^{-1} and a wavelength of 635 nm.

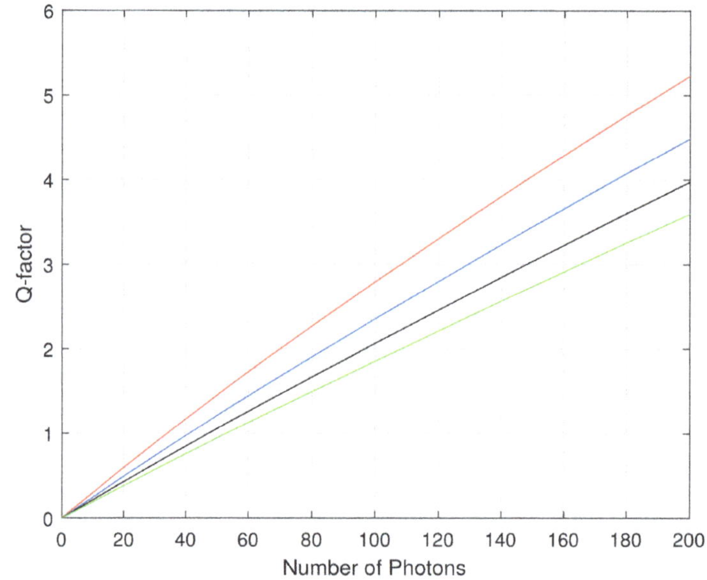

Figure 6.33. Q-factor of inverter integrator in 0.18 μm PIN-photodiode CMOS ($W_p = 2.0 \times W_n$, green: $W_n = 0.88$ μm, black: $W_n = 0.66$ μm, blue: $W_n = 0.44$ μm, red: $W_n = 0.22$ μm) without reset noise and assuming real double sampling.

Table 6.1. Results for the 0.18 μm CMOS inverter without reset noise and for real double sampling.

W_n (μm)	W_p (μm)	n_{PH} for $Q = 3$	Sensitivity (dBm)
0.88	1.76	166	−55.1
0.66	1.32	149	−55.6
0.44	0.88	130	−56.2
0.22	0.44	109	−57.0

photons was also 109 (see figure 6.20). It is indeed very surprising that the 0.18 μm CMOS technology does not reduce the necessary number of photons compared to the 0.35 μm technology, when inverter integrators are compared. Therefore, the circumstances leading to this outcome will be investigated next.

Looking at the integration efficiency (figure 6.34) for $W_n = 0.44$ μm, we find a value of 73% for $C_T = 5.7$ fF. This integration efficiency is about 50% better than that of the 0.35 μm CMOS inverter (compare figure 6.10). Therefore, we have to look for another explanation why the 0.18 μm CMOS inverter is not better than the 0.35 μm CMOS inverter.

The transconductances of the NMOS transistor with $W_n = 0.44$ μm and of the PMOS transistor with $W_p = 0.88$ μm are 130.4 μS and 83.8 μS, respectively. These values are smaller than those of the 0.35 μm inverter in section 6.1. Furthermore, the output conductances are $g_{DS,n} = 2.46$ μS and $g_{DS,p} = 1.74$ μS. These values are also

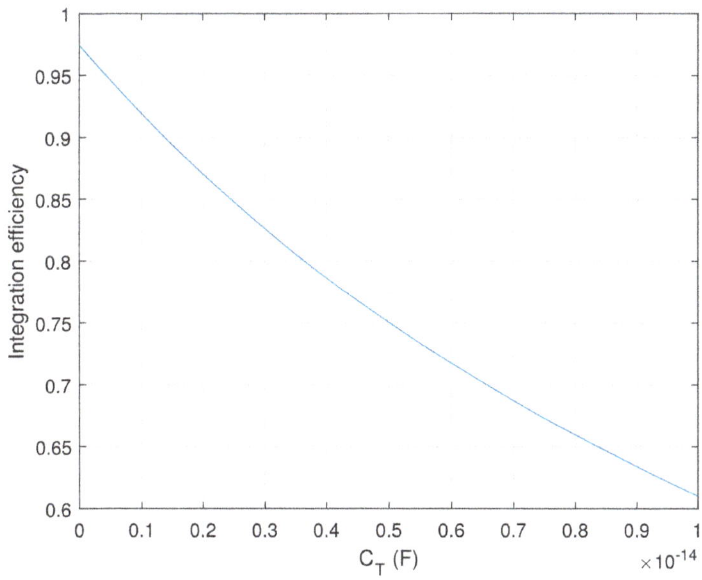

Figure 6.34. Integration efficiency for $A = 38$ and $C_{\text{int}} = 0.43$ fF.

Table 6.2. Rms electronic output noise voltage for the 0.18 μm CMOS inverter.

W_n (μm)	W_p (μm)	rms noise voltage (mV)
0.88	1.76	3.3
0.66	1.32	3.7
0.44	0.88	4.6
0.22	0.44	6.5

smaller than that of the 0.35 μm CMOS inverter in section 6.1 and they explain the higher gain of the 0.18 μm LP CMOS inverter. By the way, there is a certain deviation between gain from (AC) simulation and when we calculate the gain with these transconductances and output conductances.

But what is the most important difference is the noise of the inverters. The rms output noise voltage of the 0.35 μm inverter ranges from about 1.3 mV to 1.56 mV for W_n from 1 μm to 0.7 μm (see figure 6.25). For the 0.18 μm CMOS inverter we calculate the output noise voltages with equation (6.8) listed in table 6.2. The noise increased more strongly for the 0.18 μm LP CMOS inverter than the integration efficiency. This is the surprising reason, why the 0.18 μm LP CMOS inverter is not better than the 0.35 μm CMOS inverter.

But, there is a measure to improve the optical receiver with the 0.18 μm CMOS inverter. We add a load resistor R to the output of the inverter as shown in figure 6.35. The bias voltage V_{bo} has to be connected to a voltage source, which

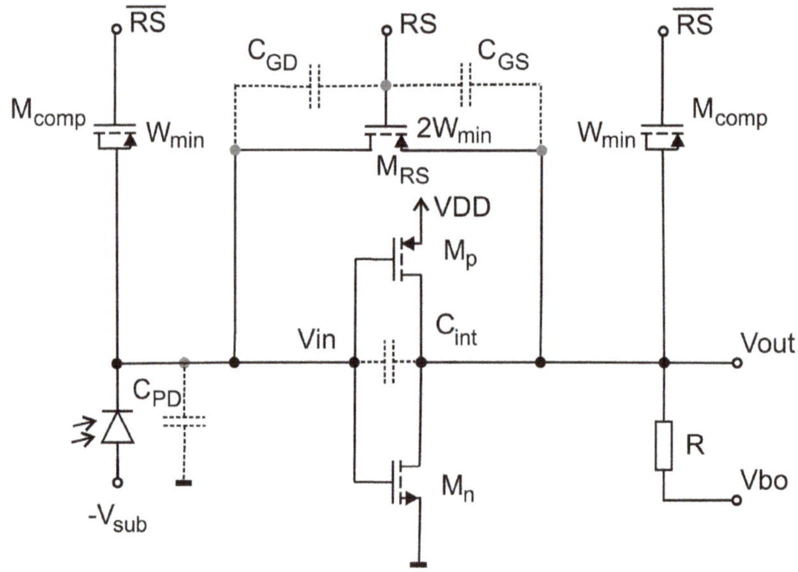

Figure 6.35. Inverter integrator with load resistor R and compensation transistors for clock feed-through of reset transistor.

delivers the same voltage as the DC operating voltage of the inverter (being present during reset).

By doing this we increase the output conductance to

$$g_{\mathrm{out}} = g_{\mathrm{DS,n}} + g_{\mathrm{DS,p}} + \frac{1}{R}. \tag{6.25}$$

Of course we also reduce the gain to

$$A_{\mathrm{inv,R}} = \frac{g_{m,\mathrm{n}} + g_{m,\mathrm{p}}}{g_{\mathrm{ds,n}} + g_{\mathrm{ds,p}} + 1/R}. \tag{6.26}$$

With a 200 kΩ resistor we obtain a gain of about 19. If we scale R inversely proportional to W_{n}, the gain of the inverter with such a load resistor is constant.

The noise current at the output of the inverter with the load resistor is

$$\overline{i_{\mathrm{n,out}}^2} = \left(\frac{8}{3}k_{\mathrm{B}}T(g_{m,\mathrm{n}} + g_{m,\mathrm{p}}) + \frac{4k_{\mathrm{B}}T}{R}\right)\Delta f. \tag{6.27}$$

With the output conductance (equation (6.25)), we can calculate the output noise voltage:

$$\overline{v_{\mathrm{n,out}}^2} = \frac{\overline{i_{\mathrm{n,out}}^2}}{g_{\mathrm{out,\,R}}^2} = \frac{\frac{8}{3}k_{\mathrm{B}}T(g_{m,\mathrm{n}} + g_{m,\mathrm{p}}) + \frac{4k_{\mathrm{B}}T}{R}}{\left(g_{\mathrm{DS,n}} + g_{\mathrm{DS,p}} + \frac{1}{R}\right)^2}\Delta f \tag{6.28}$$

Before we calculate the noise voltage values, we should also consider the integration efficiency because it will be reduced compared to the inverter without load resistor by the lower gain. Figure 6.36 plots the integration efficiency for the gain of 19 and the integration capacitance of 0.43 fF. For $C_T = 5.7$ fF, we can read an integration efficiency of 57%. Compared to that of the inverter without load resistor of 73%, it reduces only slightly with the load resistor and our hope is that the noise reduces more strongly.

But let us cosider the Q-factor first. The Q-factor of the inverter with load resistor is depicted in figure 6.37 for different gate widths of the NMOS transistor. We indeed see an improvement compared to figure 6.31. Now, 91 photons for $W_n = 0.88\mu$m, 84 photons for $W_n = 0.66\mu$m, 76 photons for $W_n = 0.44$ μm and 68 photons for $W_n = 0.22$ μm are sufficient for $Q = 3$. Table 6.3 summarizes the results of the 0.18 μm CMOS inverter with load resistor and lists the sensitivities for 118 Mb s^{-1} and 635 nm.

For considering the higher thermal noise due to real double sampling according to equation (6.23), the Q-factor is plotted in figure 6.38 and the results inclusive of rms noise voltages and sensitivities are listed in table 6.4. The rms output noise voltages are smaller than those of the inverter without load resistor listed in table 6.2. Compared to the inverter without load resistor using real double sampling (see figure 6.33) we also see an improvement in the sensitivities.

The relationships of the integrating receiver are really complex and astonishing, since the measure of adding the load resistor actually worsens the inverter but improves the USPAR receiver.

Before we investigate a USPAR integrated circuit with common-source amplifier in 0.18 μm CMOS, it may be interesting to investigate the influence of $1/f$-noise on

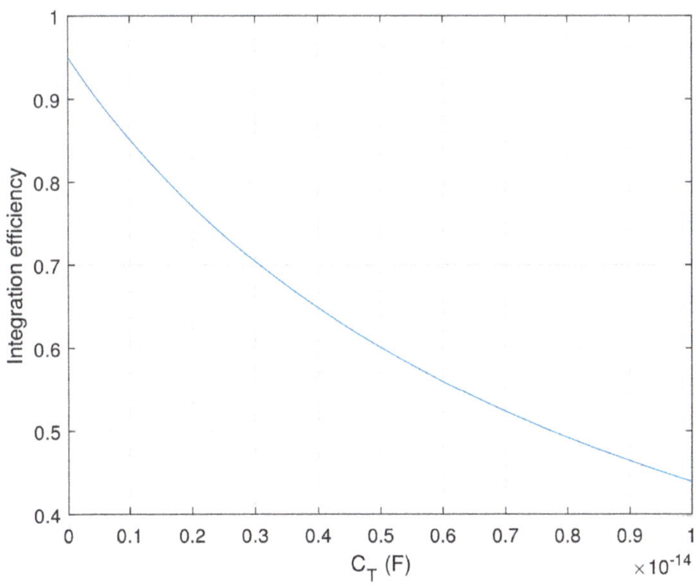

Figure 6.36. Integration efficiency for $A = 19$ and $C_{\text{int}} = 0.43$ fF.

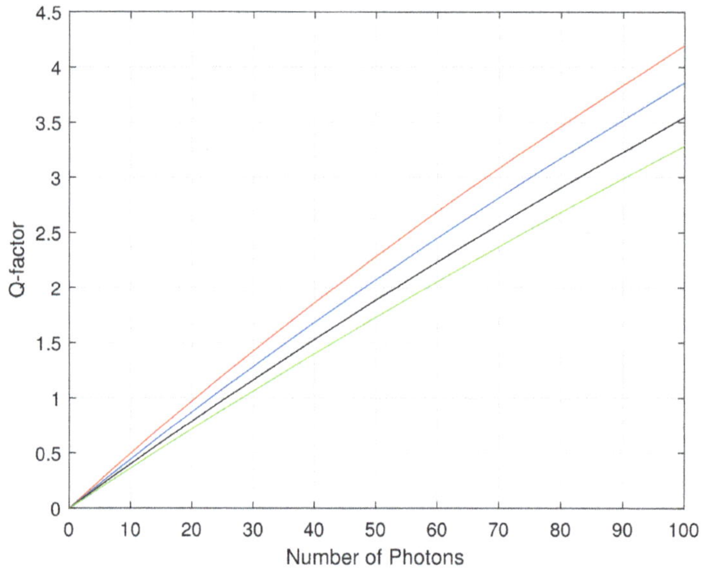

Figure 6.37. Q-factor of inverter integrator in 0.18 μm PIN-photodiode CMOS with load resistor R ($W_p = 2.0 \times W_n$, green: $W_n = 0.88$ μm, black: $W_n = 0.66$ μm, blue: $W_n = 0.44$ μm, red: $W_n = 0.22$ μm; R inversely proportional to W_n) assuming perfect double sampling, i.e. without reset noise.

Table 6.3. Results for the 0.18 μm CMOS inverter with load resistor R assuming perfect double sampling.

W_n (μm)	W_p (μm)	R (kΩ)	rms noise voltage (mV)	n_{PH} for $Q = 3$	Sensitivity (dBm)
0.88	1.76	100	1.5	91	−57.7
0.66	1.32	133	1.7	84	−58.1
0.44	0.88	200	2.1	76	−58.5
0.22	0.44	400	3.0	68	−59.0

the total noise. For this purpose, the model derived in section 3.9 for the dependence of the output noise voltage of the integrator on the photodiode capacitance is used. For the 0.18 μm HV CMOS technology the noise power spectral density $S_i(f)$ was determined for an inverter with $W_n = 4.52$ μm and $W_p = 4.92$ μm:

$$S_i(f) = \frac{3.81 \times 10^{-16} A^2}{f^{1.15}} + \frac{5.29 \times 10^{-16} A^2}{f^{0.95}} + 2.66 \times 10^{-23} A^2/\text{Hz}, \tag{6.29}$$

where the first term represents the $1/f$-noise of the NMOS transistor, the second term defines the $1/f$-noise of the PMOS transistor and the third term determines the thermal noise of both transistors in the inverter.

The load capacitance was 40 fF and no load resistor was connected. The input capacitance C_{in} of this inverter was 12.8 fF and C_F was 3.0 fF. The transconductance

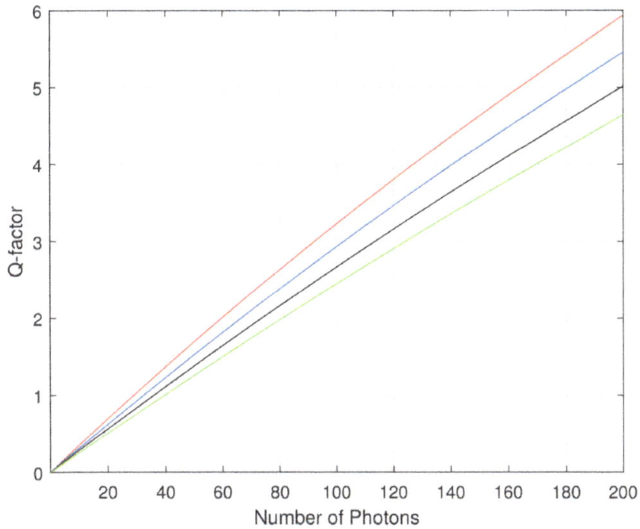

Figure 6.38. Q-factor of inverter integrator in 0.18 μm PIN-photodiode CMOS with load resistor R ($W_p = 2.0 \times W_n$, green: $W_n = 0.88$ μm, black: $W_n = 0.66$ μm, blue: $W_n = 0.44$ μm, red: $W_n = 0.22$ μm; R inversely proportional to W_n) without reset noise and assuming real double sampling.

Table 6.4. Results for the 0.18 μm CMOS inverter with load resistor R for real double sampling.

W_n (μm)	W_p (μm)	R (kΩ)	rms noise voltage (mV)	n_{PH} for $Q = 3$	Sensitivity (dBm)
0.88	1.76	100	1.5	125	−56.4
0.66	1.32	133	1.7	115	−56.7
0.44	0.88	200	2.1	103	−57.2
0.22	0.44	400	3.0	93	−57.65

of this inverter was 1.545 mS and its output resistance was 25.04 kΩ. The noise integral (see equation (3.56)) was calculated with the boundaries 1 Hz and 10 GHz. Figure 6.39 shows that the influence of the $1/f$-noise is stronger for larger photodiode capacitances.

6.4 Integrator with common-source amplifier in 0.18 μm CMOS

Another possibility to reduce the gain is to implement a common-source amplifier with a PMOS transistor as constant current load, as we did it in section 6.2. Figure 6.21 shows the circuit diagram of such an integrating receiver. In 0.18 μm LP CMOS, the gain is 19 and the integration capacitance for $W_n = 0.44$ μm is 0.146 fF. For the photodiode capacitance we assume 0.9 fF and inclusive metal connection to the amplifier input we take 2.2 fF as we did for the 0.18 μm CMOS inverter described above. W_p was two times W_n, when varying W_n from 0.88 μm to 0.22 μm. Figure 6.40 shows the Q-factor for the integrating receiver with the common-source

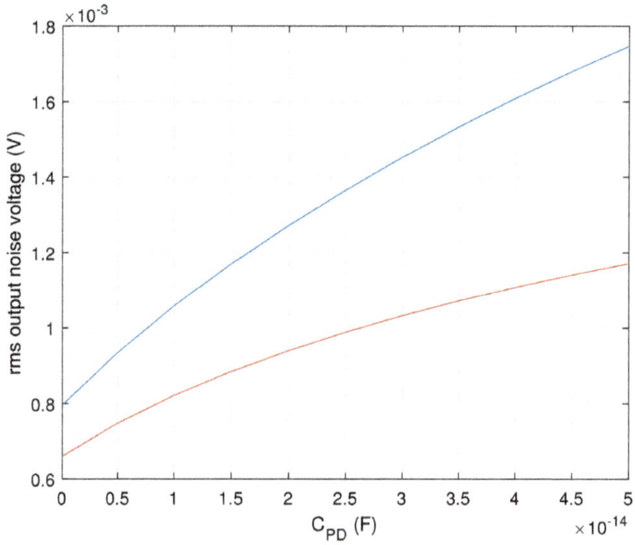

Figure 6.39. RMS output noise voltage of inverter integrator in 0.18 μm PIN-photodiode CMOS ($W_n = 4.52\,\mu$m, $W_p = 4.92\,\mu$m, blue: with thermal noise and $1/f$-noise, red: with thermal noise only).

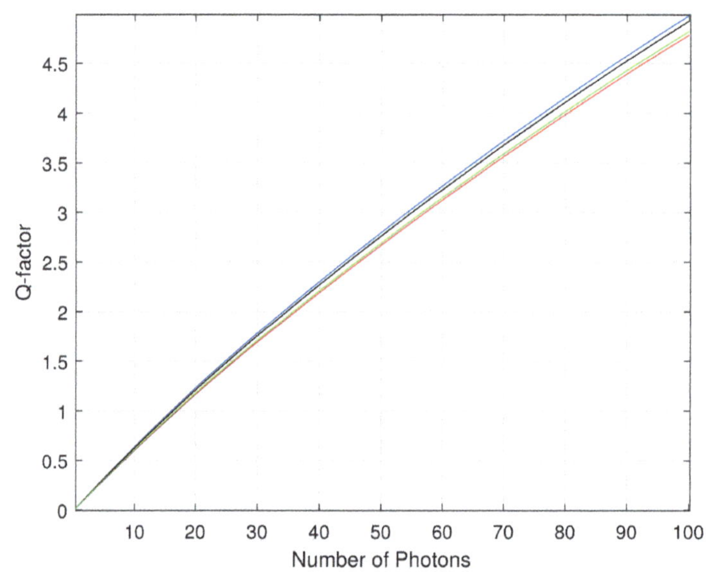

Figure 6.40. Q-factor of common-source integrator in 0.18 μm PIN-photodiode CMOS with load resistor R ($W_p = 2.0 \times W_n$, green: $W_n = 0.88\,\mu$m, black: $W_n = 0.66\,\mu$m, blue: $W_n = 0.44\,\mu$m, red: $W_n = 0.22\,\mu$m) assuming perfect double sampling, i.e. without reset noise.

amplifier. The bandwidth was held constant at 160 MHz. The quantum efficiency of the photodiode was chosen as 90% again.

The blue and black curves ($W_n = 0.44\,\mu$m and $W_n = 0.66\,\mu$m, respectively) are located left of the red and green curves ($W_n = 0.22\,\mu$m and $W_n = 0.88\,\mu$m, respectively). The common-source amplifier with $W_n = 0.44$ μm needs the lowest number of photons (55) for $Q = 3$, but all four versions are close together (55 to 58 photons). Table 6.5 summarizes the results. The sensitivity was calculated for a data rate of 118 Mb s^{-1} and a wavelength of 635 nm. It should be mentioned that to reach the sensitivities listed in this table in practice will be extremely difficult or even impossible, because the simple analytical model does not consider many parasitic effects.

For real double sampling, the dependence of the Q-factor on the photon number is displayed in figure 6.41 and table 6.6 lists the necessary numbers of photons for $Q = 3$ as well as the corresponding sensitivities at 118 Mb s^{-1} and 635 nm. The larger thermal noise stemming from double sampling worsens the sensitivity by about 1.2 dB compared to table 6.5.

Table 6.5. Results for the 0.18 μm CMOS receiver with common-source amplifier assuming perfect double sampling (without reset noise).

$W_n(\mu m)$	$W_p(\mu m)$	n_{PH} for $Q = 3$	Sensitivity (dBm)
0.88	1.76	57	−59.8
0.66	1.32	56	−59.85
0.44	0.88	55	−59.9
0.22	0.44	58	−59.7

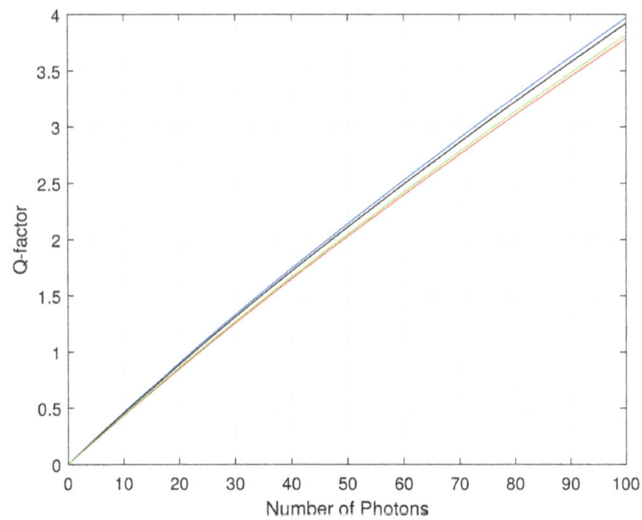

Figure 6.41. Q-factor of common-source integrator in 0.18 μm PIN-photodiode CMOS with load resistor R ($W_p = 2.0 \times W_n$, green: $W_n = 0.88$ μm, black: $W_n = 0.66$ μm, blue: $W_n = 0.44$ μm, red: $W_n = 0.22$ μm) assuming real double sampling for eliminating reset noise.

Table 6.6. Results for the 0.18 μm CMOS receiver with common-source amplifier assuming real double sampling.

$W_n(\mu m)$	$W_p(\mu m)$	n_{PH} for $Q = 3$	Sensitivity (dBm)
0.88	1.76	77	−58.5
0.66	1.32	74	−58.64
0.44	0.88	73	−58.7
0.22	0.44	77	−58.5

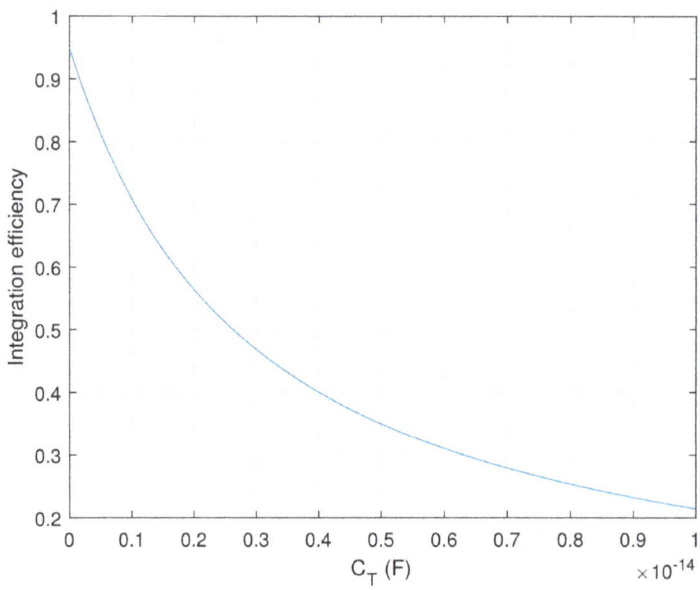

Figure 6.42. Integration efficiency for $A = 19$ and $C_{int} = 0.146$ fF.

To complement the view of the receiver with the common-source amplifier, we should show its integration efficiency. Figure 6.42 reveals that its integration efficiency for the input-node capacitance $C_T = 4.35$ fF is 38%, which is about two times larger than that of the common-source integrator in 0.35 μm CMOS.

The conclusion of this investigation by simulations is that the common-source integrating receiver in 0.18 μm CMOS technology has the best sensitivity. However, the sensitivity is not much better than that of the 0.35 μm common-source integrating receiver. For a further improvement, it will be necessary to reduce the input-node capacitance—a quite difficult task.

A receiver using a common-source amplifier in the integrator was designed and fabricated in 0.18 μm HV CMOS [11]. The PIN photodiode of type 4 with the UV window (opto window) was implemented. This photodiode is depicted in figure 5.20(b) and described in section 5.2. The P/N-junction capacitance of this PIN photodiode was 1.5 pF at 20 V reverse bias. For the 30 μm long metal line in level 3, the capacitance extracted from the layout was 1.5 fF. So, the total capacitance of photodiode and metal connection to the input transistor was 3.0 fF. The responsivity of PIN4 was 0.39 A W^{-1} at 635 nm, which corresponds to the quantum efficiency of 0.7634 [11]. Comparing to the detection efficiencies of SPADs and SPAD arrays, the quantum efficiency of the PIN photodiode is higher. The 10%-to-90% rise/fall times of the PIN4 photodiode at 20 V reverse voltage were 0.73 ns and 0.92 ns, respectively. These values should enable data rates of up to 500 Mb s^{-1}.

The receiver circuit using an input transistor M_1 in common-source (CS) configuration is shown in figure 6.43. All transistors had minimum gate lengths and the 1.8 V transistors from the low-power process module were used. M_1 consists of two minimum-width gate fingers and the PMOS constant-current load element M_2 has four minimum-width gate fingers. The inverter formed by M_3 and M_4 with twice the gate widths served for the generation of the gate–source voltage of M_2. In addition, the reference voltage V_{Ref} for the differential post amplifier is delivered by this inverter.

Although the circuit is used as a photo-charge integrator, there is no integration capacitor visible in figure 6.43. The reason for this is that the photo-charges are integrated on a parasitic capacitance of transistor M_1—its gate–drain overlap capacitance. This is the smallest possible capacitance being usable. For the implemeted transistor, C_{GD} is 0.145 fF [11]. This is much smaller than the smallest poly–poly or metal-insulator-metal (MIM) capacitors of 10 fF available in the design kit of the process used. By using C_{GD} as integration capacitor, the parasitic capacitance of metal lines from the transistor to the integration capacitor are avoided, which is most advantageous.

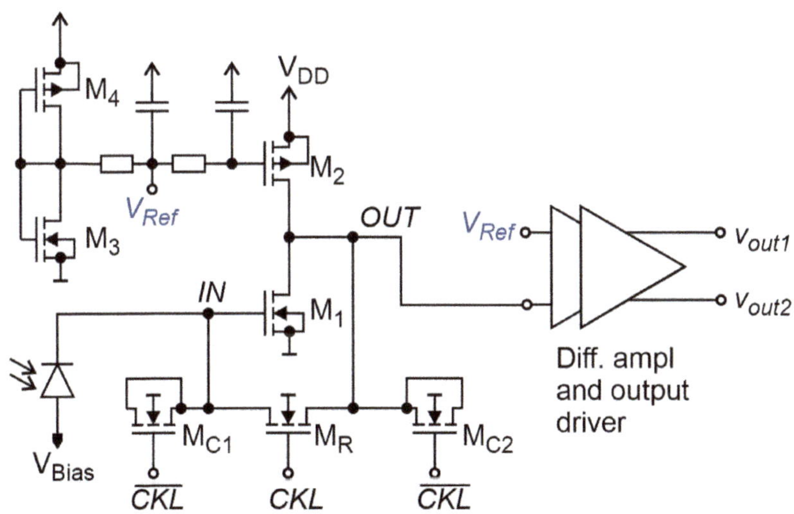

Figure 6.43. Circuit diagram of the fabricated CS-integrator receiver, where actually the gate–drain overlap capacitance of M_1 is exploited as integration capacitor. Reproduced from [11] CC BY 4.0.

In order to examine the sensitivity to process variations (i.e. to fabrication tolerances), Monte Carlo simulations were performed for the core receiver circuit formed by M_1–M_4. The top part of figure 6.44 presents the input-to-output transfer curve of the CS amplifier. In the lower part of this figure, the gain curve with a FWHM of about 35 mV is depicted. According to these Monte Carlo results, the DC operating point shifts somewhat, but the gain behaves well. The histogram shown in figure 6.45 reveals that the gain stays within the range from −38 to −30 with a mean value of −33.5 and a standard deviation of 1.5 [11]. So, the fabricated core receiver circuit should work well enough.

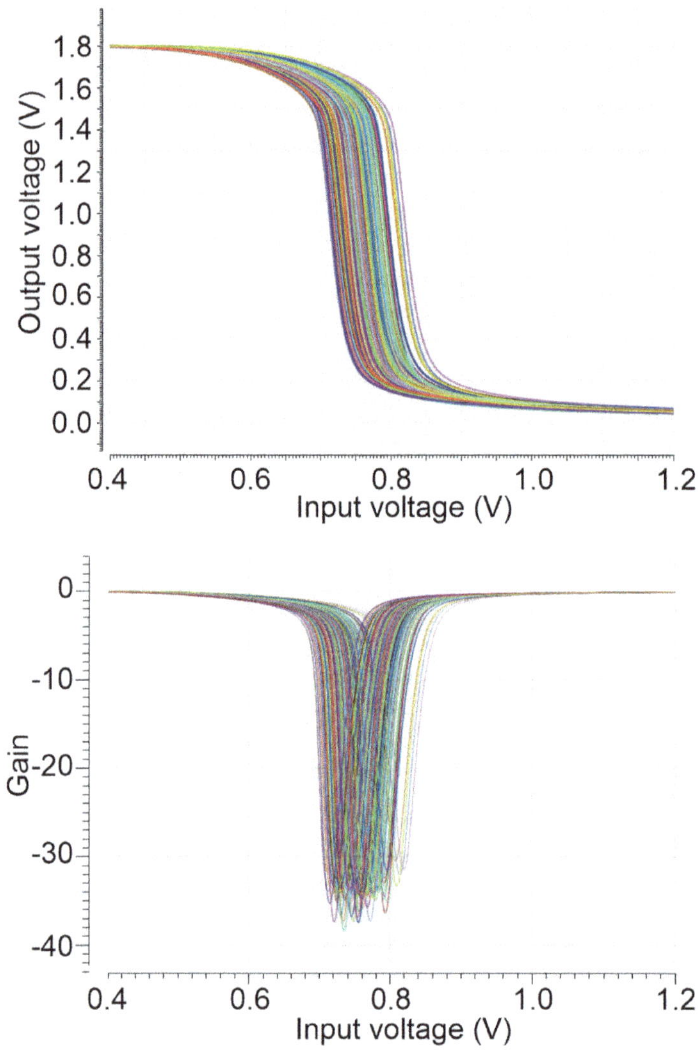

Figure 6.44. Transfer characteristics (top) and gain (bottom) of CS-amplifier resulting from 200 Monte Carlo simulation runs. Reproduced from [11] CC BY 4.0.

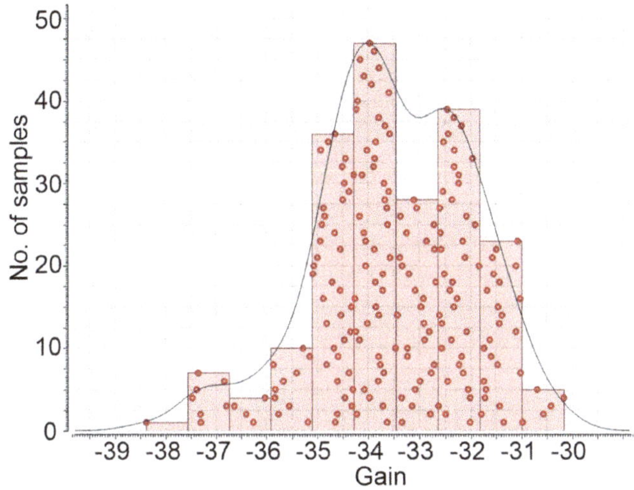

Figure 6.45. Histogram of the gain A of the CS-amplifier resulting from 200 Monte Carlo simulation runs. Reproduced from [11] CC BY 4.0.

The reset of the integration capacitor is done with transistor M_R. However, due to its gate–drain and gate–source overlap capacitances charges are injected for both rising and falling clock edges. These injection charges can be compensated with the MOS capacitors M_{C1} and M_{C2}. Keeping in mind that both overlap capacitances C_{GS} and C_{GD} of the MOS capacitors are connected in parallel and these have to compensate one overlap capacitance (C_{GS} or C_{GD}) of M_R, M_{C1} and M_{C2} must have minimum gate width and M_R's gate width must be twice the minimum gate width. Therefore, the charge injection of M_R with a clock step height of 1.8 V corresponds to 1640 electron charges for an overlap capacitance of 0.145 fF ($W_{MR} = 2 \cdot W_{min}$). This amount of charges is much larger than the photo-charge aimed at in USPAR. Consequently, it is necessary that M_R and the compensating MOS capacitors match well enough. Fortunately, due to the fact that the bits will be sampled before and after integration to determine the bit value (double sampling), some remaining signal distortion due to mismatch and imperfect reset can be tolerated to a certain extent.

The minimum gain (determined by Monte Carlo simulations, see above) of 30 (we use its magnitude here), $C_{int} = 0.356$ fF (considering extracted parasitics, see below), and $C_T = 5.4$ fF lead to an integration efficiency of 0.65 [11]. Therefore, the integration of photo-charges on a feedback capacitor results in a much larger output signal than integrating on the photodiode (more exactly: on the input node capacitance without the amplifier's input capacitance C_{in}, see section 1.4 and figure 1.4). This enhancement of the signal voltage is $\eta_{int} C_{PD}/C_{int} = 5.5$ for $\eta_{int} = 0.65$, $C_{PD} = 3$ fF, and $C_{int} = 0.356$ fF [11].

Further interesting results from circuit simulations were a gain of 11 of the differential post amplifier and output buffer, the power consumption of 24.15 mW including that of the output driver of 21.6 mW, the current consumption of the CS-amplifier of 13.7 μA, and the current consumption of the dummy inverter of about 27 μA [11].

The reset is not so simple as one might think or expect. Therefore, we have to investigate the reset and its consequences in detail. After the reset transistor M_R is switched off to enable a new integration, charge injection or an imperfect reset can lead to an error voltage V_{NO} or V_{NI} at the output or at the input, respectively. If these error voltages remain constant within a measurement cycle, double sampling can cancel the errors out fully. However, an exponential settling behavior to a final value with the time constant τ occurs, which will cause an error in the difference of the two sampling values. To determine the dynamics and the time constant, Kirchhoff's current law can be utilized for the input and output nodes of the integrator [11]:

$$I_{PD} = C_{int}\left(\frac{dV_{OUT}}{dt} - \frac{dV_{IN}}{dt}\right) - C_T\frac{dV_{IN}}{dt} \tag{6.30}$$

$$-g_{m,n}V_{IN} - g_o V_{OUT} = C_L\frac{dV_{OUT}}{dt} + C_{int}\left(\frac{dV_{OUT}}{dt} - \frac{dV_{IN}}{dt}\right) \tag{6.31}$$

In these equations, V_{IN} and V_{OUT} are the error voltages at the input and output of the integrator, respectively, both related to the operating point. C_L represents the total load capacitance at the output node of the integrator. The transconductance $g_{m,n}$ belongs to transistor M_1 and g_o stands for the total output conductance of M_1 and M_2. I_{PD} is the photocurrent. Equations (6.30) and (6.31) have to be split into an equation for V_{IN} and an equation for V_{OUT}:

$$\begin{aligned} &- g_o I_{PD} - (C_{int} + C_L)\frac{dI_{PD}}{dt} = (g_o(C_{int} + C_T) \\ &+ g_{m,n}C_{int})\frac{dV_{IN}}{dt} - (C_T C_{int} + C_T C_L + C_L C_{int})\frac{d^2V_{IN}}{dt^2} \end{aligned} \tag{6.32}$$

$$\begin{aligned} &- g_{m,n}I_{PD} + C_{int}\frac{dI_{PD}}{dt} = (g_o(C_{int} + C_T) \\ &+ g_{m,n}C_{int})\frac{dV_{OUT}}{dt} - (C_T C_{int} + C_T C_L + C_L C_{int})\frac{d^2V_{OUT}}{dt^2} \end{aligned} \tag{6.33}$$

The solution for an error voltage $V_{IN}(0) = V_{NI}$ when $V_{OUT}(0) = 0$ being caused by, e.g. charge injection at the input node either by M_R or M_{C1} is after setting $I_{PD} = 0$ [11]:

$$V_{IN}(t) = V_{NI} + V_{NI}\frac{g_{m,n}C_{int}}{g_o(C_{int} + C_T) + g_{m,n}C_{int}}(e^{-t/\tau} - 1) \tag{6.34}$$

$$V_{OUT}(t) = V_{NI}\frac{g_{m,n}(C_{int} + C_T)}{g_o(C_{int} + C_T) + g_{m,n}C_{int}}(e^{-t/\tau} - 1) \tag{6.35}$$

For charge injection to the output node, the error voltage $V_{OUT}(0) = V_{NO}$ results. Using $I_{PD} = 0$ gives [11]:

$$V_{IN}(t) = V_{NO}\frac{C_{int}}{C_{int} + C_{T}}e^{-t/\tau} \tag{6.36}$$

$$V_{OUT}(t) = V_{NO}e^{-t/\tau} \tag{6.37}$$

For a constant photocurrent I_{PD} and the initial conditions $V_{OUT}(0) = V_{IN}(0) = 0$, the solution is [11]:

$$V_{IN}(t) = I_{PD}\frac{C_{int}}{C_{int} + C_{T}}K(e^{-t/\tau} - 1) - I_{PD}\frac{g_{o}}{g_{o}(C_{int} + C_{T}) + g_{m,n}C_{int}}t \tag{6.38}$$

$$V_{OUT}(t) = I_{PD}K(e^{-t/\tau} - 1) + I_{PD}\frac{g_{m,n}}{g_{o}(C_{int} + C_{T}) + g_{m,n}C_{int}}t \tag{6.39}$$

with

$$K = \frac{(C_{int} + C_{T})\left[(C_{L} + C_{int})g_{m,n} + g_{o}C_{int}\right]}{\left[g_{o}(C_{int} + C_{T}) + g_{m,n}C_{int}\right]^{2}}. \tag{6.40}$$

Of course superpositions of these solutions are also solutions because of their linearity. All above listed solutions contain parts with exponential settling being determined by the time constant τ [11]:

$$\tau = \frac{C_{T}C_{int} + C_{L}C_{int} + C_{L}C_{T}}{g_{o}(C_{int} + C_{T}) + g_{m,n}C_{int}} \tag{6.41}$$

Because of the exponential settling with τ, sampling should be done at least about 4τ after the end of the reset. For the circuit presented in [11], τ was 298 ps for the ideal operating point. However, due to an imperfect reset, the operating point of the integrator shifts from detection cycle to detection cycle since the error voltage depends on the last bit. In particular, a long series of the same bits can cause a large shift of the operating point voltage and the integrator leaves the original small-signal regime at the ideal operating point and a new set of small-signal parameters would be needed.

Since the simulation results were quite promising, the integrating receiver was fabricated. The chip microphotograph of this USPAR chip with a size of $670 \times 670\ \mu m^2$ is presented in figure 6.46. The most interesting core circuit is zoomed in figure 6.47. The layout corresponding to this zoomed core is depicted in figure 6.48. The active area of the core circuit (excluding post amplifier and output driver) is about 0.015 mm^2 [11].

Before we can consider measured results, the experimental set-up for characterization of this USPAR chip has to be explained. A high-extinction-ratio 635 nm light source [12] was modulated with the return-to-zero PRBS7 signal generated by a

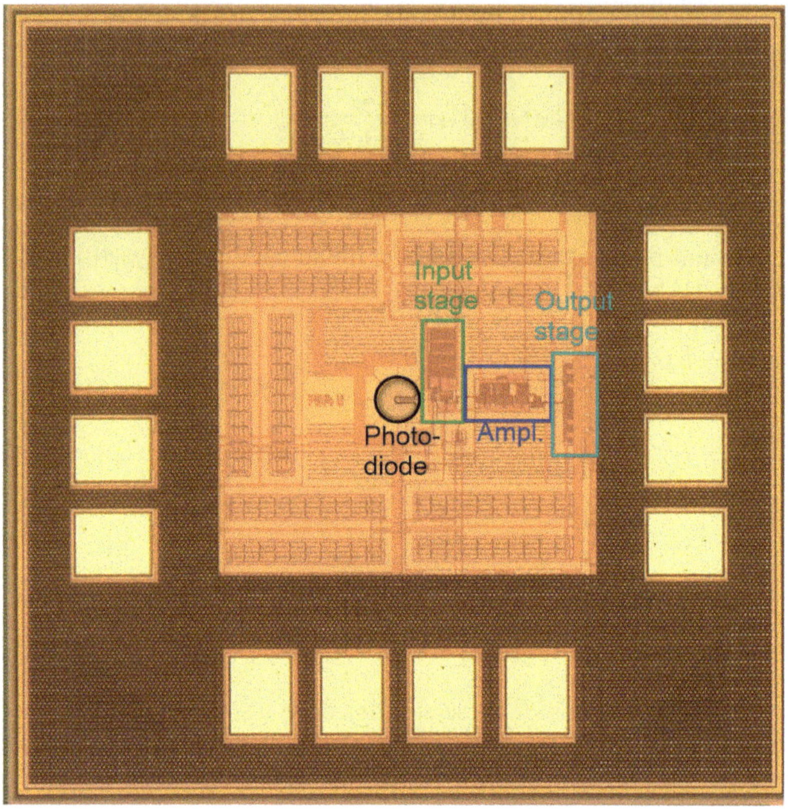

Figure 6.46. Chip photo of the receiver with CS-integrator. Reproduced from [11] CC BY 4.0.

programmable Sympuls BPG 40 G–128 Mbit pattern generator. The duration of the laser pulse was set to 5.3 ns (26.5% of a bit). A 50 Mb s^{-1} clock provided by a Centillax TG1C1-A clock synthesizer was fed into the bit pattern generator as well as into a second bit pattern generator BMG 12GIG, which divided the clock period into 64 time slots in order to define the duration and the position of the reset signal inside the data bits [11]. The output stream of the integrating receiver with a length of more than 900 000 bits and the reset signal were stored with a sampling rate of 20 GS s^{-1} and a bandwidth of 7 GHz in a Keysight mixed-signal oscilloscope MSOV204A. To save storage, only the differential output signal ($v_{\text{out1}} - v_{\text{out2}}$, see figure 6.43) was stored.

The stored receiver output signals were compared to the PRBS7 signal in MATLAB. The signals[4] were synchronized during the evaluation in MATLAB. Considering the reset signal with the start at $t = 0$, the two sampling instants t_{s1} and t_{s2} were optimized in MATLAB for lowest bit error ratio. t_{s1} had to be long enough after the end of the reset and before the laser pulse (in the case of a logical '1'). t_{s2} had

[4] There was a quite long signal propagation time in electrical cables and in the optical fiber.

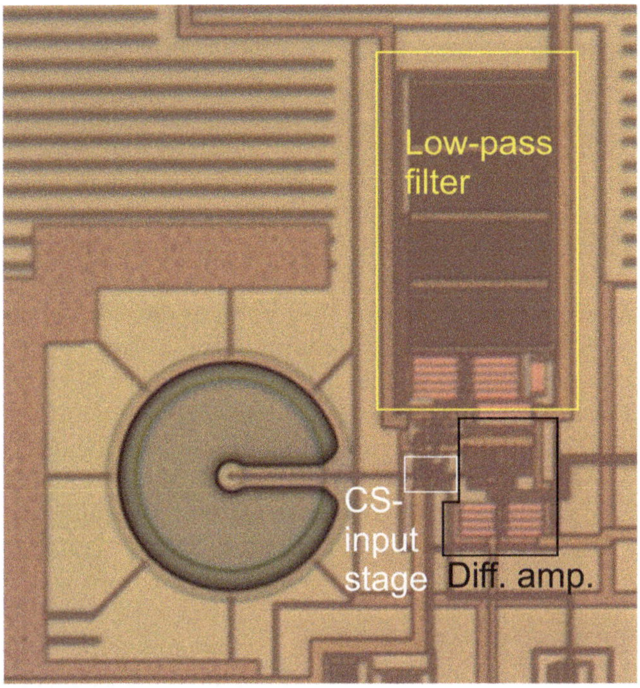

Figure 6.47. Zoomed part of the chip photo with photodiode and core circuit of the receiver with CS-integrator. Reproduced from [11] CC BY 4.0.

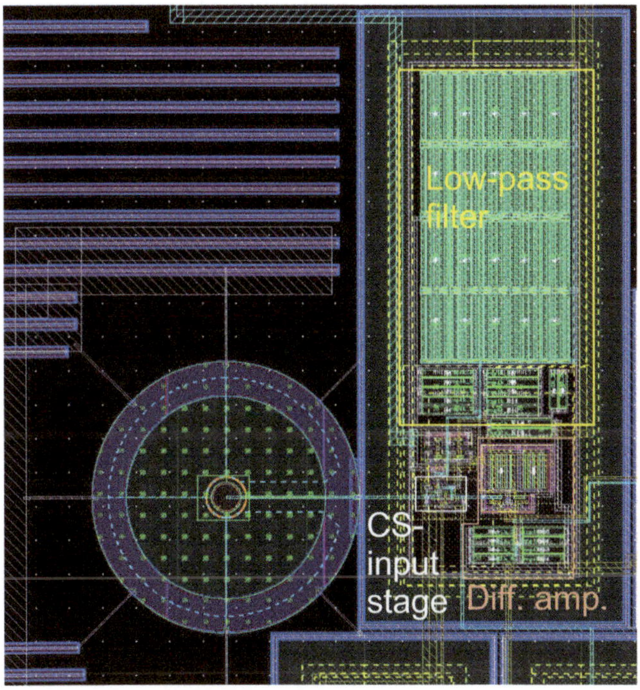

Figure 6.48. Layout of zoomed part with photodiode and core circuit. Reproduced from [11] CC BY 4.0.

to be after the laser pulse and before the reset of the next bit. Depending on v_t being larger or smaller than a threshold $v_{out1} - v_{out2}$ the decision for a logical '1' or logical '0' was done. The bit error ratio was optimized by variation of t_{s1}, t_{s2} and v_t in MATLAB. For $t_{s1} = 4.8$ ns and $t_{s2} = 14.5$ ns, the lowest BER was obtained.

A typical section of the integrating receiver's output signal is shown in figure 6.49. The blue vertical lines in this figure constitute the first sampling point after reset and before integration. The second sampling point after the integration is illustrated by the green vertical lines. Unfortunately, the DC operating point of the integrator wanders. It depends on the last bits. The reset transistor obviously is not strong enough [11]. The output voltage due to integration of photo-charges amounts up to about 170 mV, whereas the peaks due to charge injection range up to about 300 mV. It seems that the compensation of charge injection with the MOS capacitors is not good enough. Channel-charge injection [13] may play a role in addition to clock feed-through. It is also visible in figure 6.49 that because of the misfunctioning reset and in turn moving operating point of the integrator, the post amplifier and output driver saturate for many negative voltage spikes and especially around 200 ns as indicated by many cut or compressed negative spikes stemming from charge injection.

Despite all these problems, the receiver achieved a BER = 1.76×10^{-3} determined by MATLAB at 50 Mb s^{-1} with an average optical input power of −56.4 dBm at a wavelength of 635 nm and a pseudo-random bit sequence with the length $2^7 - 1$ [11]. The BER increased to 4.5×10^{-3} for −57.0 dBm average optical input power. Figure 6.50 contains the measured BER values. The sensitivity of −56.4 dBm corresponds to an average optical power of 2.3 nW or 147 photons at 635 nm inside each bit. Since half of the bits are '0s' and '1s', 294 photons are necessary in a '1'-bit. If a more advanced error correction, as described in [14] can be used, a BER of 6.5×10^{-3} is possible and the sensitivity improves to −57.0 dBm.

The experimentally found sensitivity is worse than the theoretically expected one. When we compare the number of electrons 1640 injected by the reset transistor having twice the minimum gate width to the number of signal electrons 223 (294 photons times the quantum efficiency of the photodiode of 0.7634), the ratio of injected to signal electrons is 7.4. The clock injection visible in figure 6.49 is

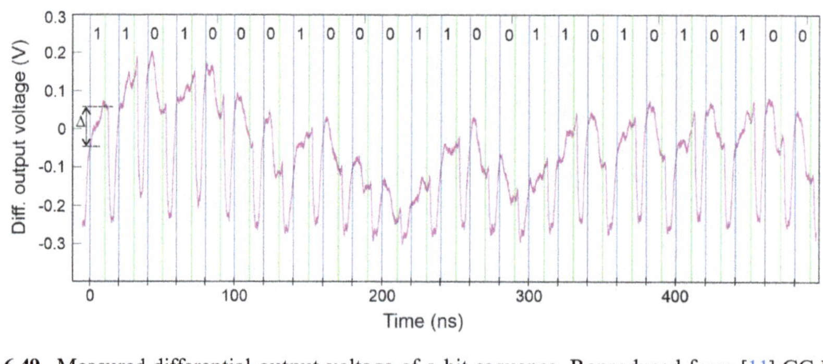

Figure 6.49. Measured differential output voltage of a bit sequence. Reproduced from [11] CC BY 4.0.

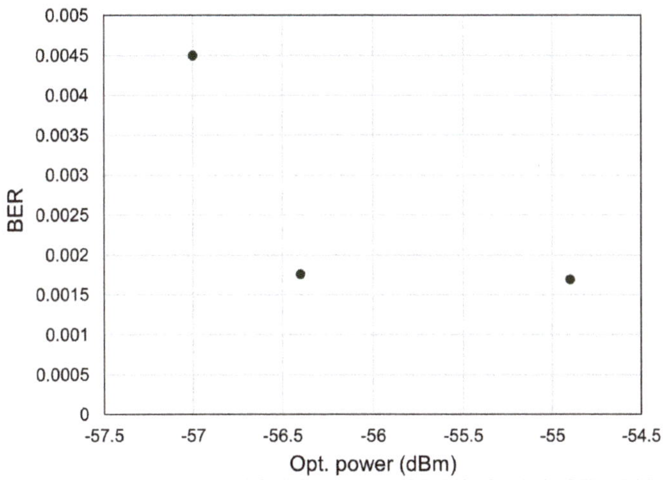

Figure 6.50. Bit error ratios versus average optical input power at 50 Mb s^{-1} and 635 nm. Reproduced from [11] CC BY 4.0.

approximately two times the maximum signal voltage. The compensation capacitors M_{C1} and M_{C2} reduce the clock injection, however, not as effectively as expected. These MOS capacitors obviously mismatch strongly with the reset transistor or there are parasitic capacitances not found or determined correctly in the extraction from the layout. Additionally, figure 6.49 reveals a movement of the operating point in dependence on the last bits before. The different heights of the integration swing and of the negative voltage spikes due to clock injection follow from this moving operating point.

Although there were severe imperfections, the fabricated ultra-sensitive integrating receiver achieved a sensitivity of −56.4 dBm, which is a little bit better than the sensitivity of −55.7 dBm of a receiver exploiting four SPADs [12]. The sensitivity of the first USPAR integrated circuit is compared to the state of the art in figure 6.51. The remaining gap to the quantum limit of the first USPAR integrated circuit is about 18 dB. The gated SPAD receiver of [16] showed a sensitivity of −57 dBm at 50 Mb s^{-1} leaving a gap of 15.7 dB to the quantum limit. This gated 1-SPAD receiver based on sub-bits is only 2.3 dB better than the first USPAR integrated circuit. Table 6.7 summarizes the properties of these and further published highly sensitive optical receivers. Higher data rates can be handled with multi-SPAD receivers as for instance with the 64 × 64 SPAD receiver [18]. Also the gap to the quantum limit is smaller for this multi-SPAD receiver. However, the circuit complexity and the power consumption (230 pJ per bit inclusive an ADC function for 4-PAM) are high. The active area of this multi-SPAD receiver is 1.8 mm^2. In contrast, without output driver the power consumption of the ultra-sensitive PIN-photodiode receiver is 51 pJ per bit and its active area is only 0.0145 mm^2. Further SPAD-based receivers being appropriate for higher data rates are so-called silicon photomultiplier (SiPM) receivers, as suggested in [19], where 5676 passively quenched SPADs are present.

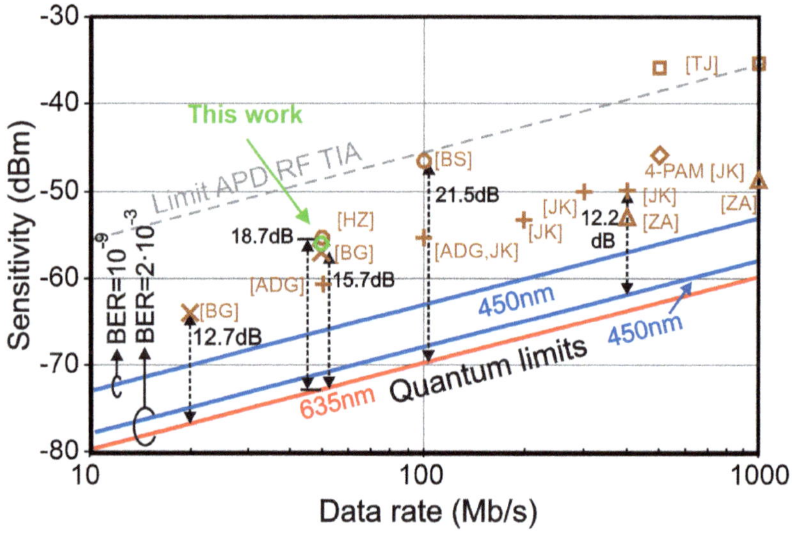

Figure 6.51. Comparison to sensitivities of APD and SPAD receivers (BER = 2×10^{-3}, SPAD-based: [ADG]: [15], [BG]: [16], [HZ]: [12], [BS]: [17], [JK]: [18]; [ZA]: [19], [19] reports a SiPM receiver). The gray dashed line represents the limit of resistive-feedback TIAs with 200 μm diameter APDs in linear mode (BER = 10^{-9}, APD based: [TJ]: [20]). Reproduced from [11] CC BY 4.0.

Table 6.7. Comparison to the state of the art of the integrating 0.18 μm CMOS receiver with common-source amplifier (S = sensitivity, DR = data rate, λ = wavelength). Reproduced from [11] CC BY 4.0.

Reference	Detector type	S (dBm)	DR (Mb s^{-1})	λ (nm)	BER	Active area (mm^2)	Power cons.[0]	Technology (μm)
[16]	1 Gated SPAD	−64	20	635	2×10^{-3}	0.66	19.2mW	0.35 CMOS
[16]	1 Gated SPAD	−57	50	635	2×10^{-3}	0.66	41 mW	0.35 CMOS
[12]	4-SPAD	−55.7	50	635	2×10^{-3}	0.1	19.1 mW[a]	0.35 CMOS
[17]	4-SPAD	−46.3	100	635	2×10^{-3}	0.08[b]	24.8 mW[c]	0.35 CMOS
[18]	64 × 64 SPAD	−49.9	400	450	1.8×10^{-3}	1.8	230 pJ/bit	0.13 CMOS
[15]	64 × 64 SPAD	−55.2	100	450	2×10^{-3}	1.8	230 pJ/bit	0.13 CMOS
[19]	SiPM	−58.4	400	405	10^{-3}	9.42[d]	—	Special[e]
[19]	SiPM	−49	1000	405	10^{-3}	9.42[d]	—	Special[e]
[20]	200 μm APD	−35.8	500	675	10^{-9}	0.1	250 mW[f]	0.35 BiCMOS
[20]	200 μm APD	−35.5	1000	675	10^{-9}	0.1	250 mW[f]	0.35 BiCMOS
[20]	200 μm APD	−32.2	2000	675	10^{-9}	0.1	250 mW[f]	0.35 BiCMOS
[11]	PIN	−56.4	50	635	2×10^{-3}	0.0145	51 pJ/bit	0.18 CMOS

[0]Without outpout driver. [a]380 pJ/bit. [b]Only SPADs + 4 quenchers. [c]248 pJ/bit. [d]SiPM only. [e]Not integrated. [f]Incl. output driver.

The sensitivity of this SiPM receiver is −58.4 dBm at 400 Mb s^{-1} and −49 dBm at 1 Gb s^{-1}. The SiPM, however, was fabricated in a special technology and does not allow integration in CMOS chips together with transistors.

The comparison of the first USPAR IC to the APD receiver of [20] also should be done. The linear-mode APD receiver in 0.35 μm BiCMOS using resistive feedback has sensitivities of −35.8 dBm at 500 Mb s^{-1}, of 35.5 dBm at 1 Gb s^{-1} and of −32.2 dBm at 2 Gb s^{-1} with a BER of 10^{-9} and at a wavelength of 675 nm [20]. This APD receiver left a gap to the quantum limit (for a BER of 10^{-9}) of about 20 dB. In this sense, the first integrating ultra-sensitive PIN-photodiode receiver leaves with 18 dB a smaller gap to the quantum limit. It should be mentioned that the gray line in figure 6.51 representing the sensitivity limit of 200 μm diameter APD receivers in the linear mode with resistive feedback will shift down a little bit, if small-diameter APDs like in USPAR are considered. This, however, does not have any influence on the gap to quantum limit of the USPAR receiver. Of course, the APD receiver is appropriate for much higher data rates.

Summarizing and concluding, the first integrating ultra-sensitive PIN-photodiode receiver outperforms linear-mode APD and some SPAD receivers with respect to the remaining gap to the quantum limit without making use of impact ionization. And what is more, the potential of the capacitive-feedback technique is not exhausted yet. A redesign of the reset and compensation circuitry should allow to improve the sensitivity. However, a disadvantageous aspect is the strong increase of the integration capacitor due to parasitic capacitances in the layout. These issues pointed us to the following investigation of the possibility to live without reset.

6.5 Reset-less integrating receiver

Since it will be difficult to compensate the charges injected by the reset transistor (clock feed-through, see section 6.1) perfectly due to mismatch of the reset transistor and compensation transistors, a reset-less receiver would be highly desirable. In addition, reset noise ceases to exist and double sampling is not necessary. Furthermore, non-return-to-zero coding becomes possible. We have, however, to be aware that photocurrents never can change the direction. A reset-less receiver therefore needs a compensating current to subtract the photocurrent pulses again in order to keep the operating point of the optical receiver. Fortunately a switch, which delivers or switches a compensating current on or off within nanoseconds, is not necessary. This is because there are equal numbers of logical zeros and ones within not too long bit sequences (in so-called DC-free encoding). Consequently, the half of the logical-one photocurrent can be applied as compensating DC current by a constant current PMOS transistor (M_{comp} in figure 6.52). The constant-current transistor M_{comp} needs the control signal V_{cont} from a control loop to be able to compensate the average photocurrent.

Now only C_{DB} and C_{GD} of the minimum-size PMOS transistor M_{comp} contribute in addition to the input capacitance of the inverter to the total input-node capacitance C_T. Since M_{RS} (see figure 6.3) was of two times the minimum size and the MOS capacitor M_{comp} in figure 6.3 was of minimum size, the photocurrent compensating constant current transistor M_{comp} saves two thirds of the parasitic capacitance contributions of M_{RS} and M_{comp} in figure 6.3. The integration efficiency of the reset-less integrating receiver increases therefore to 0.56 (C_T = 8.5 fF instead of about 10 fF for the inverter in 0.35 μm CMOS with $W_n = 1$ μm and $W_p = 2.5$ μm).

But there are not only advantages of the reset-less receiver. The constant current transistor adds noise to the input node. For simplicity, we assume that the control loop does not add any noise. So, the noise added by the constant current PMOS

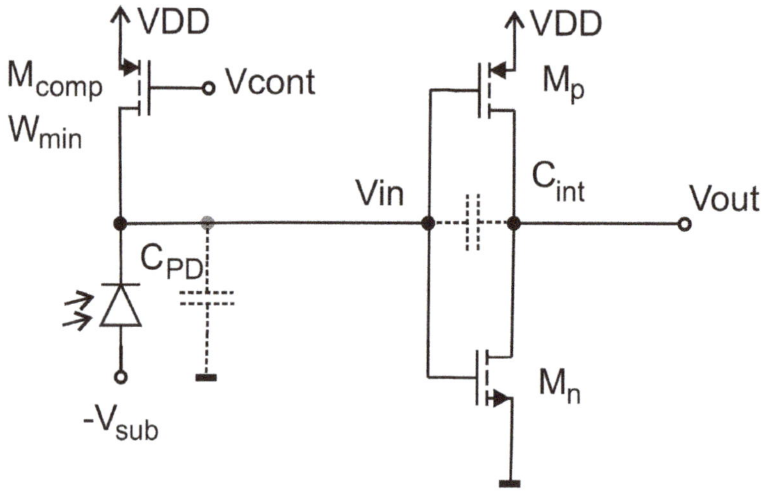

Figure 6.52. Inverter integrator with average photocurrent compensation by M_{comp} (reset-less integrating receiver).

transistor is only its drain current noise. However, here we cannot apply equation (3.8) because the constant current transistor operates in the nanoampere range, i.e. in the subthreshold regime (weak inversion) [21, 22]. The (one-sided) spectral power density S_{id} in the WI noise model [23] is

$$S_{\text{id}} = 2qI_{\text{D}}\left(1 + e^{-\frac{V_{\text{DS}}}{U_{\text{th}}}}\right), \tag{6.42}$$

where U_{th} is the thermal voltage. In saturation, the contribution from the e-function is negligible and

$$S_{\text{id}} = 2qI_{\text{D,sat}}, \tag{6.43}$$

with the drain saturation current $I_{\text{D,sat}}$, which is also obtained in [24]. This is shot noise. In this sense, the MOSFET in the subthreshold regime behaves like a bipolar transistor, which has collector current shot noise. Indeed in the subthreshold regime, the drain current depends exponentially on the gate–source voltage like the collector current of a bipolar transistor.

Instead of the one-sided spectral power density (SPD), the two-sided SPD [25]

$$S_{\text{id}} = qI_{\text{D}} \tag{6.44}$$

with the unit [A^2 Hz^{-1}] can be used for MOSFETs in the subthreshold range [22]. Then the integration has to be done from $-\infty$ to $+\infty$. But here we will use the one-sided SPD and can integrate from 0 to ∞.

Making the bandwidth Δf (for integration of the noise SPD) equal to the data rate as done in [26] is not really physical. Furthermore, the signal-to-noise ratio was optimized in [27] for the amplifier bandwidth equal to 0.9 times the bit rate. Therefore, we use here

$$\Delta f = mB = m\frac{1}{\Delta t}, \tag{6.45}$$

where B is the bit rate or data rate and Δt is the duration of a bit. So, we obtain for the shot noise of the constant current transistor:

$$\overline{i_{\text{en}}^2} = 2qI_e\Delta f = 2q\frac{n_e q}{\Delta t}\Delta f, \tag{6.46}$$

where I_e is the constant current through M_{comp}, which is formed by the number of electrons n_e flowing in a bit period Δt, i.e. $I_e = n_e q/\Delta t$. Therefore, we obtain

$$\overline{i_{\text{en}}^2} = 2q^2 n_e m\frac{1}{(\Delta t)^2} = 2q^2 n_e m B^2. \tag{6.47}$$

The rms noise current is

$$i_{\text{en}}^{\text{rms}} = qB\sqrt{2\,mn_e} \tag{6.48}$$

and in a next step we integrate it on C_{int} (where we must not forget the integration efficiency η_{int}), to obtain the corresponding rms output noise voltage:

$$u_{\text{out, en}}^{\text{rms}} = \frac{i_{\text{en}}^{\text{rms}}\Delta t\eta_{\text{int}}}{C_{\text{int}}} = \frac{q\eta_{\text{int}}\sqrt{2\,mn_e}}{C_{\text{int}}}. \tag{6.49}$$

Now, we recall that the constant current I_e is half the logical 'one' photocurrent and we can use

$$n_e = \frac{\eta_d n_{\text{PH}}}{2}, \tag{6.50}$$

where η_d is the quantum efficiency of the photodiode.

Finally, we obtain the mean square noise voltage at the output caused by the shot noise of the constant compensation current:

$$\overline{u_{\text{out, en}}^2} = m\eta_d n_{\text{PH}}\frac{q^2\eta_{\text{int}}^2}{C_{\text{int}}^2}. \tag{6.51}$$

Since the compensation current is a DC current, we have to consider this mean square noise voltage at the output for the logical 'one' and also for the logical 'zero' to compute the Q-factor:

$$Q = \frac{\eta_d\eta_{\text{int}}n_{\text{PH}}q/C_{\text{int}}}{\sqrt{v_{\text{n,out}}^2 + n_{\text{PH}}\eta_d^2\frac{q^2\eta_{\text{int}}^2}{C_{\text{int}}^2} + mn_{\text{PH}}\eta_d\frac{q^2\eta_{\text{int}}^2}{C_{\text{int}}^2}} + \sqrt{v_{\text{n,out}}^2 + mn_{\text{PH}}\eta_d\frac{q^2\eta_{\text{int}}^2}{C_{\text{int}}^2}}} \tag{6.52}$$

This noise distance Q was calculated for the reset-less receiver with the above described inverter integrator in 0.35 μm CMOS with $W_p = 2.5 \times W_n$. The results are plotted in figure 6.53 and compared to the necessary photon numbers of the resetted inverter integrator using real double sampling in table 6.8.

Despite of the shot noise of the constant current transistor, the reset-less receiver needs about 10 photons less than the resetted receiver (using real double sampling) with the same inverter integrator. Fortunately, the difference between both receivers is not smaller since the photon noise is less than 1.5 mV, as visible in figure 6.54. Note that the noise from the constant compensating current is 'half' of the photon noise but that it is also present in the logical 'zero'. Concluding, the reset-less receiver approach is advantageous compared to the resetted receiver since the shot noise from the constant current transistor contributes less noise than the implementation of double sampling.

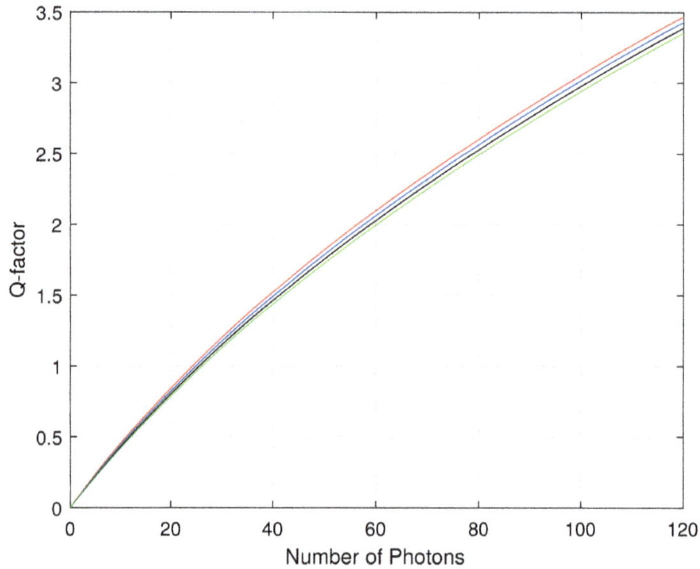

Figure 6.53. Q-factor of reset-less inverter integrator in 0.35 μm PIN-photodiode CMOS ($W_p = 2.5 \times W_n$, $m = 1$, green: $W_n = 1.0$ μm, black: $W_n = 0.9$ μm, blue: $W_n = 0.8$ μm, red: $W_n = 0.7$ μm).

Table 6.8. Results for the 0.35 μm reset-less CMOS receiver with $m = 1$ compared to the inverter receiver with reset from figure 6.3.

W_n (μm)	W_p (μm)	n_{PH} for $Q = 3$ reset-less RX from figure 6.53	n_{PH} for $Q = 3$ RX with reset and double sampling from figure 6.20
1.0	2.5	104	115
0.9	2.25	102	113
0.8	2.0	100	111
0.7	1.75	98	109

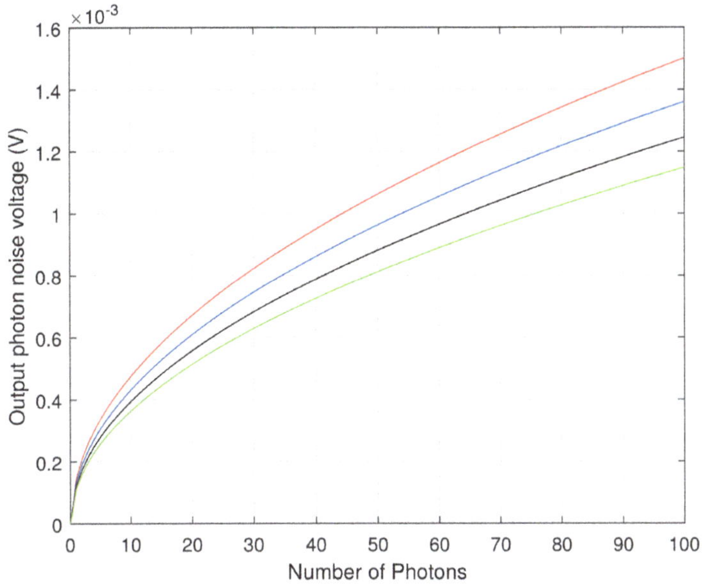

Figure 6.54. Rms output noise voltage of inverter integrator in 0.35 μm PIN-photodiode CMOS due to photon noise ($W_p = 2.5 \times W_n$, green: $W_n = 1.0$ μm, black: $W_n = 0.9$ μm, blue: $W_n = 0.8$ μm, red: $W_n = 0.7$ μm).

Let us investigate next, how the reduction of the integration capacitance by reducing the gate width of the PMOS transistor in the inverter influences the performance. Here we make both gate widths equal to each other and obtain a DC operating voltage of about 1.2 V, which is even good because the dynamic output voltage range is then about 2 V (be aware that the supply voltage in 0.35 μm CMOS is 3.3 V).

The inverter with $W_n = W_p = 0.7$ μm in 0.35 μm CMOS achieved a gain of about 17 and a bandwidth of about 140 MHz according to circuit simulations. The simulations were performed for a data rate of 100 Mb s^{-1}. The photodiode capacitance inclusive connection to the amplifier input was 3.5 fF. The extracted (parasitic) capacitances for these gate widths according to circuit simulations were 2.6 fF inverter input capacitance and 4.09 fF contribution of the reset and compensation transistors to the input node. The inverter input capacitance was scaled with the gate widths. The contribution of the current compensation transistor of the reset-less receiver to the input node capacitance was 1.11 fF. With these values the Q-factor calculations with equation (6.23) for real double sampling determine photon numbers from 92 to 93 for $Q = 3$ in the case of the reset inverter receiver using double sampling. For the reset-less inverter receiver 84 to 87 photons were necessary.

The extraction from the layout of the reset receiver revealed a capacitance of 0.23 fF between input and output of the inverter, which unfortunately adds to the integration capacitance and therefore reduces the signal output voltage. For the reset-less receiver with the same inverter, the additional extracted capacitance between input and output was 0.156 fF. These extracted capacitances were added to the integration capacitance and the results for the Q-factor considering reset with real double sampling, respectively, for the reset-less case are presented in figures 6.55 and 6.56.

Table 6.9 lists the numbers of necessary photons and compares them for the reset-less and the reset integrating receivers. The reset-less receiver achieves a better sensitivity. The smaller input-node capacitance and the relatively low shot noise added to the input node by the current compensating transistor are responsible for this. However, a perfect control loop is required with the reset-less receiver, which will need very good noise filtering. Another advantage of the reset-less receiver is that the additional (parasitic) capacitance being present in the layout is smaller, because there is no reset transistor, which lies between input and output of the integrating inverter.

A complete reset-less integrating receiver was investigated in 0.35 μm PIN-photodiode CMOS. The block diagram of this optical receiver is introduced in figure 6.57 [28]. In fact, two versions were considered. The first one implements an NMOS transistor M_n for compensation of the average photocurrent. M_n actually works as a source follower, which seems to be unusual. The second version exploits a PMOS transistor M_p as current source for compensation of the average photocurrent. It should be mentioned that the PMOS version suffered from a long settling time after turning on. Therefore, the NMOS transistor sinking the current I_{st} (see figure 6.57) was added to increase the current I_{co} delivered by M_p. With $I_{st} = 4$ nA, the current from the PMOS transistor M_p was increased to 6 nA for an average photocurrent of 2 nA. In such a way the start-up time was shortened.

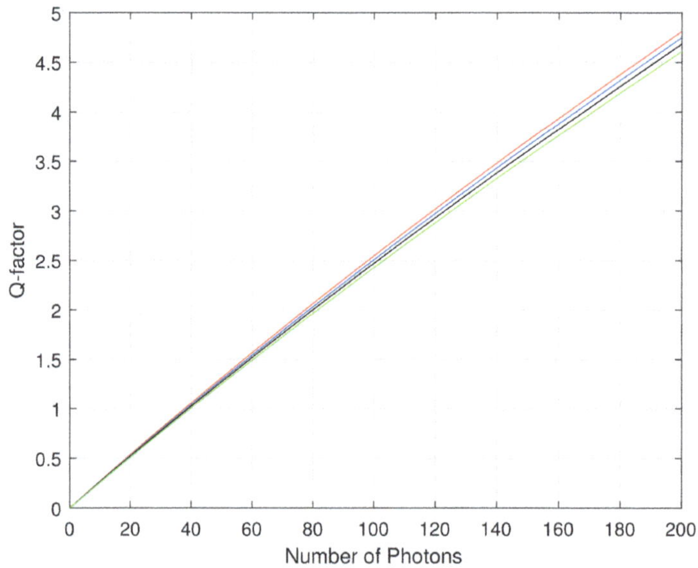

Figure 6.55. Q-factor of reset inverter integrator in 0.35 μm PIN-photodiode CMOS after extraction from the layout ($W_p = 1.0 \times W_n$, green: $W_n = 1.0$ μm, black: $W_n = 0.9$ μm, blue: $W_n = 0.8$ μm, red: $W_n = 0.7$ μm).

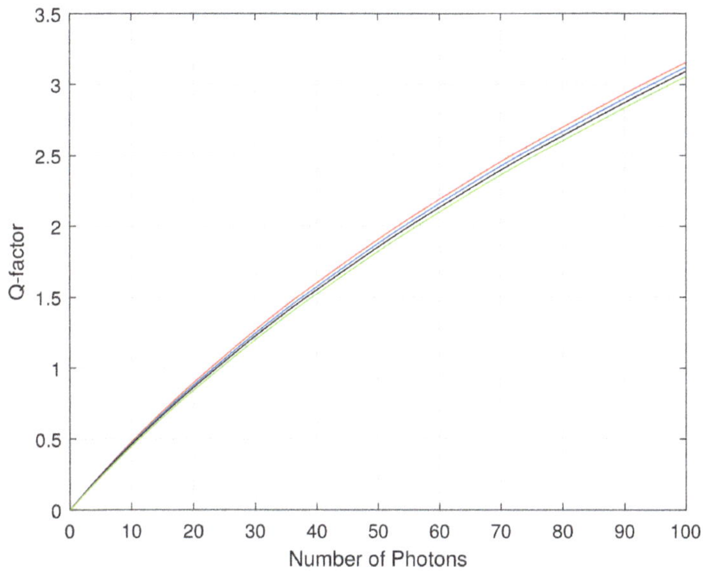

Figure 6.56. Q-factor of reset-less inverter integrator in 0.35 μm PIN-photodiode CMOS after extraction from the layout ($W_p = 1.0 \times W_n$, $m = 1$, green: $W_n = 1.0$ μm, black: $W_n = 0.9$ μm, blue: $W_n = 0.8$ μm, red: $W_n = 0.7$ μm).

Table 6.9. Results for the 0.35 μm reset-less ($m = 1$) and reset CMOS inverter receiver with $W_n = W_p$, both considering extraction of parasitic capacitances from the layouts.

W_n (μm)	W_p (μm)	n_{PH} for $Q = 3$ reset-less RX	n_{PH} for $Q = 3$ RX with reset
1.0	1.0	98	126
0.9	0.9	96	124
0.8	0.8	95	122
0.7	0.7	93	120

Since the transfer function of an integrator increases to infinity, if the frequency decreases to zero, a differentiator follows behind a source follower (SF). The combination of the frequency response of integrator and differentiator is flat. Towards the output of the optical receiver, a three-stage inverter driver with a 50 Ω output is integrated [28].

A PI (proportional integral) controller adjusts the gate voltage of the current compensating NMOS or PMOS transistor. The DC operating point output voltage of the integrator is generated with a matching replica inverter whose output is connected to its input. A dummy capacitance emulating the capacitance of the PIN photodiode is connected to the input of the replica inverter. In such a way integrating inverter and replica inverter see the same substrate noise. This pseudo-differential principle was very often successfully exploited with resistive-feedback

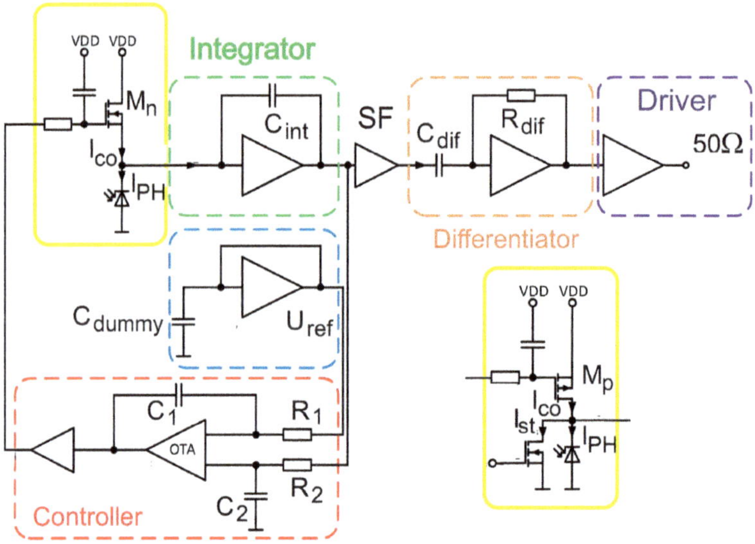

Figure 6.57. Reset-less optical receiver with NMOS transistor (M_n, in upper left yellow frame) and PMOS transistor (M_p, in lower right yellow frame) for the compensation of the average photocurrent [28].

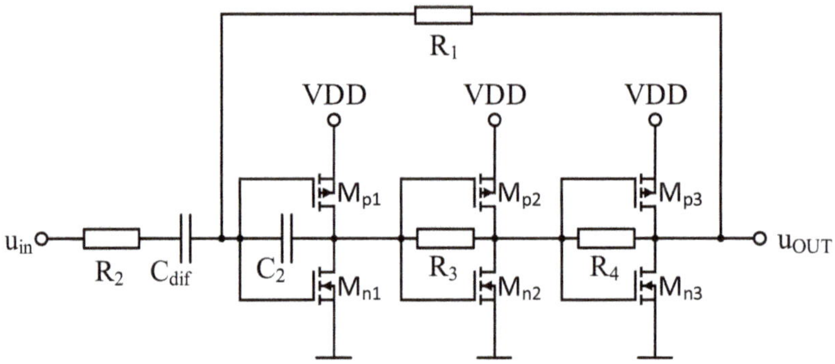

Figure 6.58. Circuit of differentiator using three inverters [28].

TIAs [29–32]. The PI controller introduces a lower cut-off frequency. An R–C filter reduces the noise at the gate of the current compensating transistor (M_n or M_p).

The differentiator consists of three inverters with several feedback paths (see figure 6.58) to cope with process tolerances and temperature influences [28]. The transfer function of the differentiator decreases to zero, if the frequency decreases to zero. Together with the integrator, therefore, a flat frequency response follows. The controller, however, causes a decrease towards low frequencies.

We have to be aware that the input voltage range of an inverter for a high and approximately constant gain is quite narrow, as figure 6.59 shows for an inverter in 0.35 μm PIN-photodiode CMOS technology. The so-called trip-point, where the output voltage is equal to the input voltage, is at 1.65 V. This would be the operating point of a resetted inverter integrator. The maximum gain is 16.6. The transfer characteristics are

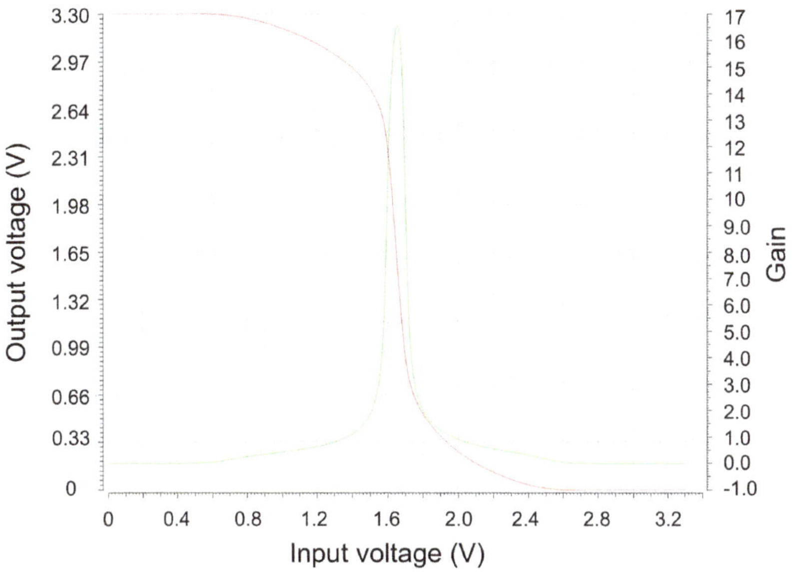

Figure 6.59. Transfer characteristics (red) and gain (green) of an inverter [28].

about linear for output voltages between 0.99 V and 2.31 V. But the gain drops already considerably at the boundaries of this interval. For these boundaries of the output voltage, the width of the input voltage range is less than 100 mV.

Although, the expected signal voltage on the integration capacitance is in the millivolt range for a '1'-bit, the bit sequence must be short. Usually, there are equal numbers of '0's and '1s' in a bit sequence. For a 128-bit sequence, there can be up to 64 '1'-bits and up to 64 '0'-bits following in series. Then, the signal voltage is in the range of hundreds of millivolts. Considering the gain of the inverter (10 to 16), the change in the inverter's input voltage is in the range of tens of millivolts for a 128-bit sequence. For the circuit simulations, a 2^7-1 pseudo-random bit sequence (PRBS) was used. Figure 6.60 shows the photocurrent for a 100 Mb s^{-1} NRZ sequence. The photocurrent takes values of 0 and 4 nA leading to an average photocurrent of 2 nA. The rise and fall time of the photocurrent was 1 ns (the duration of a bit at 100 Mb s^{-1} is 10 ns in NRZ). It should be mentioned that for the 128-bit sequence at 100 Mb s^{-1}, the minimum signal frequency is 390.6 kHz [28]. So, the frequency response of the receiver has to be flat from 390 kHz to at least about 70 MHz. The simulated frequency response is indeed flat from about 200 kHz to several 10 MHz with an upper −3 dB cut-off frequency of about 70 MHz (see figure 6.61). The total transimpedance in the middle of the flat range is 41.7 MΩ for the PMOS version and 42.2 MΩ for the NMOS version.

Figure 6.62 presents the result of circuit simulations for the integrator output voltage obtained with the bit sequence of figure 6.60. It is obvious that detecting the '0's and '1's would be quite difficult. Furthermore, we see that the output voltage of the integrator changes in a range of about 120 mV for a peak photocurrent of 4 nA and a 128-bit sequence at 100 Mb s^{-1}.

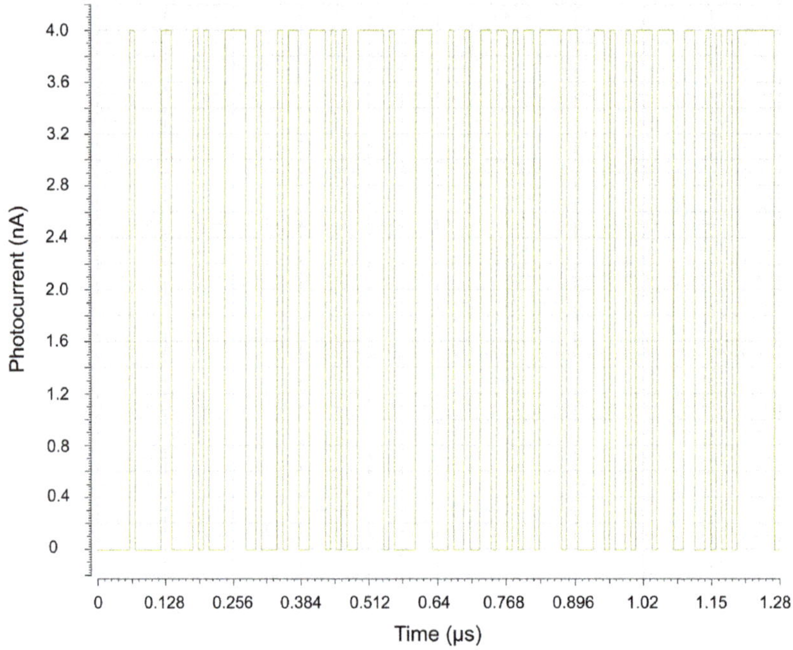

Figure 6.60. Photocurrent in an NRZ 128-bit sequence at 100 Mb s^{-1} (2 nA average photocurrent) [28].

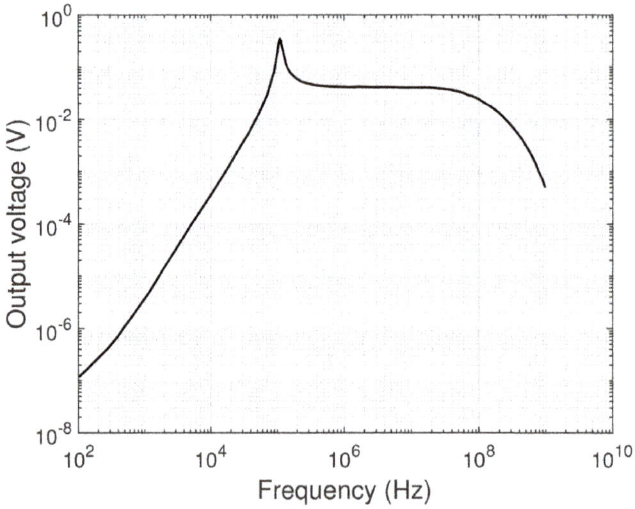

Figure 6.61. Frequency response of the output voltage of the differentiator for a current compensating PMOS transistor [28].

The differentiator was implemented to make the frequency response of the receiver flat in a broad bandwidth. Figure 6.63 validates this behavior, since the transient response after the differentiator corresponds well to the digital bit sequence of figure 6.60. The simulated eye diagram after the differentiator (see figure 6.64)

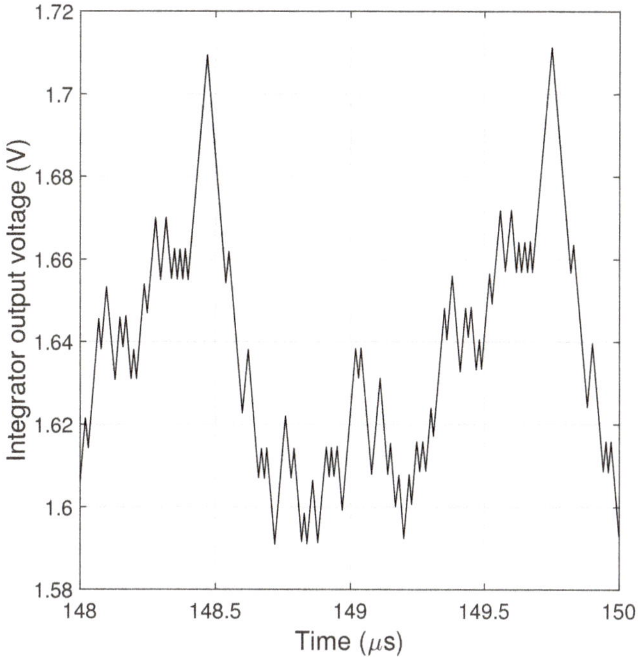

Figure 6.62. Output voltage of the integrator for 2 nA average photocurrent in a 128 NRZ bit sequence at 100 Mb s^{-1} [28].

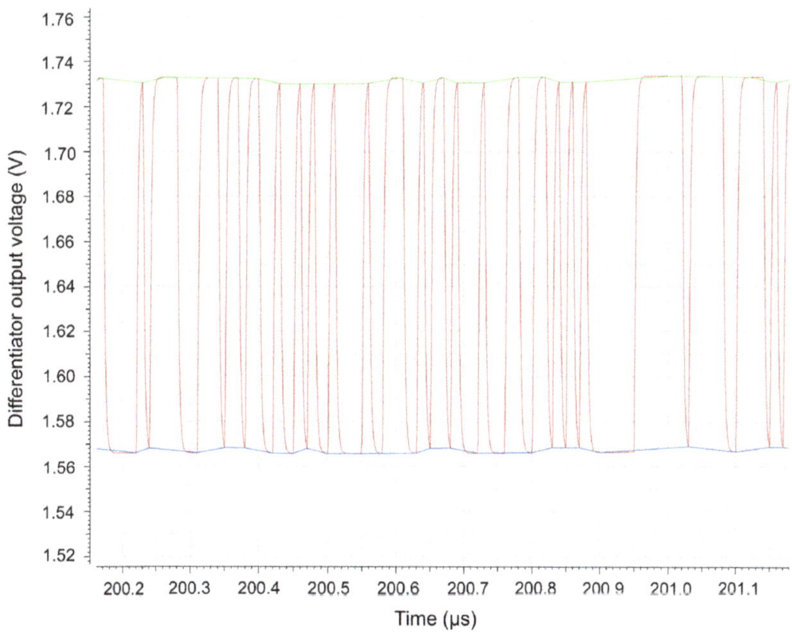

Figure 6.63. Transient response at the output of the differentiator for a current compensating PMOS transistor [28].

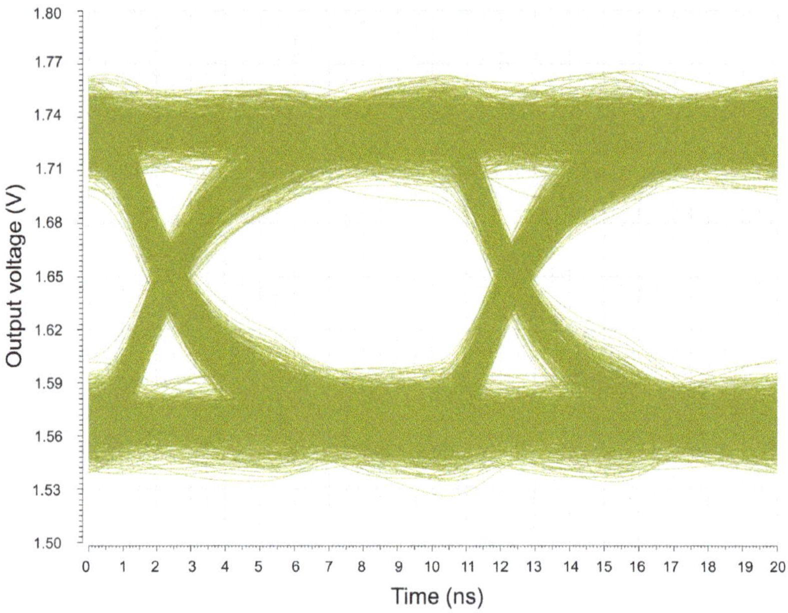

Figure 6.64. Eye diagram with transient noise at the output of the differentiator for a current compensating PMOS transistor [28].

also confirms the desired behavior and that the transmitted digital values can be recovered well.

Another important property of an integrated circuit is its behavior for process variations. The output voltages of the differentiator for 100 Monte Carlo runs considering process variations and mismatch are depicted in figure 6.65 [28]. For the transient noise, a minimum frequency of 1 Hz and a maximum frequency of 100 MHz were used. We see, that the operating point shifts but that all bits reach the output correctly. The receiver version with the NMOS transistor for photocurrent compensation was also successful in such a Monte Carlo analysis.

An important aspect of the reset-less integrating optical receiver, is the peak in the frequency response (see figure 6.61) at about 100 kHz. The consequence of this peak, which rises about a factor of 10 over the output voltage value in the flat range from 300 kHz to 30 MHz, is that it increases the noise of the receiver. This peak is less pronounced for the NMOS version [28]. The reason for this peak is the phase response of an integrating receiver. The integrator has a phase lag of 90°, which does not leave enough phase lag for the operational transconductance amplifier (OTA) in the control loop. The same control loop worked well with resistive-feedback TIAs for background-light suppression [29–32], i.e. without a peak in the receiver's frequency response. The resistive-feedback TIAs have less phase lag than integrating, capacitive-feedback TIAs. The peak is a matter of principle for reset-less integrating receivers.

The noise of this rest-less optical receiver was also investigated. The PMOS version showed an equivalent input noise current of 377.9 pA in a band from 70 kHz (lower −3 dB cut-off freuqency) to 100 MHz. The equivalent input noise current of

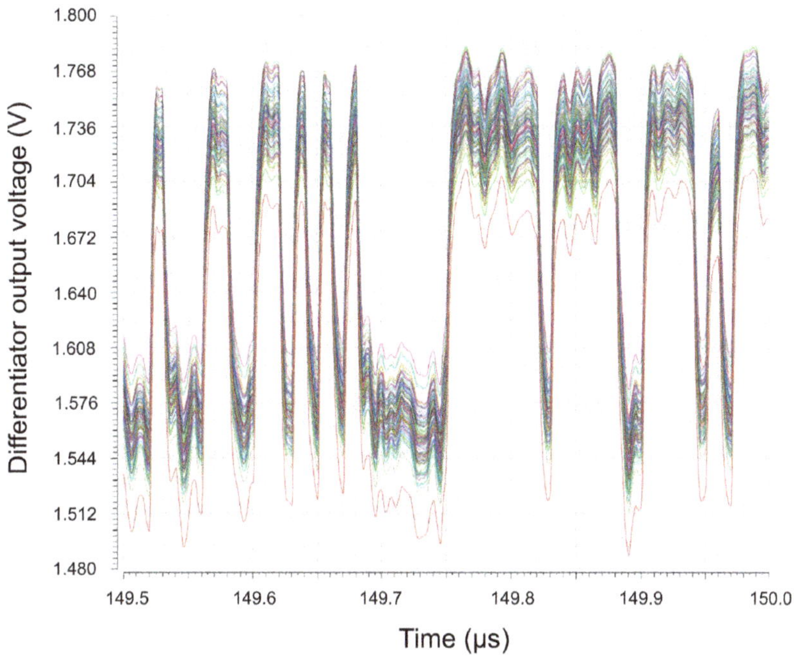

Figure 6.65. Transient response at the output of the differentiator inclusive transient noise with a current compensating PMOS transistor obtained by 100 Monte Carlo simulations [28].

the NMOS version was 313.8 pA from its lower −3 dB cut-off freuency of 150 kHz to 100 MHz. The corresponding Q-factors of the PMOS and of the NMOS versions were 6.90 and 8.68, respectively [28]. When switching off the start-up transistor in the PMOS version, its Q-factor improved to 8.78 making the PMOS version to the better reset-less receiver. But the peak in the frequency response alerts us to a stability issue.

Furthermore, a simulation of the output noise of the differentiator resulted in 11.9 mV rms for the PMOS version and in 9.59 mV rms for the NMOS version. Comparing with the inverter output noise in section 6.1, shows that the control loop and the differentiator add a considerable amount of noise.

6.6 Reset-less integrating receiver with punch-through device

There is a possibility to save some parasitic capacitance of the constant current transistor used in the last section above, mainly the drain-bulk capacitance of this transistor (see figure 6.52). For this purpose, the punch-through effect known from bipolar transistors and MOSFETs can be exploited. In bipolar transistors with open base, the electric field can reach through the base for high collector-emitter voltages and lead to breakdown. If the drain-source voltage in MOSFETs is too large, the electric field extends from the drain to the source, a punch-through current flows and breakdown can happen.

PTDs were used to replace the reset transistor in quantum image sensors because they reduced the capacitance of the storage node and because reset noise can be avoided [33–36]. Unfortunately, it is not possible to use a PTD instead of the reset transistor in USPAR ICs, because they are too slow for Mbit data rates and—what is more important—the operating point does not fit. Here, we will exploit the punch-through effect to replace the constant current transistor in a reset-less USPAR IC by a special PTD as shown in figure 6.66. By doing this, there is the possibility to reduce the input-node capacitance. However, there is no chance to get rid of the shot noise of the constant compensating current, since the PTDs involve a potential barrier [35].

The PTD for the reset-less integrating receiver must be optimized for a low capacitance and a low punch-through voltage between both cathodes. The device consists therefore of two small N+ cathodes with an ideally low P-type doping beween them. Figure 6.67 introduces such a PTD being located in a low-doped P-epitaxial layer. To achieve the lowest capacitance, we assume the 24 μm thick P-layer of the 0.18 μm PIN-photodiode HV CMOS technology. Cathode 1 of this PTD is connected to the input of the integrator and cathode 2 is controlled by V_{cont}. The circular structure has the advantage that cathode 1 will not be influenced by surrounding doping regions and their potentials. Nevertheless, the radial extension of the simulation region was chosen as 10 μm.

First, no shallow-trench isolation (STI) was implemented. The diameter of the central cathode 1 was chosen as 0.88 μm and the width of the cathode 2 N+ region was 1 μm with a gap d_n of 2 μm or 3 μm between cathode 1 and cathode 2. A gap of 1 μm was too small because the two N+ dopings overlapped due to the low epitaxial layer doping. For the 2 μm and 3 μm gaps, the capacitances obtained by device simulation with ATLAS [37] are about 0.7 fF and 0.5 fF, respectively, at the voltage differences needed for a current of about 1 nA. These capacitances can be considered as too large for application in the reset-less integrating receiver.

Therefore, an STI in the gap between the cathodes and around the cathode 2 ring was implemented in the device structure used in device simulation. The thickness of

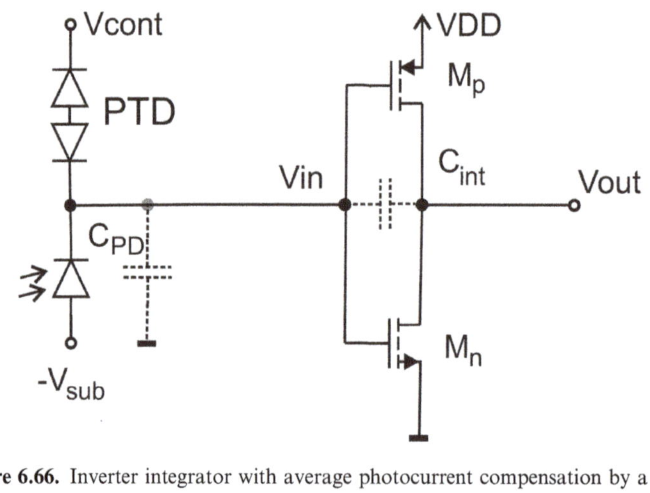

Figure 6.66. Inverter integrator with average photocurrent compensation by a PTD.

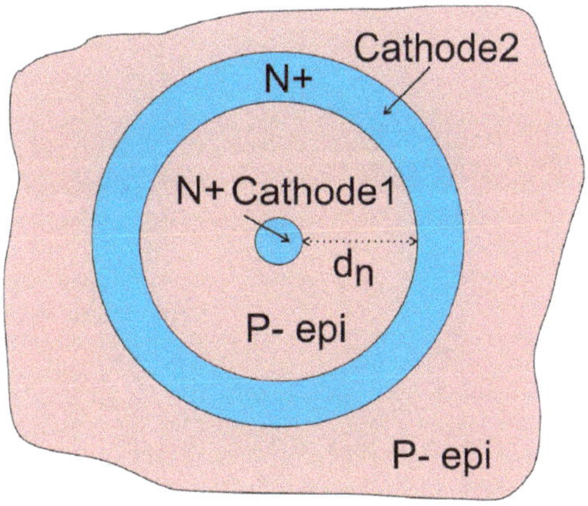

Figure 6.67. Top view of a PTD located in a low-doped P-epitaxial layer.

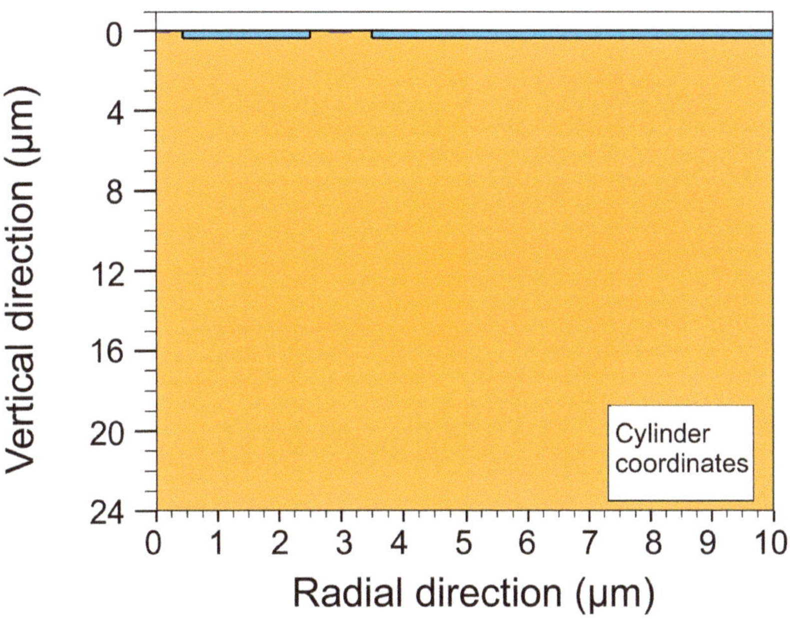

Figure 6.68. Simulation region of the PTD located in a low-doped P-epitaxial layer and using an STI for definition of a 2 μm gap between the two cathodes.

the STI in the 0.18 μm process is 0.4 μm. The structure (radial cross-section) of the device is shown in figure 6.68. The STI is represented in blue color in this figure. The anode is at the bottom and it is connected to 0 V in the device simulations.

The capacitance of the PTD with 3 μm gap between the two cathodes is about 0.2 fF (from cathode 1 to cathode 2) plus 0.06 fF (from cathode 1 to anode) at the voltage of 8.5 V at cathode 1 (see figure 6.69), where the current is about 1 nA (see figure 6.70). It should be mentioned that the difference voltage between cathode 2

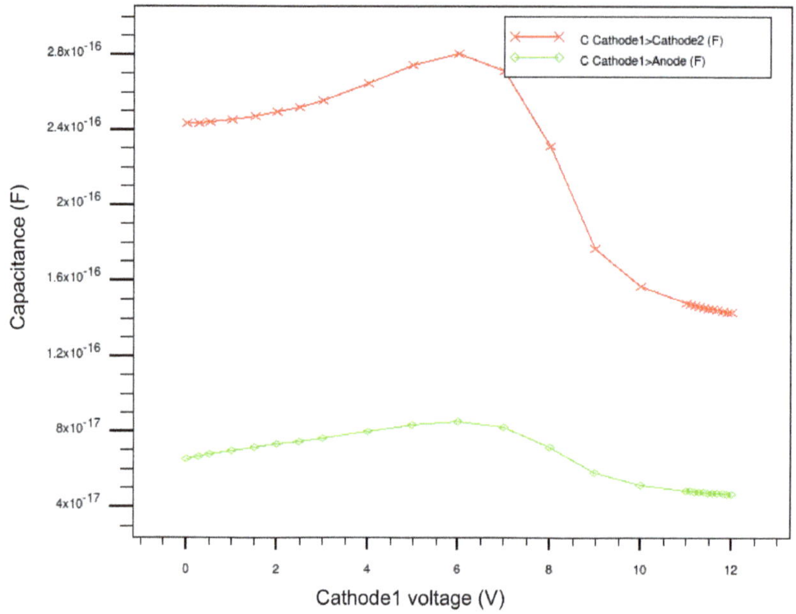

Figure 6.69. Simulated capacitance of the PTD located in a low-doped P-epitaxial layer and using an STI for definition of a 3 μm gap between the two cathodes (cathode 2 voltage = 12 V, the anode is at 0 V).

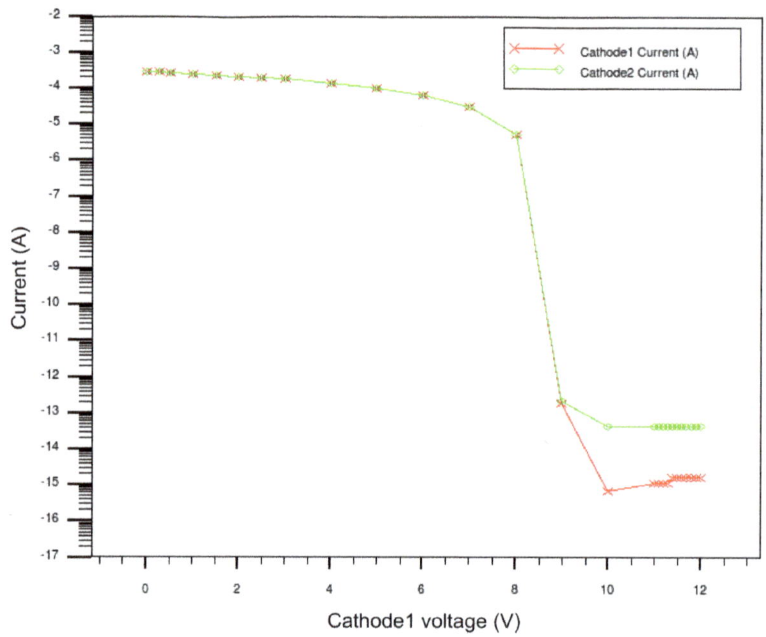

Figure 6.70. Simulated current of the PTD located in a low-doped P-epitaxial layer and using STI for definition of a 3 μm gap between the two cathodes (cathode 2 voltage = 12 V, the anode is at 0 V).

and cathode 1 is 3.5 V for the cathode 1 voltage of 8.5 V. Therefore, a control circuit with a supply voltage in excess of 3.5 V will be needed to regulate the current into the 1 nA range.

In order to reduce the necessary difference voltage of the two cathodes, the gap between them should be reduced. This was done and figure 6.71 shows the capacitance for a gap of 2 μm width. At 9.1 V cathode 1 voltage, where the current is about 1 nA (see figure 6.72). The capacitance is about 0.2 fF (from cathode 1 to cathode 2) plus about 0.055 fF (from cathode 1 to anode). Therefore, the difference voltage of both cathodes is still 2.9 V, which is larger than the supply voltage of the 0.18 μm CMOS technology of 1.8 V for circuits with the low-voltage high-speed MOS transistors.

Therefore, simulations with a gap of 1 μm were performed for the device using STI. Indeed, the STI prevents the two N+ cathodes from flowing together as is visible in figure 6.73. The STI is violet in this plot. In the gaps between the STI, the N+ doping regions (donor) are present. Towards the bottom (below about 20 μm), the P-type doping increases to the P+ bulk substrate doping (acceptor). The green region is the low-doped P- epitaxial layer.

Figure 6.74 depicts the simulated results for the capacitance of the PTD with STI and 1 μm gap between the N+ cathodes in dependence on the voltage. The capacitance at 9.7 V, where the current is about 1 nA (see figure 6.75), is 0.24 fF (from cathode 1 to cathode 2) plus 0.02 fF (from cathode 1 to anode). Figure 6.75 presents the simulated results for the current of the PTD with STI and 1 μm gap between the N+ cathodes in dependence on the voltage. The necesssary voltage

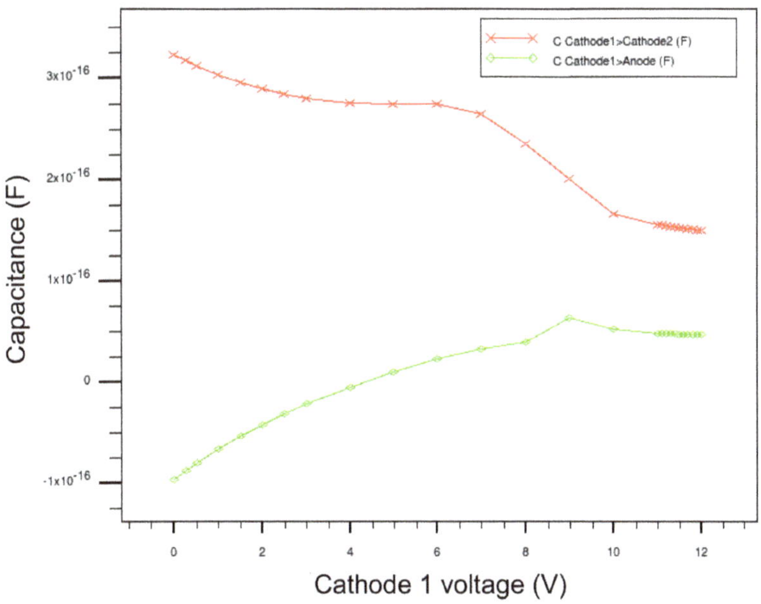

Figure 6.71. Simulated capacitance of the PTD located in a low-doped P- epitaxial layer and using STI for definition of a 2 μm gap between the two cathodes (cathode 2 voltage = 12 V).

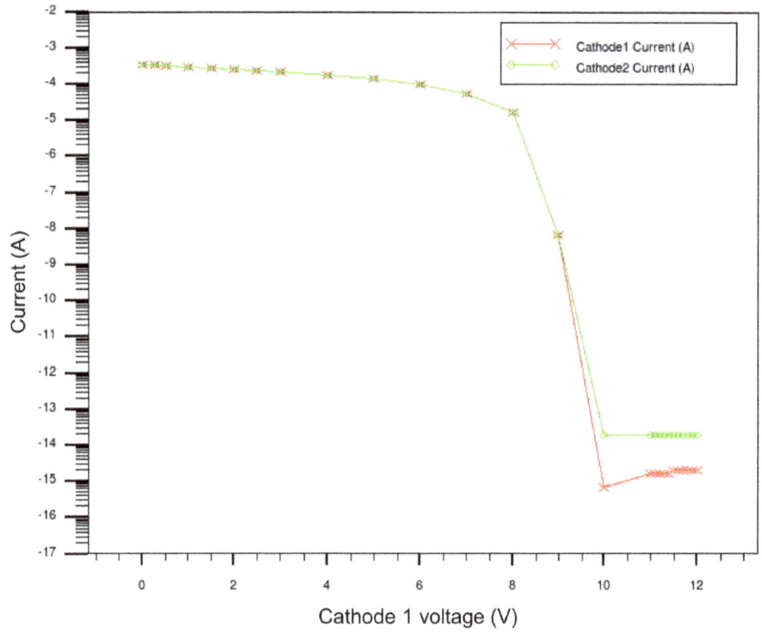

Figure 6.72. Simulated current of the PTD located in a low-doped P-epitaxial layer and using STI for definition of a 2 μm gap between the two cathodes (cathode 2 voltage = 12 V).

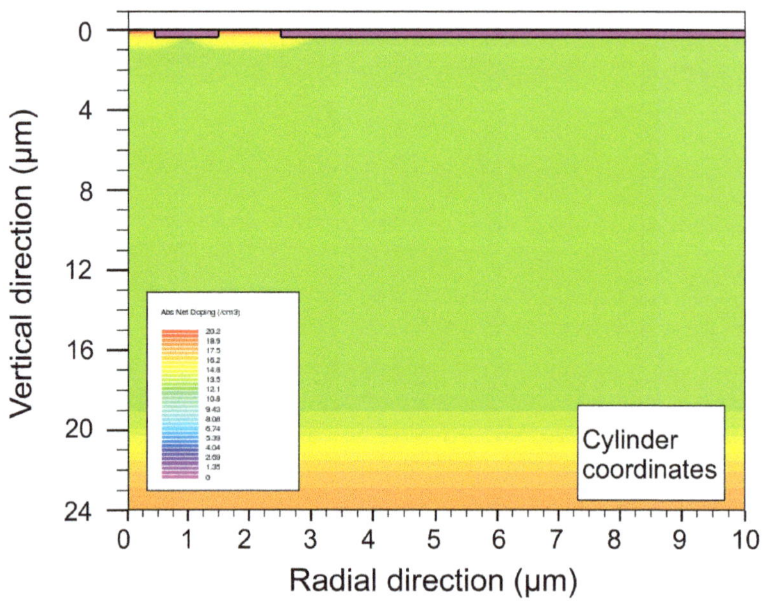

Figure 6.73. Simulation region including net doping of the PTD located in a low-doped P-epitaxial layer and using STI for definition of a 1 μm gap between the two cathodes.

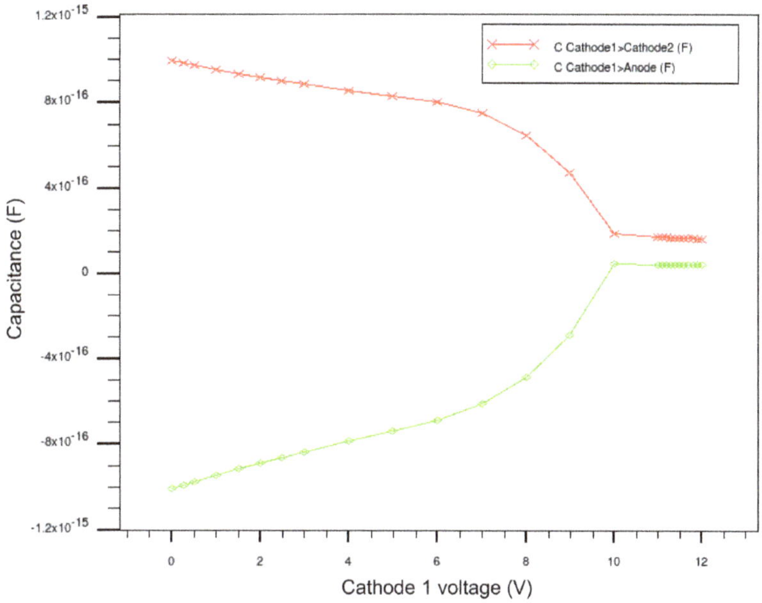

Figure 6.74. Simulated capacitance of the PTD located in a low-doped P- epitaxial layer and using STI for definition of a 1 μm gap between the two cathodes (cathode 2 voltage = 12 V).

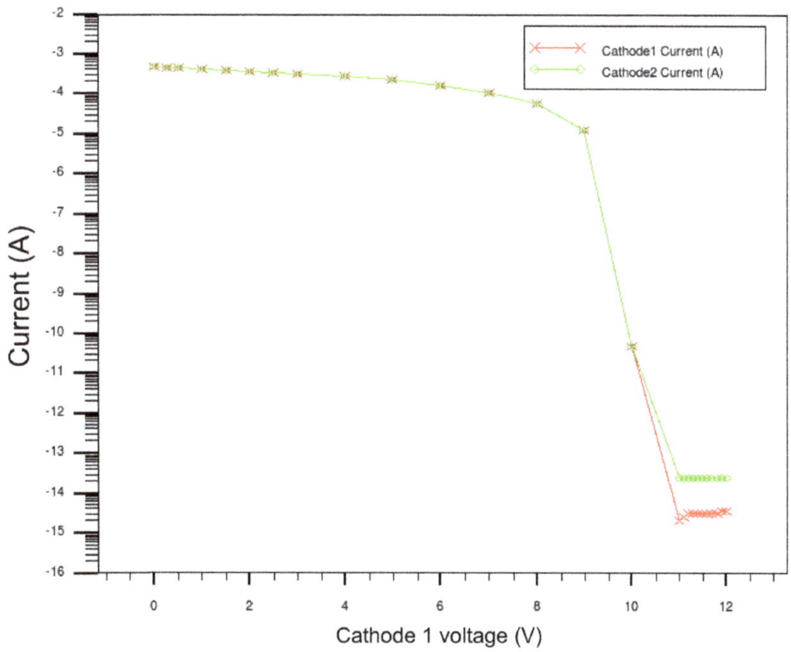

Figure 6.75. Simulated current of the PTD located in a low-doped P-epitaxial layer and using STI for definition of a 1 μm gap between the two cathodes (cathode 2 voltage = 12 V).

difference between the two cathodes for a current of about 1 nA is still 2.3 V. In order to get along with a controller circuit having a supply voltage of 1.8 V, the gap should be reduced further below 1 μm. Then, however, the two cathode N+ doping regions will flow together.

Summarizing, the circular PTD inside the low-doped P-epitaxial layer with STI and 1 μm gap needs a difference voltage of 2.3 V. When we assume an input voltage of the integrating amplifier of 0.9 V (remember that cathode 1 should be connected to the amplifier input), cathode 2 of the PTD should be on a potential of 3.2 V, which is larger than the 1.8 V supply of the integrator in 0.18 μm CMOS. Another deep N-well will be needed to isolate the higher-voltage controller circuit needed for this PTD. This is possible and the other deep N-well should be far away from the photodiode, since it would draw photogenerated current away from the cathode of the photodiode otherwise (see figure 5.15). However, the PTD cannot be placed far away from the photodiode, because the capacitance of the metal line from the cathode of the photodiode or from the receiver input to the PTD would destroy the capacitance advantage of the PTD over the constant current transistor. But if the PTD is close to the photodiode, cathode 2, which is at 3.2 V (much more positive as the deep N-well of the integrator at 1.8 V), will draw photocurrent away from the cathode of the photodiode. Concluding: unfortunately, the circular PTD inside the low-doped P-epitaxial layer cannot be implemented in an USPAR receiver.

The next idea was to investigate a PTD with two short N+ stripes located in the P-epitaxial layer. Figure 6.76 shows the structure of such a PTD. The simulation region had a width (in horizontal direction, x-axis) of 38 μm and a height (in vertical direction, y-axis) of 24 μm. The device simulator calculates currents and capacitances for a thickness (in the third z-direction) of the simulation region of 1 μm. With a gap $d_n = 1$ μm, the two N+ regions flow together, because the gap between them has the low P-type doping of about 10^{13} cm^{-3}. This problem can be solved with an STI between the two N+ regions. But the difference voltage needed for a current in the order of 1 nA between the two cathodes is about 2 V. Therefore, only the results for the PTD with a gap of 2 μm are shown. The net doping of this PTD is shown in figure 6.77. The two N+ regions are well separated.

Figure 6.78 presents the results of current–voltage device simulations. The cathode 1 current of 1 nA is obtained at the cathode 1 voltage of 11.67 V, which

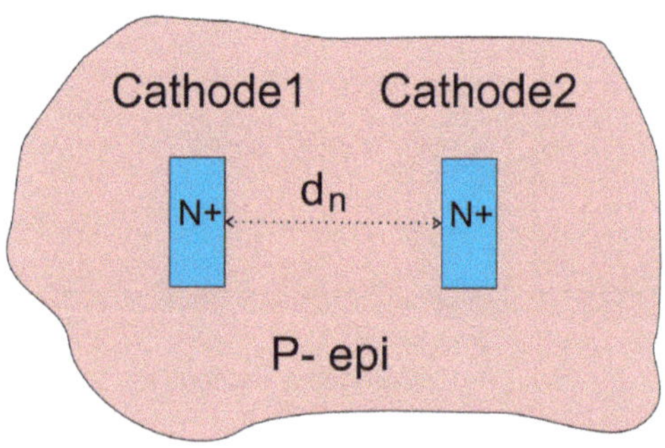

Figure 6.76. Top view of the PTD with two short N+ stripes located in a low-doped P-epitaxial layer.

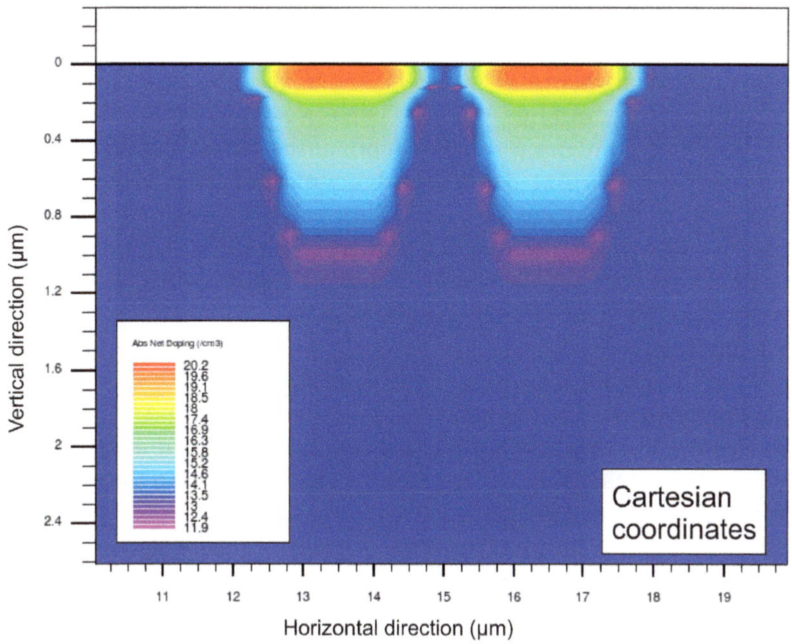

Figure 6.77. Zoomed cross-section of the PTD with two short N+ stripes located in a low-doped P-epitaxial layer (gap $d_n = 2\ \mu m$). The device is surrounded by STI.

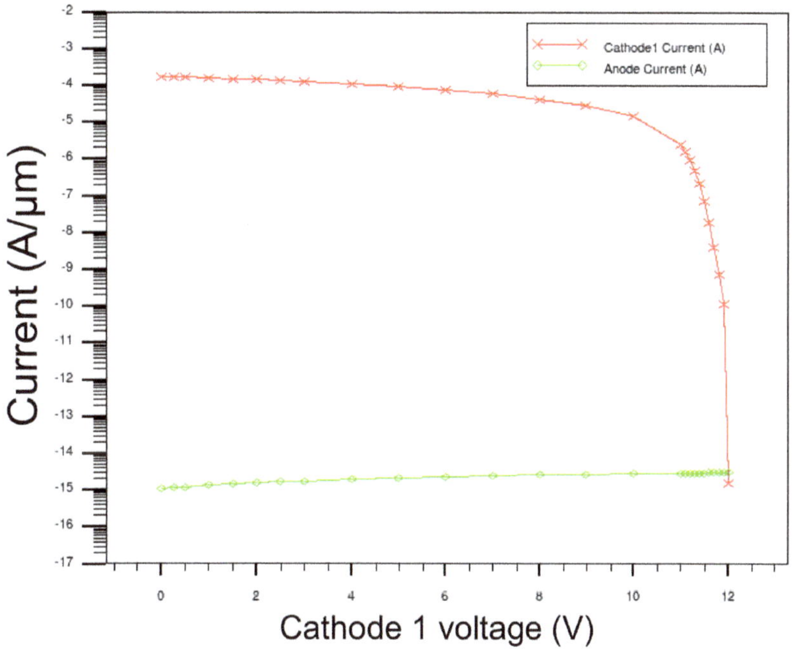

Figure 6.78. Current of the PTD with two short N+ stripes located in a low-doped P-epitaxial layer (gap $d_n = 2\ \mu m$, cathode 2 voltage = 12 V).

means the difference voltage of the two cathodes is only 0.33 V to reach the current of 1 nA between the two cathodes. This is a very interesting and good result. The capacitance of this Cartesian PTD with the gap d_n of 2 μm is plotted in figure 6.79. At the cathode 1 voltage of 11.67 V this PTD has a capacitance of about 0.1 fF μm^{-1} (note that the device simulator assumes a depth of the simulation region (thickness in the third z-direction) of 1 μm). Since the minimum dimension of the N+ of the 0.18 μm CMOS process is 0.44 μm, the possibility exists to reduce the capacitance to about 0.05 fF. This is also a very advantageous result for the capacitance. The Cartesian PTD with a gap of 2 μm and without STI between the two cathodes needs a 1.8 V-only controller, which can be placed inside the same deep N-well as the integrating amplifier.

However, in the simulations more than 10 μm distance to P-wells or other N-type regions are presupposed. The implementation of the PTD in an USPAR receiver will require a longer metal line between photodiode and integrator input. But there still should be a gain in capacitance for the input node compared to the constant current transistor in the reset-less USPAR IC.

Nevertheless, a PTD inside the deep N-well, which isolates the integrating amplifier, would save some chip area and therefore the following device was investigated. Figure 6.80 shows the top view of this PTD. The deep N-well was not added in the device simulations, because it should not have any influence on the characteristics of the PTD. The size of the two N+ regions is 1 μm times 0.44 μm.

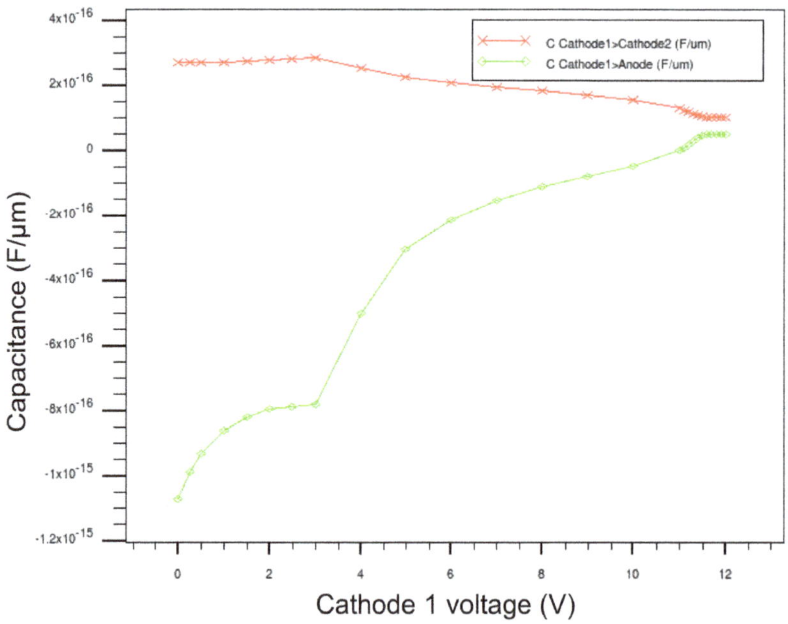

Figure 6.79. Capacitance of the PTD with two short N+ stripes located in a low-doped P-epitaxial layer (gap $d_n = 2$ μm, cathode 2 voltage = 12 V).

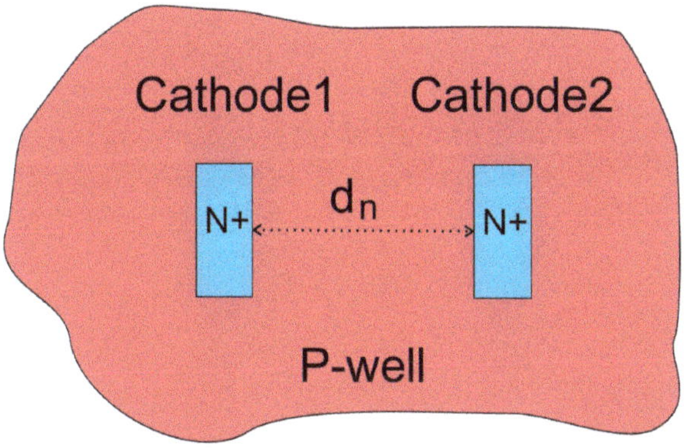

Figure 6.80. Top view of the PTD with two short N+ stripes located inside a P-well.

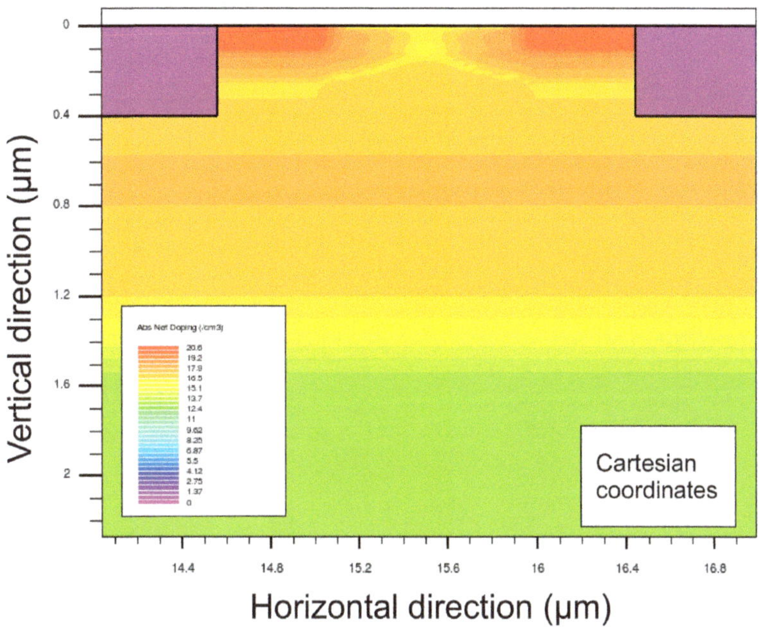

Figure 6.81. Zoomed cross-section of the PTD with two short N+ stripes located inside a P-well.

The PTD is surrounded by STI to isolate it better from the ambient. Figure 6.81 shows the cross-section of the device inclusive net doping. Thanks to the rather high P-well doping of about 10^{17}cm^{-3}, the two N+ regions do not flow together although the gap between them is only 1 μm.

Figure 6.82 presents the results of device simulations for the current–voltage characteristics of the PTD inside the P-well. Due to the high P-well doping concentration, a difference voltage between the two cathodes of about 5 V is necessary for a cathode 1 current of 1 nA. For such a difference voltage, the capacitance of this PTD is approximately 0.4 fF μm^{-1} (see figure 6.83). Therefore,

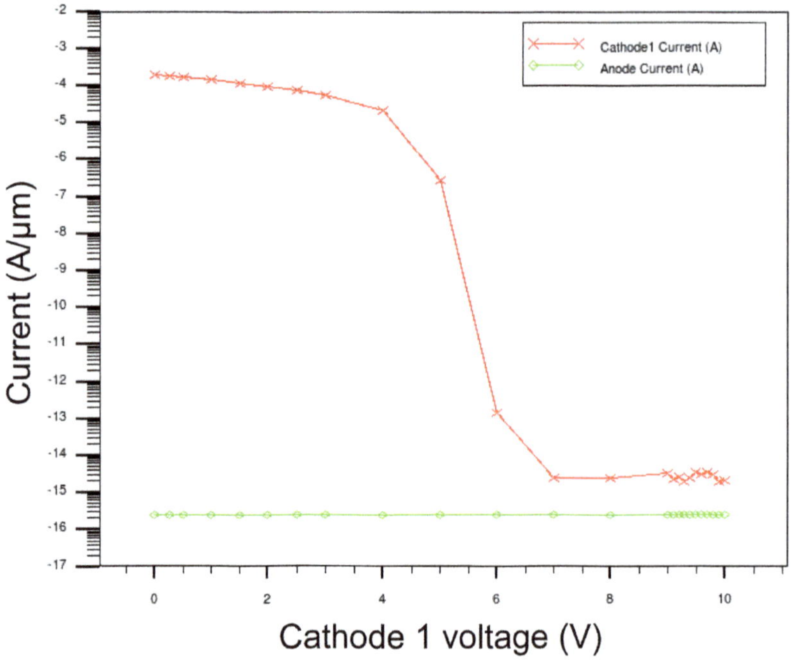

Figure 6.82. Current–voltage characteristics of the PTD with two short N+ stripes located inside a P-well (cathode 2 voltage = 10 V).

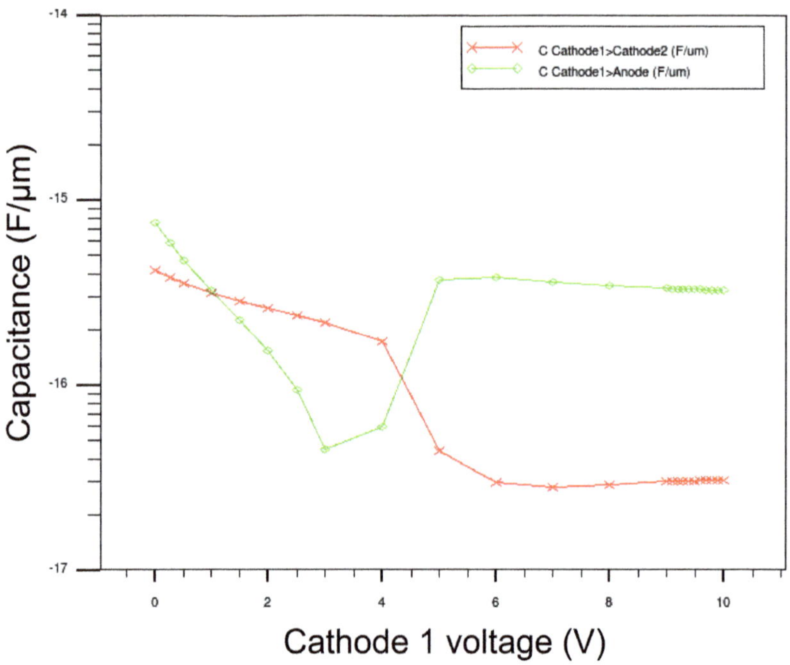

Figure 6.83. Capacitance–voltage characteristics of the PTD with two short N+ stripes located inside a P-well (cathode 2 voltage = 10 V).

despite the rather high P-well doping, a PTD having a capacitance of 0.2 fF is possible when the two cathodes are cut to extensions of 0.5 μm times 0.44 μm. Such a PTD can be placed inside a rather small P-well, which is located inside the same deep N-well as the integrator of an USPAR integrated circuit. Of course, an HV controller is necessary for the high difference voltage of the two cathodes. However, this controller can be placed into another deep N-well at a large distance from the photodiode to avoid current flow between the photodiode and this second deep N-well.

Summarizing, two solutions for the replacement of the constant-current transistor of a reset-less USPAR integrated circuit are available: the Cartesian PTD inside the low-doped epitaxial layer and the PTD inside a P-well. Both should result in a reduction of the input-node capacitance, which should increase the integration efficiency of a reset-less USPAR integrated circuit.

References

[1] Förtsch M and Zimmermann H 2003 Low-offset CMOS OEIC for optical storage systems *IEEE Trans. Consumer Electronics* **49** 1125–8

[2] Schneider K, Zimmermann H and Wiesbauer A 2004 Optical Receiver in Deep-sub-μm CMOS with −28.2 dBm Sensitivity at 1.25 Gb s^{-1} *Electron. Lett.* **40** 262–3

[3] Schneider K and Zimmermann H 2006 Three-stage burst-mode transimpedance amplifier in deep-sub-μm CMOS technology *IEEE Trans. on Circuits and Systems I: regular Papers* **53** 1458–67

[4] Schneider K and Zimmermann H 2006 *Highly Sensitive Optical Receivers* (Berlin, Heidelberg: Springer)

[5] Davidovic M, Seiter J, Hofbauer M, Gaberl W, Schidl S and Zimmermann H 2014 64 × 48 tof sensor in 0.35 μm CMOS with high ambient light immunity *IET Electron. Lett.* **50** 1375–7

[6] Davidovic M, Seiter J, Hofbauer M, Gaberl W and Zimmermann H 2014 Dual-lock-in sensor IC with integrated PIN photodetectors *Analog Integr. Circ. Sig. Process* **81** 797–804

[7] https://www.cadence.com/en_US/home/tools/custom-ic-analog-rf-design/circuit-simulation/spectre-simulationplatform.html, accessed 16.09.2022

[8] X-Fab Semiconductor Foundries 2019 Process & Device Specification XO035—0.35 μm Modular CMOS for Fast Optical Applications

[9] Zimmermann H 2004 *Silicon Optoelectronic Integrated Circuits* (Berlin, Heidelberg: Springer)

[10] Säckinger E 2018 *Analysis and Design of Transimpedance Amplifiers for Optical Receivers* (Hoboken, NJ: Wiley)

[11] Schneider-Hornstein K, Goll B and Zimmermann H 2023 Ultra-sensitive pin-photodiode receiver *IEEE Photonics J.* **15** 1–9 7201409

[12] Zimmermann H, Steindl B, Hofbauer M and Enne R 2017 Integrated fiber optical receiver reducing the gap to the quantum limit *Sci. Rep.* **7** 2652

[13] Sheu B and Hu C 1984 Switch-induced error voltage on a switched capacitor *IEEE J. Solid-State Circuits* **19** 519–25

[14] ITU-T 2004 G.975.1: forward error correction for high bit-rate dwdm submarine systems *Telecommunication Standardization Sector*

[15] Griffiths A D, Herrnsdorf J, Almer O, Henderson R K, Strain M J and Dawson M D 2019 High-sensitivity free space optical communications using low size, weight and power hardware arXiv:1902.00495 https://arxiv.org/pdf/1902.00495

[16] Goll B, Hofbauer M, Steindl B and Zimmermann H 2018 A fully integrated SPAD-based CMOS data-receiver with a sensitivity of −64 dBm at 20 Mb s^{-1} *IEEE Solid-State Circuits Lett.* **1** 2–5

[17] Steindl B, Hofbauer M, Schneider-Hornstein K, Brandl P and Zimmermann H 2018 Single-photon avalanche photodiode based fiber optic receiver for up to 200 mb s^{-1} *IEEE J. Sel. Top. Quantum Electron.* **24** 1–8

[18] Kosman J, Almer O, Abbas T A, Dutton N, Walker R, Videv S, Moore K, Haas H and Henderson R 2019 29.7 a 500 mb s^{-1} −46.1 dbm CMOS SPAD receiver for laser diode visible-light communications *2019 IEEE Int. Solid-State Circuits Conf.—(ISSCC)* 468–70

[19] Ahmed Z, Singh R, Ali W, Faulkner G, O'Brien D and Collins S 2020 A SIPM-based VLC receiver for gigabit communication using OOK modulation *IEEE Photonics Technology Lett.* **32** 317–20

[20] Jukic T, Steindl B, Enne R and Zimmermann H 2016 200 μm APD OEIC in 0.35 μm BiCMOS *IET Electron. Lett.* **52** 128–30

[21] Chan L H K, Yeo K S, Chew K W J, Ong S N, Loo X S, Boon C C and Do M A 2012 MOSFET drain current noise modeling with effective gate overdrive and junction noise *IEEE Electron Device Lett.* **33** 1117–9

[22] Tian H, Fowler B and Gamal A 2001 Analysis of temporal noise in CMOS photodiode active pixel sensor *IEEE J. Solid-State Circuits* **36** 92–101

[23] Tsividis Y 1999 *Operation and Modeling of the MOS Transistor* (Boston, MA: McGraw-Hill)

[24] Blanter Y M, Büttiker M, Weihong Y, Qianqian L, Rong C and Qin H 2000 Shot noise in mesoscopic conductors *Phys. Rep.* **336** 1–166

[25] van der Ziel A 1986 *Noise in Solid State Devices and Circuits* (New York: Wiley)

[26] https://web.archive.org/web/20181024162550/ http://www.fis.unipr.it/gigi/dida/strumentazione/harvard_noise.pdf Thermal and Shot Noise. Appendix C. accessed 09.05.2022

[27] Stevens A E 1995 An integrate-and-dump receiver for fiber optic networks *PhD Thesis* Columbia University

[28] Ribisch C 2021 Untersuchung kapazitiver Transimpedanzverstärker für optische Empfänger *Master Thesis* TU Wien

[29] Swoboda R, Knorr J and Zimmermann H 2005 A 5-gb/s OEIC with voltage-up-converter *IEEE J. Solid-State Circuits* **40** 1521–6

[30] Brandl P, Jukic T, Enne R, Schneider-Hornstein K and Zimmermann H 2016 Optical wireless APD receiver with high background-light immunity for increased communication distances *IEEE J. Solid-State Circuits* **51** 1663–73

[31] Jukic T, Steindl B, Enne R and Zimmermann H 2016 400 μm diameter APD OEIC in 0.35 μm BiCMOS *IEEE Photon. Technol. Lett.* **28** 2004–7

[32] Milovancev D, Jukic T, Vokic N, Brandl P, Steindl B and Zimmermann H 2021 VLC using 800 μm diameter APD receiver integrated in standard 0.35 μm BiCMOS technology *IEEE Photonics J.* **13** 7900513

[33] Ma J, Masoodian S, Starkey D A and Fossum E R 2017 Photon-number-resolving megapixel image sensor at room temperature without avalanche gain *Optica* **4** 1474–81

[34] Ma J and Fossum E-R 2017 Analytical modeling and TCAD simulation of a quanta image sensor jot device with a JFET source-follower for deep sub-electron read noise *J. Electron Devices Society* **5** 69–78

[35] Seo M-W, Kawahito S, Kagawa K and Yasutomi K 2015 A 0.27e rms read noise 220 μV/e- conversion gain reset-gate-less CMOS image sensor with 0.11 μm CIS process *IEEE Electron Device Lett.* **36** 1344–7

[36] Seo M-W *et al* 2017 A 0.44e rms read-noise 32 fps 0.5 Mpixel high-sensitivity RG-less-pixel CMOS image sensor using bootstrapping reset *2017 IEEE Int. Solid-State Circuits Conf. (ISSCC)* 80–1

[37] Silvaco Atlas user's manual https://www.silvaco.com accessed: 23 March 2023

Chapter 7

Ultra-sensitive avalanche photodiode receivers

In this chapter, a low-noise three-inverter transimpedance amplifier (TIA) with resistive feedback and avalanche photodiode (APD) will be introduced. Then, this TIA topology is equipped with a negative capacitance for sensitivity and/or bandwidth enhancement. Furthermore, the boost of APDs compared to PIN photodiodes in integrating, i.e. capacitive-feedback TIA, receivers will be investigated.

7.1 Three-inverter TIA with resistive feedback

Wireless headphones with near infra-red optical free-space transmission were used about five decades ago, but went out of fashion then, being uncomfortable because of heavy batteries and not being robust enough for head movements. Several years ago, I had the idea to use stationary APD receivers for digital free-space audio transmission, e.g. from a notebook to the stereo equipment. It should replace an optical fiber between notebook (exactly from a USB audio adapter) and optical TOSLINK input of the stereo amplifier. The APD receiver then has to deliver a digital signal to the electrical S/PDIF input of the stereo amplifier or to a digital-to-analog converter for an analog input of the stereo amplifier.

In the following, the APD receiver with a digital output is described. The circuit of this light-to-logic APD receiver, which was fabricated in a 0.35 μm PIN-photodiode CMOS with a supply voltage VDD of 3.3 V, is presented in figure 7.1. A resistive-feedback TIA with three inverters was used. The avalanche photodiode had a diameter of 800 μm, since the large photosensitive area increases the possible free-space transmission distance (see for instance [2] and references therein). Although this APD receiver is not a typical USPAR receiver (because it uses resistive and not capacitive feedback), it achieves a rather high sensitivity. Furthermore, this receiver has to be introduced before the next APD receiver can be described below.

doi:10.1088/978-0-7503-5437-0ch7

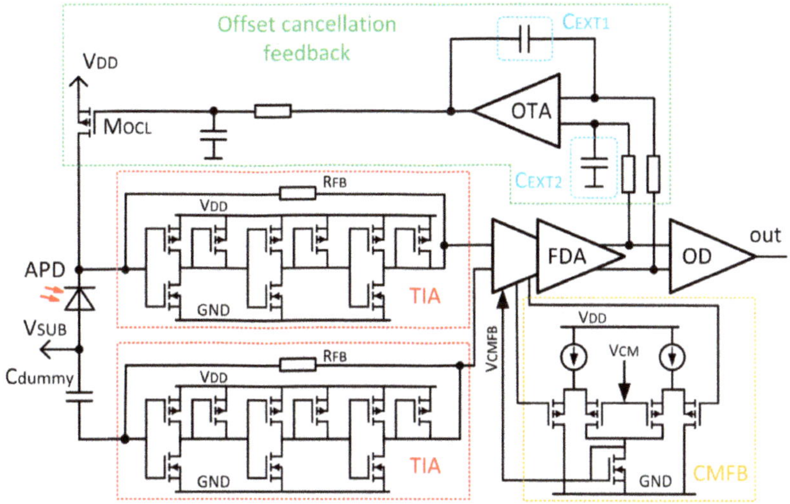

Figure 7.1. APD receiver using a three-inverter TIA. Reproduced from [1] with permission from SPIE, copyright (2021).

In contrast to the 2 Gb s^{-1} APD receiver of [2], however, the APD did not exploit modulation doping [1]. For the audio application the extension of the APD's bandwidth was not needed and modulation doping did not make sense, since it increases the excess noise of the APD [3]. The breakdown voltage of the APD was 35.1 V, which is much lower than the breakdown voltages U_{BD} of the commercial APDs AD500-8 ($U_{BD,typical}$ = 100 V) [4] and SAE500 ($U_{BD,typical}$ = 200 V) [5], both with 500 μm diameter. Due to the 800 μm diameter, the capacitance of the APD was 7 pF. Nevertheless, because of the rather low bandwidth for digital audio transmission, the feedback resistor R_{FB} was 120 kΩ, enabling the low rms input noise current of 774 pA in a bandwidth of 4 MHz [1]. Three-stage inverter TIAs are advantageous because of their high gain [6, 7]. The chip photo of this APD receiver is shown in figure 7.2. The pseudo-differential receiver architecture exploiting a second TIA (dummy or replica TIA) allows cancellation of substrate and power supply noise. The capacitance C_{dummy} has to match the capacitance of the APD to achieve best substrate noise suppression. The interested reader is referred to [8] for a detailed description of the pseudo-differential topology.

Despite the high feedback resistor value and the APD multiplication gain, the signal voltage at the TIA output is only in the millivolt range [1]. The pseudo-differential topology allows fully differential post amplifiers (FDAs). Two FDAs were implemented. The operational transconductance amplifier (OTA) in proportional-integral (PI) configuration takes the outputs of the second FDA to read the offset (due to mismatch and background light) and to cancel the offset via the controlled current source M_{OCL}. The output driver OD consisting of three inverters boosts the output signal to full digital level. To prevent pulse width distortion of the output signal, a common-mode feedback (CMFB) loop across the first FDA was necessary.

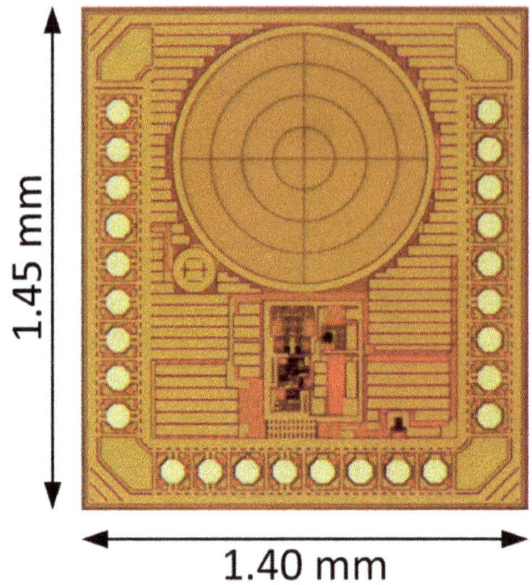

Figure 7.2. Chip photo of APD receiver using a three-inverter TIA. Reproduced from [1] with permission from SPIE, copyright (2021).

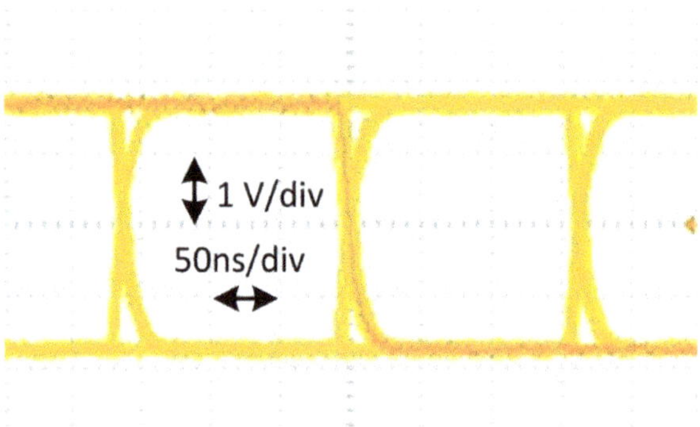

Figure 7.3. Measured eye diagram of the APD receiver using a three-inverter TIA at 6 Mb s^{-1} and −46 dBm optical power. Reproduced from [1] with permission from SPIE, copyright (2021).

The eye diagram obtained with an LED driver and a 500 μm-core-diameter plastic optical fiber (POF) at 6 Mb s^{-1} is shown in figure 7.3. The jitter is below 15 ns [1]. In order to determine the optimum operating point of the APD, its reverse voltage was varied via the substrate potential V_{SUB}. Figure 7.4 presents the resulting BER for the data rate of 5 Mb s^{-1} with four different average optical powers from −48.5 dBm to −47.6 dBm.

The BER in dependence on the average optical input power is shown in figure 7.5 for three different data rates without and with background light (BL). The BL had an intensity of 99 lx. The influence of the background light becomes stronger at lower optical signal power levels. The sensitivities of the APD receiver without

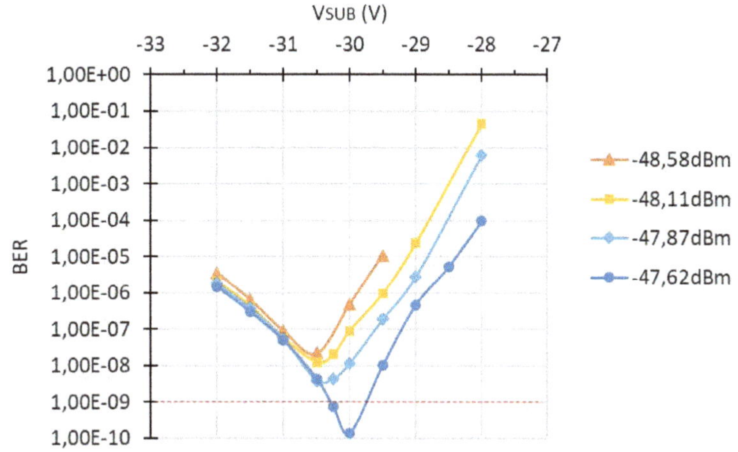

Figure 7.4. Measured BER of the APD receiver using a three-inverter TIA in dependence on substrate voltage (anode voltage of APD) at 5 Mb s^{-1}. Reproduced from [1] with permission from SPIE, copyright (2021).

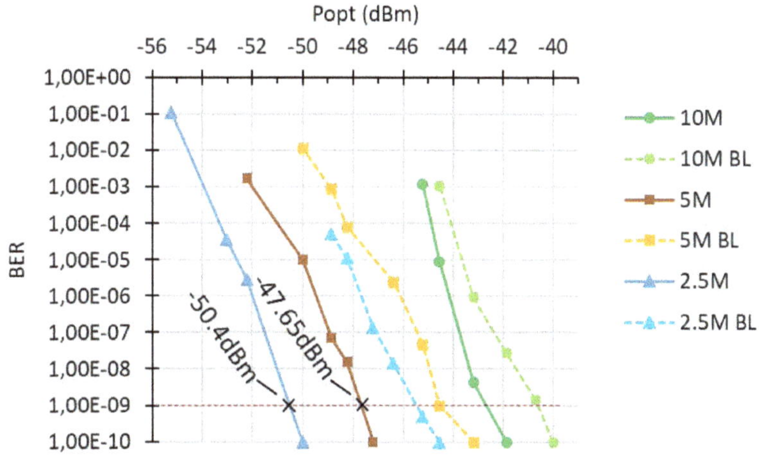

Figure 7.5. Measured BER of the APD receiver using a three-inverter TIA for different data rates (2.5 Mb s^{-1}, 5 Mb s^{-1} and 10 Mb s^{-1}) without and with BL. Reproduced from [1] with permission from SPIE, copyright (2021).

background light are −50.4 dBm at 2.5 Mb s^{-1}, −47.65 dBm at 5 Mb s^{-1} and −42.7 dBm at 10 Mb s^{-1} [1].

7.2 Three-inverter TIA with negative capacitance and resistive feedback

Now, that we know the three-stage inverter TIA, the possibility of bandwidth and noise improvement will be described. Such an improvement was suggested by implementing a negative capacitance in a three-stage inverter TIA [9]. The achieved improvements of transimpedance gain and input referred noise current were 5.7 dB and 1.8 dB, respectively. The design kit of 40 nm CMOS technology was used in [9] for the simulations. The photodiode had a capacitance of 0.5 pF. The feedback resistor across the three-inverter amplifier was 29.9 kΩ in the circuit simulations. In the reference TIA without negative capacitance, a value of the feedback resistor of only 15.5 kΩ was possible for a bandwidth of 3 GHz. Without negative capacitance, the bandwidth was 1.2 GHz with the feedback resistor of 29.9 kΩ. With the negative capacitance, the bandwidth increased to 3 GHz when keeping the 29.9 kΩ feedback resistor. But let us now observe figure 7.6 to understand the effect of the negative capacitance.

The capacitance C_{n} between output of the second inverter and the input of the amplifier constitutes a positive feedback, since the first and second inverter form a non-inverting amplifier. The gain of this two-stage amplifier is $A_1 \times A_2$. Therefore, the capacitance C_n can be transformed to a negative capacitance $C_{\mathrm{n,in}}$ between amplifier input and ground according to the Miller effect. This negative capacitance reduces the input capacitance of the amplifier, i.e. the capacitance of the input node C_T.

$$C_{\mathrm{n,in}} = (1 - A_1A_2)C_{\mathrm{n}} \approx -A_1A_2C_{\mathrm{n}}, \qquad \text{for} \quad A_1A_2 \gg 1 \tag{7.1}$$

Note that A_1 and A_2 are both negative and therefore $C_{\mathrm{n,in}}$ is negative. C_{n} poses an additional load to the second amplifier stage. The effect of this, however, should be negligible since C_{n} is rather small. Because the TIA bandwidth is inverse proportional to the total input-node capacitance [10], the positive feedback with C_{n} increases the bandwidth according to

$$f_{\mathrm{TIA,-3\,dB}} \approx \frac{\sqrt{2}A}{2\pi R_{\mathrm{F}}C_{\mathrm{T}}} \approx \frac{\sqrt{2}A_1A_2A_3}{2\pi R_{\mathrm{F}}(C_{\mathrm{T}} - A_1A_2C_{\mathrm{n}})}, \tag{7.2}$$

where A is the gain of the three-inverter TIA.

The equivalent input noise current spectral density is [10]

$$\overline{i^2_{\mathrm{n,in}}} \approx \frac{4k_{\mathrm{B}}T}{R_{\mathrm{F}}} + 4k_{\mathrm{B}}T\frac{2}{3}\frac{1 + (2\pi f R_{\mathrm{F}}C_{\mathrm{T}})^2}{g_m R_{\mathrm{F}}^2}, \tag{7.3}$$

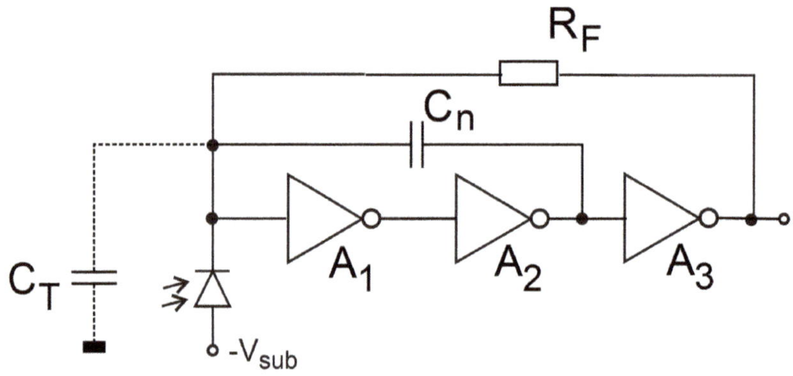

Figure 7.6. Principle of the negative capacitance needing two inverters.

if the TIA noise is mainly determined by the NMOS and PMOS transistors in the first inverter and by the feedback resistor R_F. g_m is the sum of the transconductances of the NMOS and PMOS transistors $g_{m,n}$ and $g_{m,p}$.

In [9], C_n was 12.3 fF, equation (7.2) was used to increase R_F (keeping the bandwidth of 3 GHz) and the larger R_F reduced the input noise current according to equation (7.3). The spectral rms input noise current density was reduced to 0.75 pA Hz. When integrated, the rms input noise current of 0.43 μA resulted compared to 0.53 μA of the reference TIA. The phase margin with C_n was 67.4°, which was only slightly reduced compared to the phase margin of the reference TIA of 69.2°. The stability is certainly an important issue, since simulation results are not absolutely reliable in general.

To get familiar with the negative capacitance and to judge its influence on noise and stability more reliably, we designed a new three-inverter TIA with a 800 μm diameter APD in 0.35 μm PIN-photodiode CMOS and characterized the receiver after fabrication. However, unfortunately no improvement of the sensitivity was observable. This may be because of power supply noise or substrate noise. This will need further investigation, before as a next step, the negative capacitance can be implemented in an USPAR receiver, i.e. in a capacitive-feedback TIA.

7.3 Integrating APD receivers

For the APD USPAR receiver, a model for the noise and the Q-factor will be highly useful. From chapter 3, we know that an APD in the linear mode amplifies shot noise. The shot noise can be taken from the APD operated at low electric-field strengths, where the avalanche gain M is equal to 1. (The APD works then as a PIN photodiode. Note that the reach-through APD used for the USPAR ICs is implemented in PIN photodiode CMOS; see section 5.2.) Inserting the shot noise for $\overline{i^2_{n,PIN}} = 2qI_{PIN}\Delta f$, where I_{PIN} is the photocurrent I_{PH}, into equation (3.60) results in

$$\overline{i^2_{n,\,APD}} = FM^2 2qI_{PIN}\Delta f. \tag{7.4}$$

In the following we will use $\eta_d = \eta_{PIN}$. The excess noise factor F is given by equation (3.61). $I_{PIN} = I_{PH}$ consists of $\overline{n_e}$ electrons (each with the charge q) flowing in a time interval Δt[1]:

$$I_{PIN} = \frac{q\overline{n_e}}{\Delta t} = \eta_d \overline{n_{PH}} \frac{q}{\Delta t}. \tag{7.5}$$

The APD noise current causes a noise voltage, when charging a capacitance C in the time interval Δt (Δt is the integration time.):

$$\overline{v^2_{n,\,APD}} = \frac{\overline{i^2_{n,\,APD}}\Delta t^2}{C^2} = \frac{FM^2 2q^2\overline{n_e}\Delta f\,\Delta t}{C^2}. \tag{7.6}$$

[1] I_{PIN}, $\overline{n_e}$ and $\overline{n_{PH}}$ are mean values.

Let us take the bandwidth $\Delta f = m \cdot B$, where B is the bit rate or data rate. m is usually chosen as 0.7 for resistive feedback TIAs, but is sometimes only 0.5.[2] For Δt we take $1/B$. So, $\Delta f \cdot \Delta t = m \cdot B \cdot 1/B = m$.

In the USPAR integrators we have to consider the integration efficiency η_{int}, since not the whole voltage across the capacitor C_{int} is visible as output voltage (due to the limited gain of the amplifier).[3]

$$\overline{v_{\mathrm{n,out,APD}}^2} = \frac{FM^2 2\, mq^2 \eta_{\mathrm{d}} n_{\mathrm{PH}} \eta_{\mathrm{int}}^2}{C_{\mathrm{int}}^2} \tag{7.7}$$

This output noise voltage has to be used to calculate the Q-factor of an APD USPAR integrated circuit. In equation (6.16) we have to use $\overline{u_{\mathrm{n,out,APD}}^2}$ instead of the photon noise term and we obtain

$$Q = \frac{U_{\mathrm{out}}}{\sqrt{\overline{v_{\mathrm{n,out}}^2} + \overline{v_{\mathrm{n,RS}}^2} + \overline{v_{\mathrm{n,out,APD}}^2}} + \sqrt{\overline{v_{\mathrm{n,out}}^2} + \overline{v_{\mathrm{n,RS}}^2}}}. \tag{7.8}$$

The signal output voltage U_{out} is amplified by the avalanche gain M. Instead of equation (6.18) and taking the reset noise $v_{\mathrm{n,RS}}^2$ from equation (3.69) we obtain

$$Q = \frac{M\eta_{\mathrm{d}} n_{\mathrm{PH}} \eta_{\mathrm{int}} q / C_{\mathrm{int}}}{\sqrt{v_{\mathrm{n,out}}^2 + k_{\mathrm{B}}T/C_{\mathrm{int}} + \dfrac{FM^2 2\, m\eta_{\mathrm{d}} n_{\mathrm{PH}} \eta_{\mathrm{int}}^2 q^2}{C_{\mathrm{int}}^2}} + \sqrt{v_{\mathrm{n,out}}^2 + k_{\mathrm{B}}T/C_{\mathrm{int}}}} \tag{7.9}$$

for the resetted USPAR integrated circuit or for real double sampling from equation (6.23):

$$Q = \frac{M\eta_{\mathrm{d}} n_{\mathrm{PH}} \eta_{\mathrm{int}} q / C_{\mathrm{int}}}{\sqrt{2v_{\mathrm{n,out}}^2 + \dfrac{FM^2 2\, m\eta_{\mathrm{d}} n_{\mathrm{PH}} \eta_{\mathrm{int}}^2 q^2}{C_{\mathrm{int}}^2}} + \sqrt{2v_{\mathrm{n,out}}^2}}. \tag{7.10}$$

Let us check which expression will follow from equation (7.9), if we set the (thermal) output noise of the amplifier $v_{\mathrm{n,out}}$ and the reset noise to 0:

$$Q_0 = \frac{\sqrt{\eta_{\mathrm{d}} n_{\mathrm{PH}}}}{\sqrt{F 2m}}. \tag{7.11}$$

If we set $M = 1$, which means $F = 1$, and $m = 0.5$,

$$Q_0 = \sqrt{\eta_{\mathrm{d}} n_{\mathrm{PH}}} \tag{7.12}$$

[2] Since for integrating noise, the noise bandwidth instead of the −3 dB-bandwidth should be taken and for a first-order low-pass the noise bandwidth is $\frac{\pi}{2}$ times the −3 dB-bandwidth, m can reach up to 1.1.

[3] The integration efficiency η_{int} depends on the frequency since it depends on the frequency dependent gain of the amplifier $A(\omega)$. In fact, the error should not be too large, if we use the low-frequency η_{int} for $A(\omega = 0)$. In fact, we are on the conservative side, since we overestimate the noise voltage.

is obtained. This is obviously not the same result, which follows from photon noise with a PIN photodiode (see equations (6.17) and (6.18)):

$$Q_{0,\mathrm{photonnoise}} = \frac{n_{\mathrm{PH}}}{\sqrt{n_{\mathrm{PH}}}} = \sqrt{n_{\mathrm{PH}}}. \tag{7.13}$$

So, since the photon number determines the number of photogenerated charges, the photon noise should be used in the derivation of the Q-factor for the APD integrating receiver instead of the electronic shot noise in equation (7.4).[4] The primary photocurrent in the APD i_{PIN} is

$$i_{\mathrm{PIN}} = \frac{n_e q}{\Delta t} = \frac{\eta_d n_{\mathrm{PH}} q}{\Delta t}, \tag{7.14}$$

since the number of photons n_{PH} determines the number of photogenerated electrons n_e.[5] For photon noise of $\sigma_{\mathrm{PH}} = \sqrt{n_{\mathrm{PH}}}$, we obtain

$$\overline{i_{\mathrm{n,\,PIN}}^2} = n_{\mathrm{PH}} \left(\frac{\eta_d q}{\Delta t} \right)^2 \tag{7.15}$$

and for the APD excess noise we get

$$\overline{i_{\mathrm{n,\,APD}}^2} = M^2 F n_{\mathrm{PH}} \left(\frac{\eta_d q}{\Delta t} \right)^2. \tag{7.16}$$

For the voltage noise integrated on the integration capacitance with this noise current, we get[6], if we consider the integration efficiency η_{int} of the integrator

$$\overline{v_{\mathrm{n,out,APD}}^2} = M^2 F n_{\mathrm{PH}} \left(\frac{\eta_d q \eta_{\mathrm{int}}}{C_{\mathrm{int}}} \right)^2. \tag{7.17}$$

Finally, the Q-factor follows:

$$Q = \frac{M \eta_d n_{\mathrm{PH}} \eta_{\mathrm{int}} q / C_{\mathrm{int}}}{\sqrt{\overline{v_{\mathrm{n,out}}^2} + k_B T / C_{\mathrm{int}} + \frac{F M^2 \eta_d^2 n_{\mathrm{PH}} \eta_{\mathrm{int}}^2 q^2}{C_{\mathrm{int}}^2}} + \sqrt{\overline{v_{\mathrm{n,out}}^2} + k_B T / C_{\mathrm{int}}}} \tag{7.18}$$

for the resetted USPAR integrated circuit or for real double sampling from equation (6.23):

$$Q = \frac{M \eta_d n_{\mathrm{PH}} \eta_{\mathrm{int}} q / C_{\mathrm{int}}}{\sqrt{2\overline{v_{\mathrm{n,out}}^2} + \frac{F M^2 \eta_d^2 n_{\mathrm{PH}} \eta_{\mathrm{int}}^2 q^2}{C_{\mathrm{int}}^2}} + \sqrt{2\overline{v_{\mathrm{n,out}}^2}}}. \tag{7.19}$$

---

[4] In charge-coupled devices (CCDs), where the photogenerated charges are also integrated, the photon noise determines the signal voltage [11].

[5] Note that current is the charge flowing in a time interval Δt: $I = Q/\Delta t$.

[6] $C = \frac{Q}{v}$, $v = \frac{Q}{C} = \frac{i \Delta t}{C}$.

Now, $Q_0 = \sqrt{n_{PH}}$ follows for $M = 1$ ($F = 1$) and for setting the amplifier output noise and the reset noise to 0.[7]

Next, the inverter with $W_n = 1\ \mu$m and $W_p = 2.5\ \mu$m will be used with the APD in 0.35 μm PIN-photodiode CMOS with $k = 0.0789$. This inverter had $g_m = 399\ \mu$S, $C_{int} = 0.7$ fF, $A = 16.6$ and a simulated bandwidth of 223 MHz.[8] For the quantum efficiency we take $\eta_d = 0.9$. For these parameters, the Q-factors for $M = 1$ and $M = 50$ shown in figure 7.7 are obtained. Reset noise is included. For $M = 1$, the necessary number of photons for $Q = 3$ is 164, which agrees well with the result of figure 6.5 for the PIN-photodiode USPAR integrated circuit with the same inverter as integrator.

For for an APD with $M = 50$, only 56 photons are needed for $Q = 3$. Such an APD with $k = 0.0789$ improves the sensitivity considerably over that of a PIN photodiode.

Let us compare the rms noise voltage caused by photon noise at the output of this inverter integrator for $M = 1$ (which is the case of a PIN photodiode,) and for $M = 50$. From equation (6.17) for the photon noise of a PIN integrator we calculate a rms output noise voltage of 1.07 mV for 100 photons using $\eta_{int} = 0.52$ as visible from figure 6.10 for an input-node capacitance $C_T = 10$ fF. The blue curve in

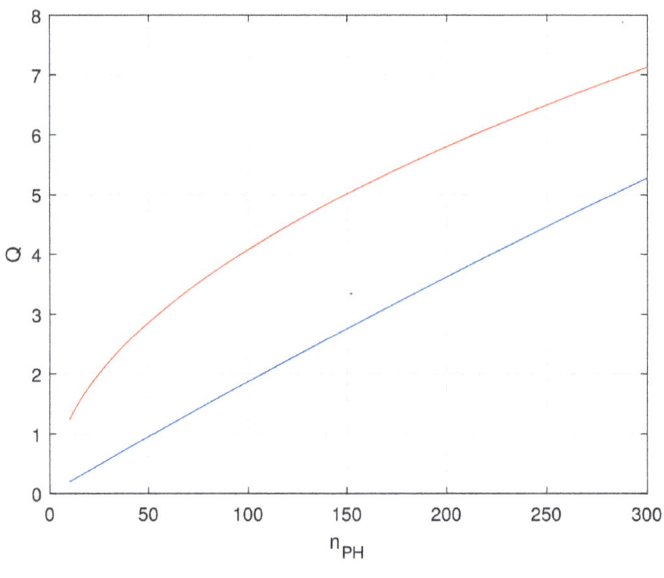

Figure 7.7. Q-factor of inverter integrator with $W_n = 1\ \mu$m and $W_p = 2.5\ \mu$m using an APD with $k = 0.0789$ (blue: $M = 1$, red: $M = 50$). Reset noise for 300 K is considered.

[7] Of course, one could say, the shot noise should be used for the PIN photodiode instead of the photon noise (equations (6.17) and (6.18)). My opinion, however, is that when we deal with a few photons, the shot noise equation is not appropriate and photon noise is straightforward. In fact, for η_d close to one, there is no difference noticeable, if $m = 0.5$. It also should be noted that the photon noise implies the optimum bandwidth of the amplifier used for integration. In this sense photon noise may be considered as optimistic.

[8] The load was a NMOS common-source amplifier with a gate width of 5 μm.

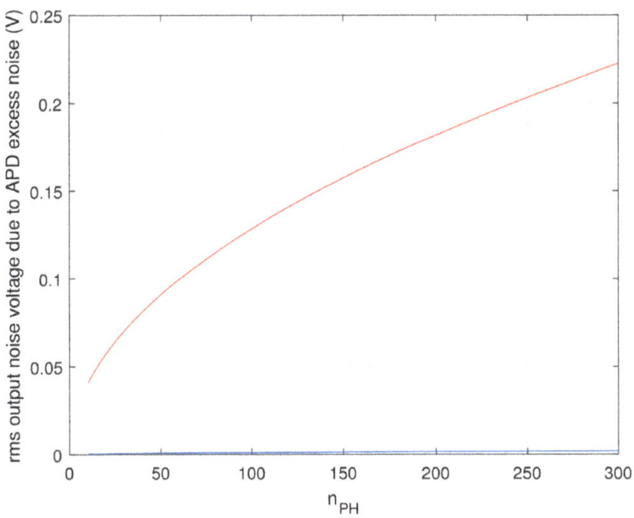

Figure 7.8. Rms output noise voltage of inverter integrator with $W_n = 1\ \mu$m and $W_p = 2.5\ \mu$m due to photon noise (blue: $M = 1$, red: $M = 50$).

figure 7.8 for $M = 1$, contains also a value of 1.07 mV for $n_{PH} = 100$. The red curve for $M = 50$, however, shows an rms output noise voltage due to photon noise of 128 mV for 100 photons. The avalanche gain M and the excess noise factor F increase the output noise voltage of the integrator due to photon noise considerably. However, the signal voltage is multiplied stronger.

Since, the ratio of the ionization coefficients k of integrated APDs varies from CMOS process to CMOS process, we should investigate the influence of k on the Q-factor. For this inverter in 0.35 μm CMOS, the dependence of the Q-factor on k is presented in figure 7.9 for $M = 50$. With increasing k, the necessary number of photons for $Q = 3$ increases from 66 ($k = 0.1$), 108 ($k = 0.2$) and 152 ($k = 0.3$) to 194 ($k = 0.4$).

In a next step, we use (real) double sampling, which requires using equation (7.10) to compare with the resetted inverter integrator. Figure 7.10 shows the results obtained with this equation for $M = 50$. The necessary photon numbers for $Q = 3$ are 64 ($k = 0.1$), 108 ($k = 0.2$), 150 ($k = 0.3$) and 194 ($k = 0.4$). There is only a very small advantage of the integrator using double sampling over the integrator with reset noise. The reason for this behavior is the dominance of the APD's excess noise over the (thermal) amplifier noise and over the reset noise.

Since it is known from resistive-feedback TIAs that there is an optimum avalanche gain for the best receiver sensitivity, equations (7.9) and (7.10) should be investigated for which avalanche gain the necessary numbers of photons are lowest for $Q = 3$. Figure 7.11 presents the Q-factor depending on M and the number of photons for $k = 0.1$. We can see that the Q-factor is largest for M around 10. From the contour plot (figure 7.12), especially from the iso-line for $Q = 3$, we can read the optimum $M = 10$ at about 36 photons.

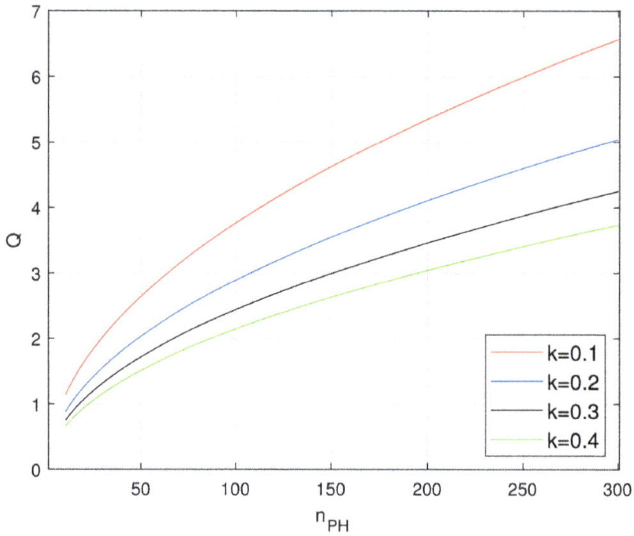

Figure 7.9. Q-factor of inverter integrator with $W_n = 1\ \mu$m and $W_p = 2.5\ \mu$m using an APD for different k-factors. Reset noise for 300 K is considered.

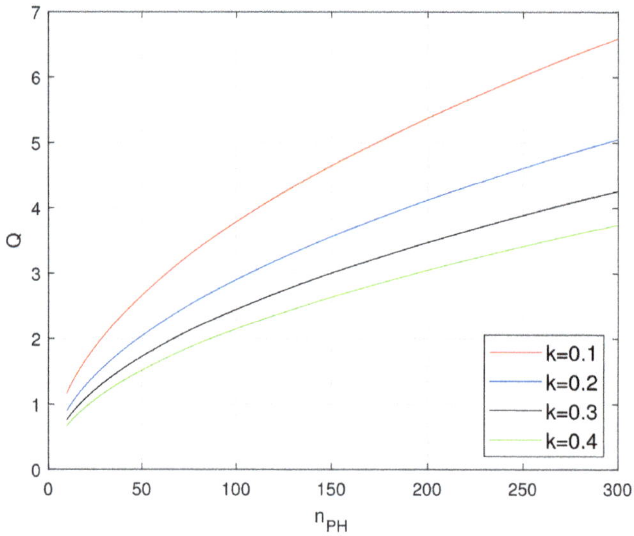

Figure 7.10. Q-factor of inverter integrator with $W_n = 1\ \mu$m and $W_p = 2.5\ \mu$m using an APD for different k-factors assuming double sampling.

Figures 7.13 and 7.14 show the Q-factor for $k = 0.2$. The optimum is shifted to lower M. For $M = 7$ to 8, the lowest photon number for $Q = 3$ of 43 is reached. For a larger ratio k of the impact ionization coefficients, the number of necessary photons for $Q = 3$ increases. Nevertheless, there is a remarkable improvement to lower photon numbers being necessary for $Q = 3$ compared to the PIN-photodiode receiver described in section 6.1 thanks to the APD.

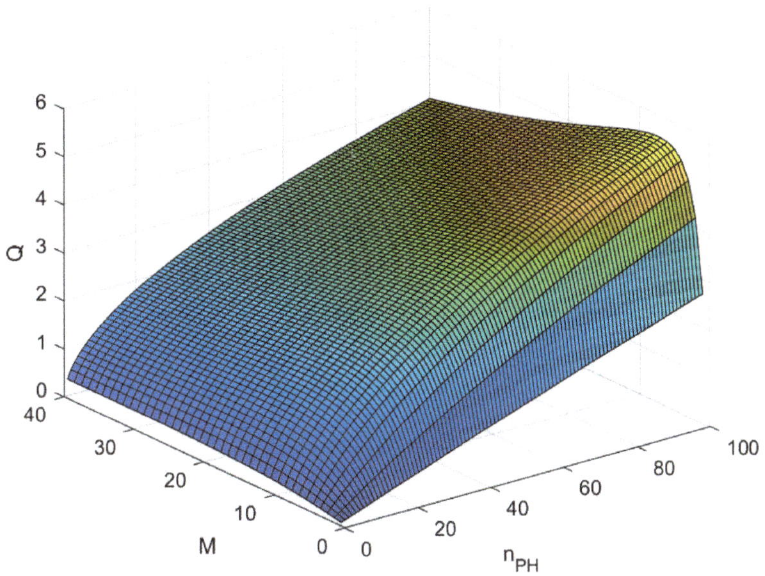

Figure 7.11. Q-factor of inverter integrator with $W_n = 1\ \mu m$ and $W_p = 2.5\ \mu m$ using an APD for $k = 0.1$ assuming double sampling in dependence on avalanche gain M and photon number n_{PH}.

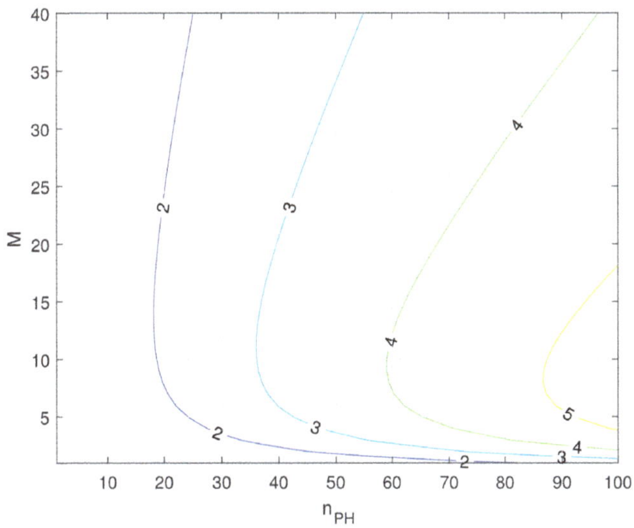

Figure 7.12. Contour plot of Q-factor of inverter integrator with $W_n = 1\ \mu m$ and $W_p = 2.5\ \mu m$ using an APD for $k = 0.1$ assuming double sampling.

Let us also compare the difference between shot noise and photon noise for $k = 0.1$, $\eta_d = 0.9$ and $m = 1$.[9] Figure 7.15 shows that the noise voltage at the

[9] For $m = 0.5$ we do not expect a pronounced difference in Q because we use $\eta_d = 0.9$.

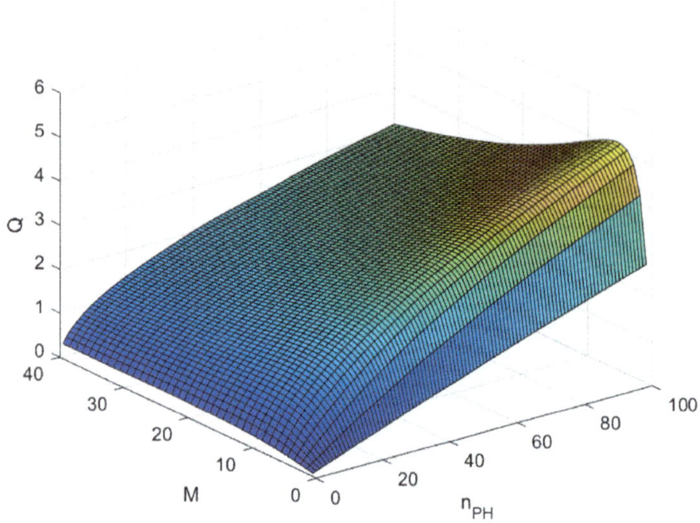

Figure 7.13. Q-factor of inverter integrator with $W_n = 1$ μm and $W_p = 2.5$ μm using an APD for $k = 0.2$ assuming double sampling in dependence on avalanche gain M and photon number n_{PH}.

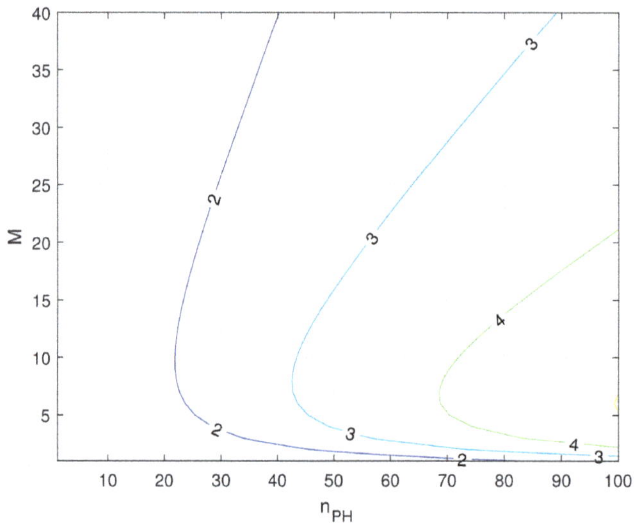

Figure 7.14. Contour plot of Q-factor of inverter integrator with $W_n = 1$ μm and $W_p = 2.5$ μm using an APD for $k = 0.2$ assuming double sampling.

integrator output caused by shot noise for $m = 1$ is about 50% larger as caused by photon noise[10].

[10] This is due to the high bandwidth, i.e. $\Delta f = B$ (B = bit rate).

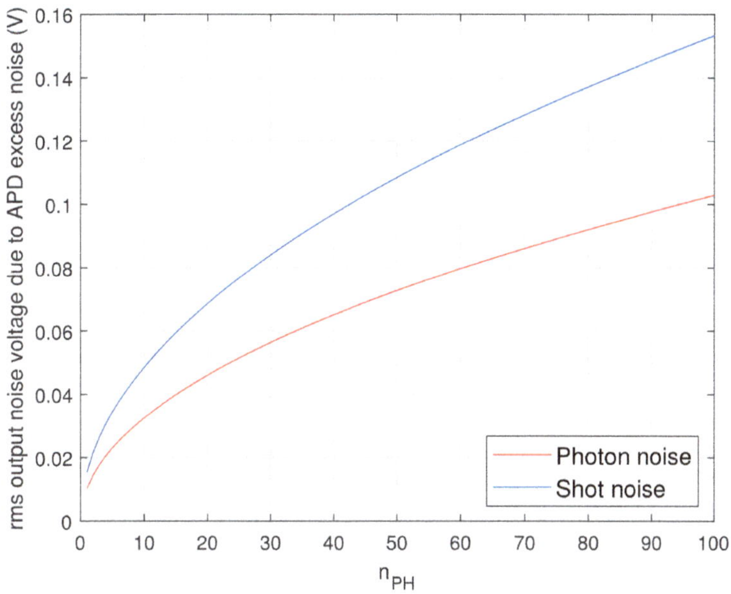

Figure 7.15. Comparison of rms output noise voltage of inverter integrator with $W_n = 1\ \mu$m and $W_p = 2.5\ \mu$m due to shot noise with $m = 1$ and due to photon noise for $M = 40$.

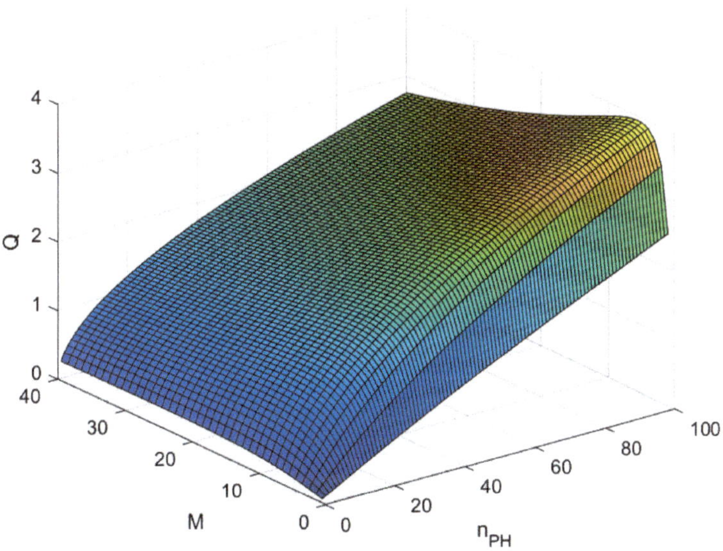

Figure 7.16. Q-factor of inverter integrator with $W_n = 1\ \mu$m and $W_p = 2.5\ \mu$m using an APD for $k = 0.1$ assuming double sampling in dependence on avalanche gain M and photon number n_{PH} calculated for shot noise with $m = 1$.

The 3D-plot of the Q-factor is depicted in figure 7.16 taking the shot noise, $k = 0.1$ and $m = 1$. The Q-factor decreased compared to the corresponding plot with photon noise for $k = 0.1$ shown above. The lowest number of photons with shot noise and $m = 1$ is 62 for $M = 6$ (see figure 7.17).

We also should be aware that the McIntyre excess noise theory and the relation between F and M is for planar APDs with homogeneous electric field.[11] The USPAR PIN and avalanche photodiodes, however, have semispherical electric field distributions and we cannot use the ratio of the impact ionization coefficients k and F does not follow the dependence on M given by equation (3.61). Meanwhile, however, the excess noise factor of a dot-APD in 0.18 μm HV CMOS was measured. Figure 7.18 compares the output noise voltage of the inverter integrator in 0.18 μm CMOS for the measured $F = 3.9$ at $M = 40$ [12] to that for the McIntyre theory with $k = 0.1$. The noise voltage at the integrator output for the measured excess noise factor is more than 10% below that obtained with the McIntyre theory for $k = 0.1$. The dot-APD in 0.18 μm CMOS obviously is characterized by a lower excess noise than the planar APDs in 0.35 μm CMOS technologies. When we insert $F = 3.9$ for $M = 40$ into equation (7.19), we obtain the Q-factor shown in figure 7.19. The USPAR integrated circuit with the dot-APD, for which $F = 3.9$ holds at $M = 40$, and with the inverter integrator in 0.18 μm CMOS needs only 41 photons for $Q = 3$ instead of 56 photons for the McIntyre theory with $k = 0.1$. When we compare with the results of the same integrator ($W_n = 0.88$ μm) with PIN-photodiode shown in figure 6.33 we see there a necessary number of photons of 166 for $Q = 3$. The

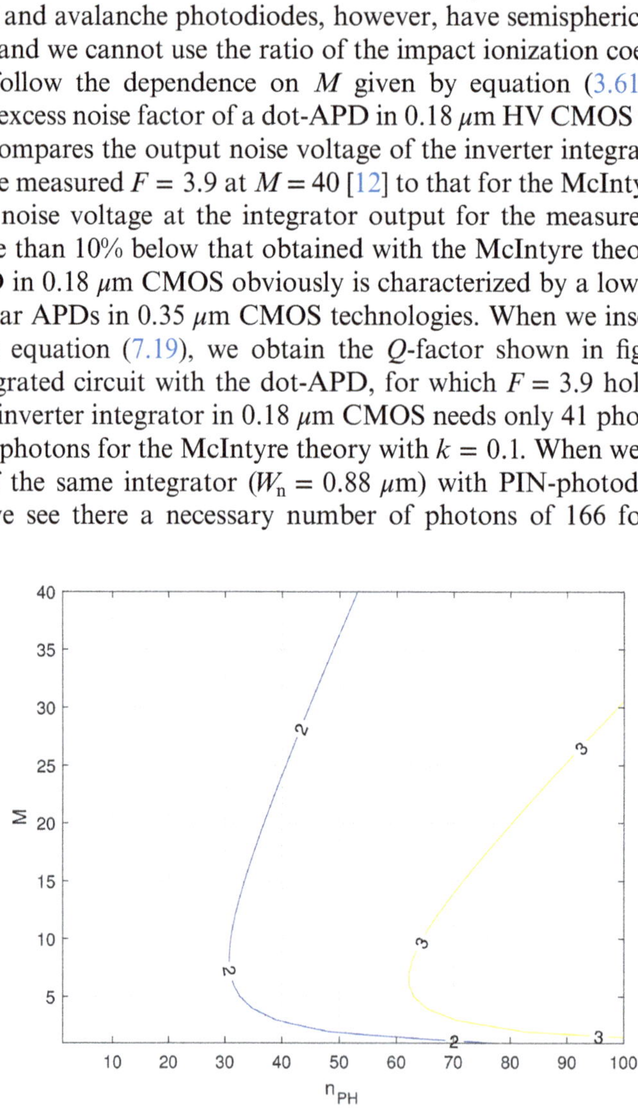

Figure 7.17. Contour plot of Q-factor of inverter integrator with $W_n = 1$ μm and $W_p = 2.5$ μm using an APD for $k = 0.1$ assuming double sampling calculated for shot noise with $m = 1$.

[11] The above estimations of Q imply a small photodiode diameter and/or a quite thick low-doped epitaxial layer to achieve the assumed photodiode capacitances of planar APDs.

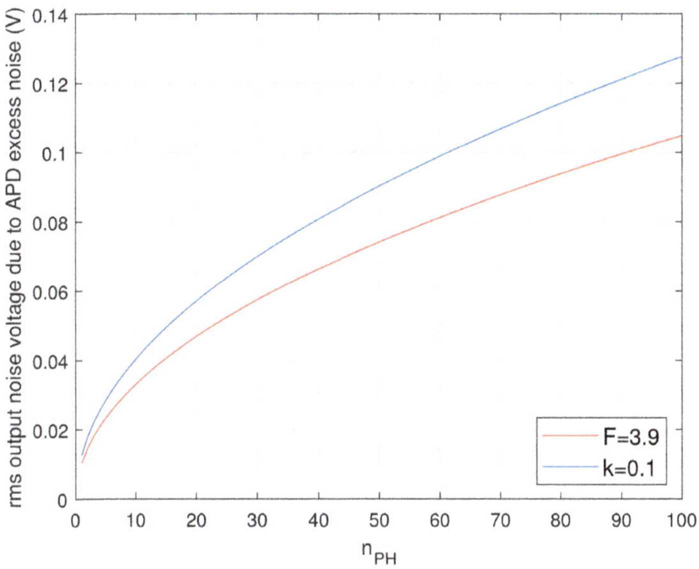

Figure 7.18. Output noise voltage of the inverter integrator in 0.18 μm CMOS with $W_n = 0.88$ μm and $W_p = 1.76$ μm using a dot-APD with $F = 3.9$ at $M = 40$ compared to the McIntyre excess noise with $k = 0.1$.

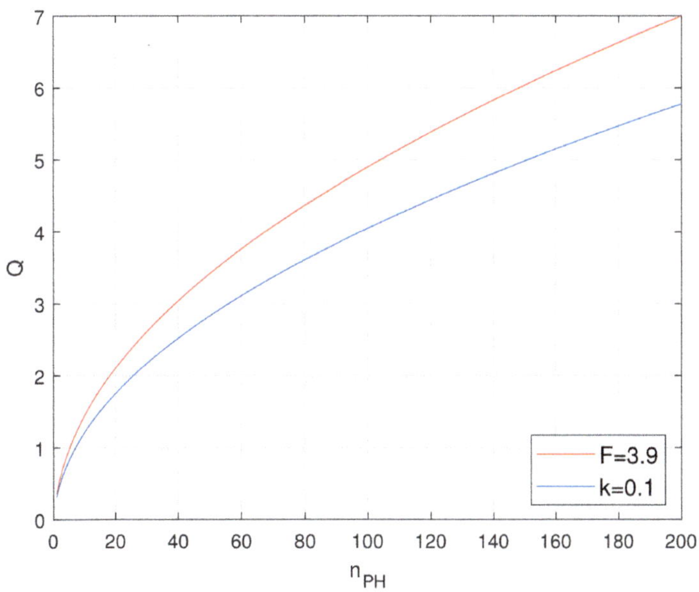

Figure 7.19. Q-factor of inverter integrator in 0.18 μm CMOS with $W_n = 0.88$ μm and $W_p = 1.76$ μm using a dot-APD with $F = 3.9$ at $M = 40$ compared to the McIntyre excess noise with $k = 0.1$ assuming double sampling.

integrating receiver with the dot-APD and $M = 40$ needs only about a third of the number of photons compared to the reference receiver with PIN photodiode.

Concluding, the usage of APDs in ultra-sensitive photocharge-integrating receivers looks quite promising. However, the Q-factors determined with photon noise may be rather optimistic and the influence of the bandwidth of the amplifier used for photocharge-integration has to be investigated in some detail.

References

[1] Gasser C and Zimmermann H 2021 Highly sensitive monolithic optoelectroni receiver with large-area avalanche photodiode for optical wireless communication at low data rates *Opt. Eng.* **60** 075109

[2] Milovancev D, Jukic T, Vokic N, Brandl P, Steindl B and Zimmermann H 2021 VLC using 800 μm diameter APD receiver integrated in standard 0.35 μm BiCMOS technology *IEEE Photonics J.* **13** 7900513

[3] Jukic T, Brandl P and Zimmermann H 2018 Determination of the excess noise of avalanche photodiodes intergated in 0.35 μm CMOS technologies *SPIE Optical Engineering* **57** 044101

[4] https://www.first-sensor.com/cms/upload/datasheets/AD5008_TO_3001349.pdf, First Sensor, First sensor APD datasheet AD500-8-TO, accessed 17.05.2022

[5] https://www.lasercomponents.com/fileadmin/user_upload/home/Datasheets/lc-apd/sae-series_redenhanced.pdf, Laser-Components, Silicon avalanche photodiode SAE-series (red-enhanced), accessed 17.05.2022

[6] Schneider K, Zimmermann H and Wiesbauer A 2004 Optical receiver in Deep-sub-μm CMOS with −28.2 dBm sensitivity at 1.25 Gb s^{-1} *Electron. Lett.* **40** 262–3

[7] Schneider K and Zimmermann H 2006 *Highly Sensitive Optical Receivers* (Berlin, Heidelberg: Springer)

[8] Brandl P, Jukic T, Enne R, Schneider-Hornstein K and Zimmermann H 2016 Optical wireless APD receiver with high background-light immunity forincreased communication distances *IEEE J. Solid-State Circuits* **51** 1663–73

[9] Li X and Yue C P 2019 3 GHz, 1 mW inverter-based TIA with capacitive feedback for enhanced gain and sensitivity in VLC applications *Electron. Lett.* **55** 469–71

[10] Säckinger E 2012 On the noise optimum of FET broadband transimpedance amplifiers *IEEE Trans. Circuits and Systems I, Reg. Papers* **59** 2881–9

[11] Nishida Y, Koike J, Ohtake H, Abe M and Yoshikawa S 1989 Design concept for a low-noise CCD image sensor based on subjective evaluation *IEEE Trans. Electron Devices* **36** 360–6

[12] Kohneh Poushi S S, Gasser C, Goll B, Hofbauer M, Schneider-Hornstein K and Zimmermann H 2023 A near-infrared enhanced field-line crowding based CMOS-integrated avalanche photodiode *IEEE Photonics J.* **15** 1–9

Chapter 8

Summary and completion

This chapter summarizes the book and draws the most important conclusions. The most relevant steps in developing the USPAR idea are specified. A clear and simple comparison of the classes of optical detectors is presented. The limitations of full-custom integrated circuit design and especially of parasitics extraction for USPAR applications are mentioned.

8.1 Recapitulation

The somehow disruptive idea of integrating the photo-charges on a small capacitor to raise the signal above the noise level of amplifiers seems to work as shown with an integrating receiver fabricated in a 0.18 μm high-voltage CMOS. This receiver exploiting a common-source amplifier and a spot PIN photodiode achieved a sensitivity of −56.4 dBm at 50 Mb s^{-1} [1], which is somewhat better than that of −55.7 dBm of a receiver using four single-photon avalanche diodes (SPADs) [2]. To make the basic idea of USPAR a success, it is necessary to integrate the photodiodes, because bondpads between photodiode and amplifier add too high capacitances. And what is more, to achieve good results, the photodiode capacitance must be minimized. It was shown that spot and dot PIN photodiodes and APDs can be designed to achieve a capacitance below 1 fF at quite large light-sensitive areas and bandwidths far into the 100 MHz regime. What would be highly advantageous, would be the development of an anti-reflection coating layer in the CMOS technology used to avoid responsivity minima for certain wavelengths.

Triggered by a publication on a quantum image sensor (QIS) [3] and inspired by an old paper on an integrating receiver [4], I developed the idea of USPAR, being aware that the shrinking of CMOS in recent decades will make it worth investigating the USPAR idea in detail. Both QIS and the integrate&dump receiver of [4] integrate the photo-charges on the photodiode's capacitance itself. This has limited the sensitivity because of the rather large integration capacitance coming with the need for a large photosensitive area. Being familiar with charge-sensitive amplifiers

doi:10.1088/978-0-7503-5437-0ch8

and integrators, the step to integrating the photo-charges on a smaller capacitance exploiting an amplifier was nearby. Just decoupling the integration capacitance from the light-sensitive area by using an amplifier with a feedback capacitor, however, was not satisfying enough since capacitors being available in ASIC design kits are from 10 fF upwards. Here, as an analog full-custom designer being aware of parasitic capacitors inside transistors, it was quite self-evident for me to choose the gate–drain overlap capacitance as integration capacitance, which can be much smaller than this value of 10 fF and smaller than the photodiode capacitance plus input capacitance of a source follower or an amplifier, if the integration is done on the photodiode. Furthermore, my experience with PIN-photodiode technology and with the reduction of the capacitance of PIN and avalanche photodiodes was another important key aspect to make the USPAR idea a success.

The USPAR idea is the product of advancing research on integrated photodiode receivers and sensors as well as 'learning' the progress of CMOS for thirty years, bringing together shrinking, photodiode physics and circuit topologies. The shrinking of CMOS led to reduction of parasitic capacitances usually exploited for rising clock frequencies and reducing power consumption. The progress of CMOS brought me to rethinking optical receivers. The challenge and fascination of research and engineering when combining a broad spectrum of know-how to an innovative solution in USPAR has accompanied me for almost four years, and I hope that sparks will awaken thirst for knowledge and curiosity in leading edge research and development in many young students.

We verified that the dark counts, the afterpulses and the optical crosstalk of SPADs can be avoided, if highly sensitive optical receivers according to the USPAR idea are realized. The circuit complexity, high power consumption and large chip area of SPAD-based receivers are not necessary, if integrating receivers following the USPAR principle are used. The imperfections of the first USPAR integrated circuit will require the implementation of some control loops to make improved USPAR integrated circuits robust. Double sampling may have to be integrated. But I am quite sure that the circuit complexity, power dissipation and chip area can be kept below those of SPAD receivers. But why should the USPAR principle be restricted to optical receivers? There are many other applications needing highly sensitive optical sensors.

8.2 Comparison

Table 8.1 compares the most important types of optical detectors. We started with integrated PIN photodiodes in the 90s, introduced PIN-photodiode receiver and PIN-photodiode sensor integrated circuits, investigated integrated APD receivers and then SPAD receivers. The USPAR idea, however, brought me back to the usage of PIN photodiodes and APDs in the linear mode because of the advantages visible from table 8.1. In my opinion, SPAD-based optical receivers are now a kind of dinosaur.[1]

[1] In many other applications, SPADs will still have advantages of course.

Table 8.1. Comparison of the most important types of optical detectors. Reproduced from [1] CC BY 4.0.

Type of detector	Detection efficiency	Internal gain	Dead time	Dark counts	Circuit effort
PIN	Up to 1	No	No	No	Low
APD	Up to 1[a]	Medium	No	No	Medium
SPAD	PDP < 1[b]	High	Several ns[d]	Yes	High[g]
SPAD array	PDE < 1[c]	High	Several ns[e]	Yes[f]	Very complex

[a] Primary quantum efficiency.
[b] PDP is the photon detection probability.
[c] PDE = PDP·FF (FF = optical fill-factor < 1).
[d] After each detection (also dark counts).
[e] After each detection per pixel, other pixels are still active.
[f] In each pixel.
[g] If active quenching or gating is implemented.

Despite of the overall success, the design of the first USPAR chip revealed the inaccuracy of parasitics extraction. The reset needed in integrating receivers also turns out as a key issue. The models for channel charge injection of the reset transistor are insufficient, which makes it impossible to develop circuits for accurate compensation.

Furthermore, USPAR chips suffer from parasitic capacitances stemming from the layout and leading to a much higher integration capacitance as expected from the pure gate–drain overlap capacitance used for charge integration. So, it is impossible to predict the sensitivity reliably and to achieve the theoretically possible sensitivity. The parasitics strongly depend on the layout. But even comparing different layout versions will leave some uncertainties due to the inaccuracy of parasitics extraction.

Because of all these difficulties, much work, research and development remains to be done in order to achieve product maturity of USPAR chips. The next years will be very exciting and we probably will see fascinating solutions and results, because circuit designers demonstrated very often really creative approaches for many applications. Of course, we also continue and the next USPAR chips are already in fabrication.

References

[1] Schneider-Hornstein K, Goll B and Zimmermann H 2023 Ultra-sensitive PIN-photodiode receiver *IEEE Photonics J.* **15** 1–9

[2] Zimmermann H, Steindl B, Hofbauer M and Enne R 2017 Integrated fiber optical receiver reducing the gap to the quantum limit *Sci. Rep.* **7** 2652

[3] Ma J, Masoodian S, Starkey D A and Fossum E R 2017 Photon-number-resolving megapixel image sensor at room temperature without avalanche gain *Optica* **4** 1474–81

[4] Jindal R P 1990 Silicon MOS amplifier operation in the integrate and dump mode for gigahertz band lightwave communication systems *J. Light. Technol.* **8** 1023–6

www.ingramcontent.com/pod-product-compliance
Ingram Content Group UK Ltd.
Pitfield, Milton Keynes, MK11 3LW, UK
UKHW051855100926
8556IPUK00005B/644